Autodesk Fusion 360 PCB Black Book (V 2.0.18719) - Colored

By
Gaurav Verma
Matt Weber
(CADCAMCAE Works)

Published by CADCAMCAE WORKS, USA. Copyright © 2024. All rights reserved. No part of this publication may be reproduced or distributed in any form or by any means, or stored in the database or retrieval system without the prior permission of CADCAMCAE WORKS. To get the permissions, contact at cadcamcaeworks@gmail.com

ISBN # 978-1-77459-137-6

NOTICE TO THE READER

Publisher does not warrant or guarantee any of the products described in the text or perform any independent analysis in connection with any of the product information contained in the text. Publisher does not assume, and expressly disclaims, any obligation to obtain and include information other than that provided to it by the manufacturer.

The reader is expressly warned to consider and adopt all safety precautions that might be indicated by the activities herein and to avoid all potential hazards. By following the instructions contained herein, the reader willingly assumes all risks in connection with such instructions.

The Publisher makes no representation or warranties of any kind, including but not limited to, the warranties of fitness for a particular purpose or merchantability, nor are any such representations implied with respect to the material set forth herein, and the publisher takes no responsibility with respect to such material. The publisher shall not be liable for any special, consequential, or exemplary damages resulting, in whole or part, from the reader's use of, or reliance upon, this material.

DEDICATION

To teachers, who make it possible to disseminate knowledge
to enlighten the young and curious minds
of our future generations

To students, who are the future of the world

THANKS

To my friends and colleagues

To my family for their love and support

Table of Contents

Preface	ix
About Authors	xi

Chapter 1 : Starting with Autodesk Fusion PCB

Overview Of Autodesk Fusion	**1-2**
Installing Autodesk Fusion (Student)	**1-2**
Starting Autodesk Fusion	**1-5**
Starting a New document	**1-6**
Starting a New Electronics Design	**1-8**
File Menu	**1-8**
Creating New Drawing	1-8
Creating New Drawing Template	1-9
Opening File	1-9
Recovering Documents	1-11
Upload	1-12
Save	1-14
Save As	1-17
Save As Latest	1-18
Export	1-18
Capture Image	1-19
Sharing File Link	1-21
View details On Web	1-22
View	1-22
Undo And Redo Button	**1-23**
User Account drop-down	**1-24**
Preferences	1-24
Autodesk Account	1-28
My Profile	1-28
Work Offline/Online	1-28
Extensions	1-29
Help drop-down	**1-29**
Search Help Box	1-29
Learning and Documentation	1-30
Quick Setup	1-31
Support and Diagnostics	1-31
Data Panel	**1-32**
Working in a Team	1-33
Cloud Account of User	1-37
Browser	**1-38**
Customizing Toolbar and Marking Menu	**1-39**
Adding Tools to Panel	1-39
Assigning Shortcut Key	1-39
Adding or Removing tool from Shortcut menu/Marking menu	1-40
Removing Tool from Panel in Toolbar	1-40
Resetting Panel and Toolbar Customization	1-40
Self Assessment	**1-41**

Chapter 2 : Introduction to Electronics Schematic

Overview of Electronics	**2-2**
Components used in Electronics	2-2
Starting New Schematic Drawing	**2-9**
Schematic Design User Interface	**2-10**
Toolbar, Tabs, and Panels	2-10
Managers	2-10
INSPECTOR Pane	2-16
SELECTION FILTER Pane	2-17
View Toolbar	2-17
Command Input Box	2-20
Sheets Tile	2-20
Workflow in Electronics Schematic Design	**2-20**
Placing Components and Blocks	**2-21**
Placing Components	2-21
Creating Pattern of Objects	2-22
Inserting Schematic in Drawing	2-23
Creating Connections	**2-24**
Creating Net	2-24
Creating Net Class	2-26
Creating Net Breakout	2-28
Drawing Bus	2-29
Breaking Out Bus	2-30
Creating Part Pin Break Out	2-31
Displaying/Editing Names	2-32
Assigning Label to Bus/Net	2-33
Creating Junction	2-33
Creating Modules	2-34
Creating Port in Module	2-35
Self Assessment	**2-37**

Chapter 3 : Modifying and Simulating Electronics Schematic

Introduction	**3-2**
Reworking on Wire/Net Segments	**3-2**
Applying Miter Wire Joints	3-2
Splitting Wire Segments/Polygon Edges	3-3
Slicing Wire Segments	3-3
Optimizing/Joining Wire Segments	3-4
Modifying Schematic Design	**3-4**
Moving Selected Objects	3-5
Rotating Objects	3-6
Mirroring Objects	3-6
Aligning Objects	3-7
Arranging Components	3-7
Copying Format	3-9
Changing Properties of Selected Object	3-9
Modifying Component Values	3-10
Invoking/Fetching Gates from Device	3-11
Replacing Part	3-12
Performing Pin Swap	3-13

Applying Gate Swap	3-14
Optimizing Wire Nets	3-14
Simulation	**3-15**
Adding Spice Model	3-15
Removing SPICE Model	3-16
Analog Source Setup	3-17
Digital Source Setup	3-18
Inserting Digital Source Component in Schematic	3-19
Applying Voltage Probe	3-21
Applying Phase Probe	3-22
Performing Simulation	**3-22**
Operating Point Simulation	3-23
DC Sweep Simulation	3-24
AC Sweep Simulation	3-26
Transient Simulation	3-27
Self Assessment	**3-29**

Chapter 4 : Documentation, Validation, and Automation

Introduction	**4-2**
Documentation Tools	**4-2**
Assembly Variant	4-2
Generating Bill of Materials	4-2
Printing Schematic	4-3
Draw Tools	**4-4**
Creating Line	4-4
Writing Text	4-5
Creating Arc	4-7
Creating Circle	4-7
Creating Rectangle	4-7
Creating Polygon Shape	4-8
Defining Part Attributes	4-9
Repositioning Attributes	4-10
Defining Document Attributes	4-11
Validation Tools	**4-12**
Synchronizing Documents	4-12
Performing Electrical Rules Check	4-13
Checking Net Classes	4-13
Automation	**4-14**
Running ULPs	4-14
Running Scripts	4-15
Library Manager	**4-15**
Updating Design from Library	4-16
Updating Design From All Libraries Used in Design	4-17
Exporting Libraries	4-17
Self Assessment	**4-19**

Chapter 5 : Introduction to PCB

Introduction	**5-2**
Starting PCB Design	**5-2**
Layers	**5-3**

Displaying Layers	5-3
Flipping Board	5-3
Single Layer/Multi Layer View	5-3
Layer Stack Manager	5-3
Creating Board Shapes	**5-9**
Creating Polyline Outline	5-9
Creating Spline Outline	5-10
Creating Arc Outline	5-10
Creating Circle Outline	5-11
Placing Objects in Drawing	**5-11**
Place Components	5-11
Creating Signal Wires	5-11
Creating Hole (NPTH)	5-12
Creating Route Manually	5-12
Creating Differential Pair Routing	5-13
Routing Multiple Airwires	5-14
Placing Vias in PCB	5-16
Meander	5-17
Quick Route Tools	**5-17**
Quick Routing Air wires	5-18
Quick Routing Signals	5-19
Quick Routing Multiple Wires	5-20
Smoothening Route	5-20
Guided Quick Routing	5-21
Performing Fanout	5-22
Autorouter	5-24
Polygons	**5-25**
Creating Polygon Pour	5-26
Creating Polygon Cutout	5-27
Creating Polygon Shape	5-28
Creating Polygon Pour from Outline	5-29
Hiding All Polygon Pour Fills	5-29
Showing All Polygon Pour Fills	5-29
Repairing Polygon Pours	5-29
Unrouting Tools	**5-30**
Unrouting	5-30
Unrouting All	5-31
Unroute Incomplete	5-31
Rework Tools	**5-31**
Rerouting	5-31
Self Assessment	**5-33**

Chapter 6 : 3D PCB Manufacturing and Design Library

Introduction	**6-2**
Performing DRC	**6-2**
Pushing PCB to 3D PCB	**6-3**
Flipping 3D PCB Component	6-4
Moving 3D PCB Components	6-5
Push to 2D PCB	6-6
Creating PCB Hole	6-6

Inserting 3D PCB in Assembly ... 6-7
    Creating Enclosure for PCB ... 6-7
    Assembly of PCB with Enclosure ... 6-11
Manufacturing (CAM) ... 6-13
    CAM Preview ... 6-13
    CAM Processor ... 6-17
    Exporting Gerber, NC Drill, Assembly, and Drawing Outputs ... 6-19
    Exporting ODB++ Output ... 6-20
    Saving IPC Netlist ... 6-21
New Electronics Library ... 6-21
    Creating New Electronics Component ... 6-22
    Creating New Symbol ... 6-27
    Creating New Footprint ... 6-29
    Creating New Package ... 6-31
Self Assessment ... 6-33

Chapter 7 : Practical and Practice

Introduction ... 7-2
Practical ... 7-2
Practice Schematics ... 7-9

Index ... I-1
    Ethics of an Engineer ... I-4
Other Books by CADCAMCAE Works ... I-5

Preface

Autodesk Fusion 360 is a product of Autodesk Inc. Fusion 360 is the first of its kind software which combines PCB Design, 3D CAD, CAM, and CAE tools in single package. It connects your entire product development process in a single cloud-based platform that works on both Mac and PC. The PCB environment of Fusion 360 provides a seamless connection between schematic design, 2D PCB and 3D PCB. Using the 2D PCB, you can create manufacturing data of design that can be sent to PCB manufacturer to get physical PCB. You can also use the 3D CAD environment of Fusion 360 to modify the orientation and placement of various electronic components in PCB design.

The **Autodesk Fusion 360 Black Book** (V 2.0.18719) is 2nd edition of our series on Autodesk Fusion 360 PCB. The book is updated on Autodesk Fusion 360 Ultimate, Student V 2.0.18719. With lots of features and thorough review, we present a book to help professionals as well as beginners in creating some of the most complex electronic design while following systematic workflow. The book follows a step by step methodology. In this book, we have tried to include related technical information for each PCB designing tool. We have tried to reduce the gap between educational use and industrial use of Autodesk Fusion 360 PCB. This edition of book, includes latest topics on Schematic design, 2D PCB design, 3D PCB design, PCB manufacturing, and electronic library management. The book covers almost all the information required by a learner to master Autodesk Fusion 360 PCB. Some of the salient features of this book are :

In-Depth explanation of concepts

Every new topic of this book starts with the explanation of the basic concepts. In this way, the user becomes capable of relating the things with real world.

Topics Covered

Every chapter starts with a list of topics being covered in that chapter. In this way, the user can easy find the topic of his/her interest easily.

Instruction through illustration

The instructions to perform any action are provided by maximum number of illustrations so that the user can perform the actions discussed in the book easily and effectively. There are about **404** small and large illustrations that make the learning process effective.

Tutorial point of view

At the end of concept's explanation, the tutorial make the understanding of users firm and long lasting. Almost each chapter of the book has tutorials that are real world projects. Moreover most of the tools in this book are discussed in the form of tutorials.

For Faculty

If you are a faculty member, then you can ask for video tutorials on any of the topic, exercise, tutorial, or concept. As faculty, you can register on our website to get electronic desk copies of our latest books, self-assessment, and solution of practical. Faculty resources are available in the `Faculty Member` page of our website (`www.cadcamcaeworks.com`) once you login. Note that faculty registration approval is manual and it may take two days for approval before you can access the faculty website.

Formatting Conventions Used in the Text

All the key terms like name of button, tool, drop-down etc. are kept bold.

Free Resources

Link to the resources used in this book are provided to the users via email. To get the resources, mail us at **cadcamcaeworks@gmail.com** with your contact information. With your contact record with us, you will be provided latest updates and informations regarding various technologies. The format to write us mail for resources is as follows:

Subject of E-mail as ***Application for resources of _____ book***.
Also, given your information like
Name:
Course pursuing/Profession:
E-mail ID:

Note: We respect your privacy and value it. If you do not want to give your personal informations then you can ask for resources without giving your information.

About Authors

The author of this book, Gaurav Verma, has authored and assisted in more than 17 titles in CAD/CAM/CAE which are already available in market. He has authored **AutoCAD Electrical Black Books** and **SolidWorks Electrical Black Books**. He has also authored books on various modules of Creo Parametric and SolidWorks. He has provided consultant services to many industries in US, Greece, Canada, and UK. He has assisted in preparing many Government aided skill development programs. He has been speaker for Autodesk University, Russia 2014. He has assisted in preparing AutoCAD Electrical course for Autodesk Design Academy. He has worked on Sheetmetal, Forging, Machining, and Casting designs in Design and Development departments of various manufacturing firms.

If you have any query/doubt in any CAD/CAM/CAE package, then you can contact the authors by writing at cadcamcaeworks@gmail.com

For Any query or suggestion

If you have any query or suggestion, please let us know by mailing us on **cadcamcaeworks@gmail.com**. Your valuable constructive suggestions will be incorporated in our books.

Page Left Blank Intentionally

Chapter 1

Starting with Autodesk Fusion PCB

Topics Covered

The major topics covered in this chapter are:

- **Overview of Autodesk Fusion**
- **Installing Autodesk Fusion (Educational)**
- **Starting Autodesk Fusion**
- **Starting a New Document**
- **File Menu**
- **Undo and Redo button**
- **User Account drop-down**
- **Help drop-down**
- **Data Panel**
- **Navigation Bar**
- **Display Bar**
- **Customizing Toolbar and Marking Menu**

OVERVIEW OF AUTODESK FUSION

Autodesk Fusion is an Autodesk product designed to be a powerful 3D Modeling software package with an integrated, parametric, feature based CAM module built into the software. Autodesk Fusion is the first of its kind 3D CAD, CAM, CAE, and PCB design tool. It connects your entire product development process in a single cloud-based platform that works on both Mac and Microsoft Windows; refer to Figure-1.

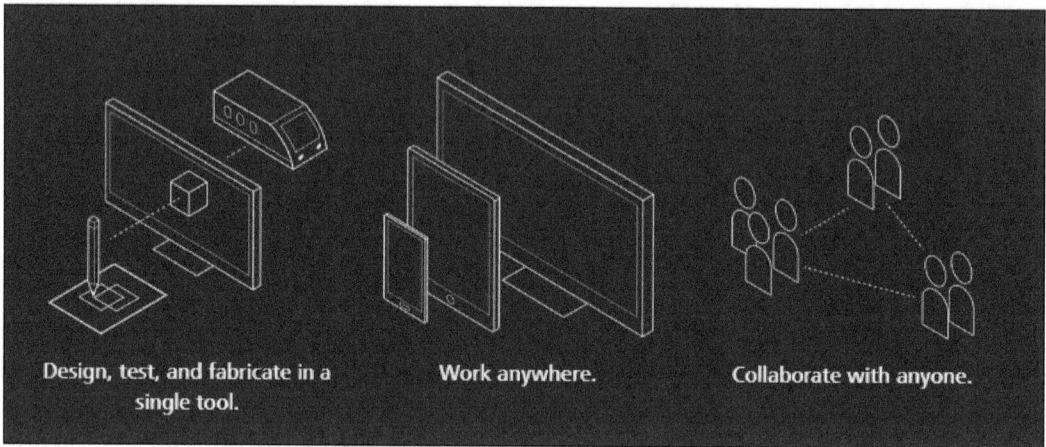

Figure-1. Overview

It combines design, collaboration, simulation, and machining in a single software. The tools in Fusion enable rapid and easy exploration of design ideas with an integrated concept to production toolset. This software needs a good network connection to work in collaboration with other team members.

This software is much affordable than any other software offered by Autodesk. To use this software, one need to pay monthly subscription of Fusion. You can work offline in this software and later save the file on Autodesk Server. User can access this software from anywhere with an internet connection. The user is able to open the saved file and also able to share files with anyone from anywhere as long as he/she has the software and good internet connection. Also, the pricing of this software is so effective that anyone can use it for manufacturing of tools and parts. Autodesk Fusion is build to work in multi body manner: both parts and assemblies build in a single file. The procedure to install the software is given next.

INSTALLING AUTODESK FUSION (STUDENT)

- Connect your PC with the internet connection and then log on to **https://www.autodesk.com/education/edu-software/overview** as shown in Figure-2.

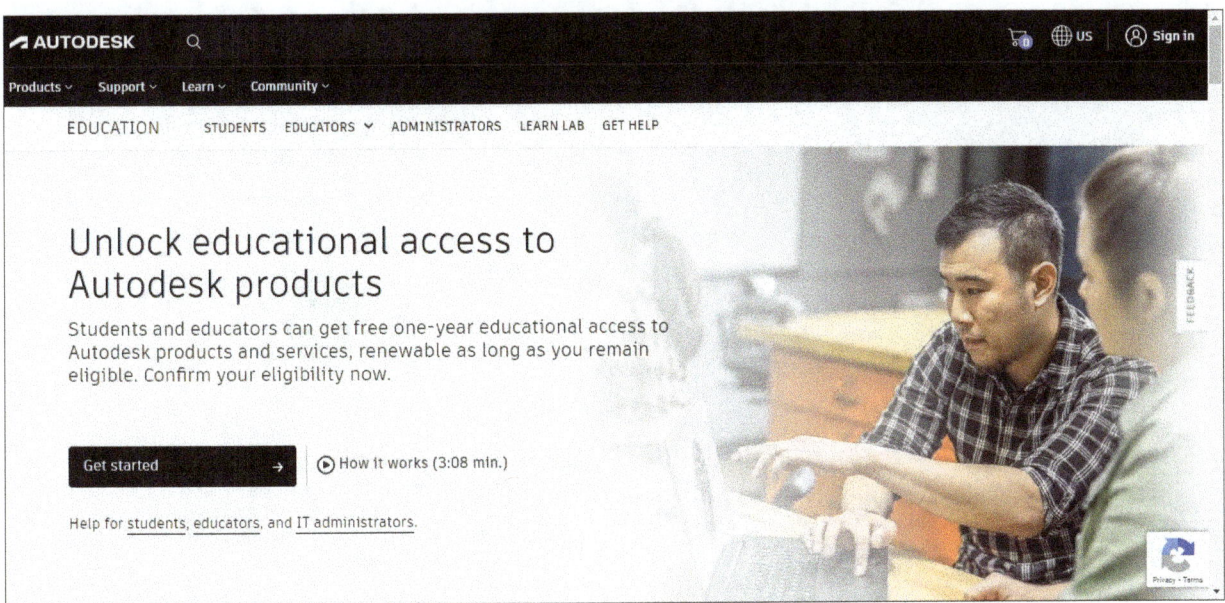

Figure-2. Autodesk website

- Click on the **GET STARTED** button, the Sign up page will be displayed; refer to Figure-3.

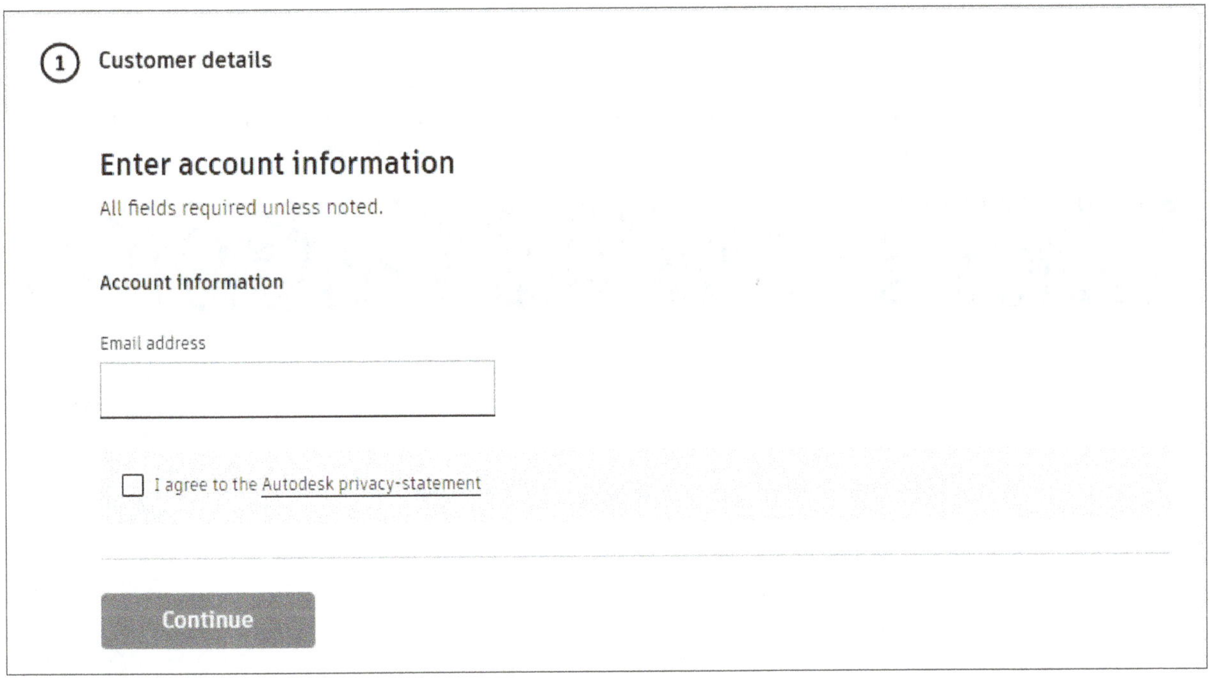

Figure-3. Sign Up page

- In this page, specify the personal details and in **Education role** drop-down, you need to select **Student** option and provide your date of birth. (There is free subscription for students with license term of 1 year which can be renewed).
- After filling the details, click on the **Submit** button from the page as shown in Figure-4. A message will be displayed confirming the account creation; refer to Figure-5.

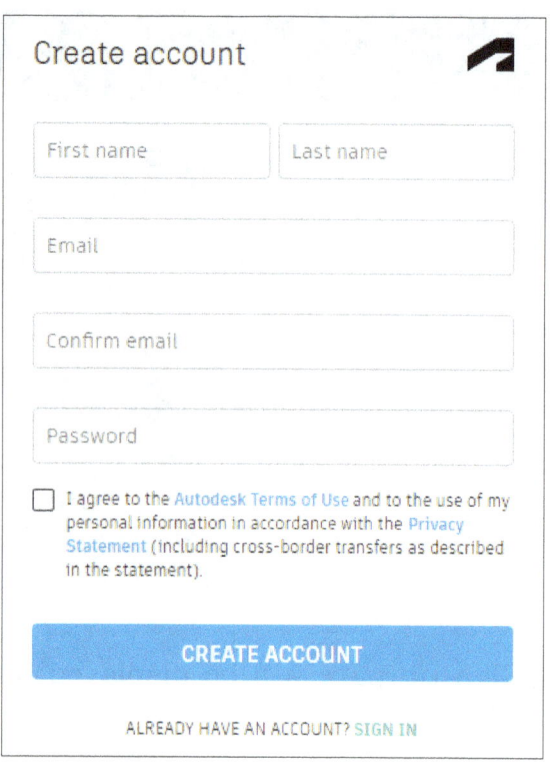

Figure-4. Creating account

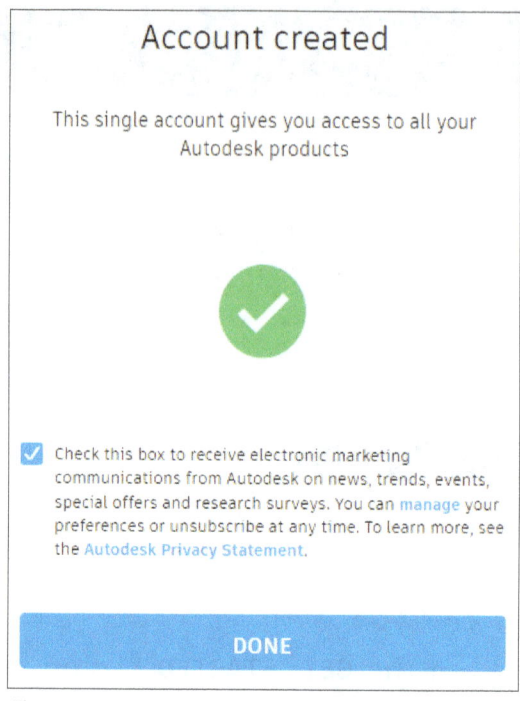

Figure-5. Account created message

- Click on the **DONE** button. You will be sent back to education page of Autodesk.
- Click on the `Get Educational Access` button from the page; refer to Figure-6. The web page to confirm your educational status will be displayed.

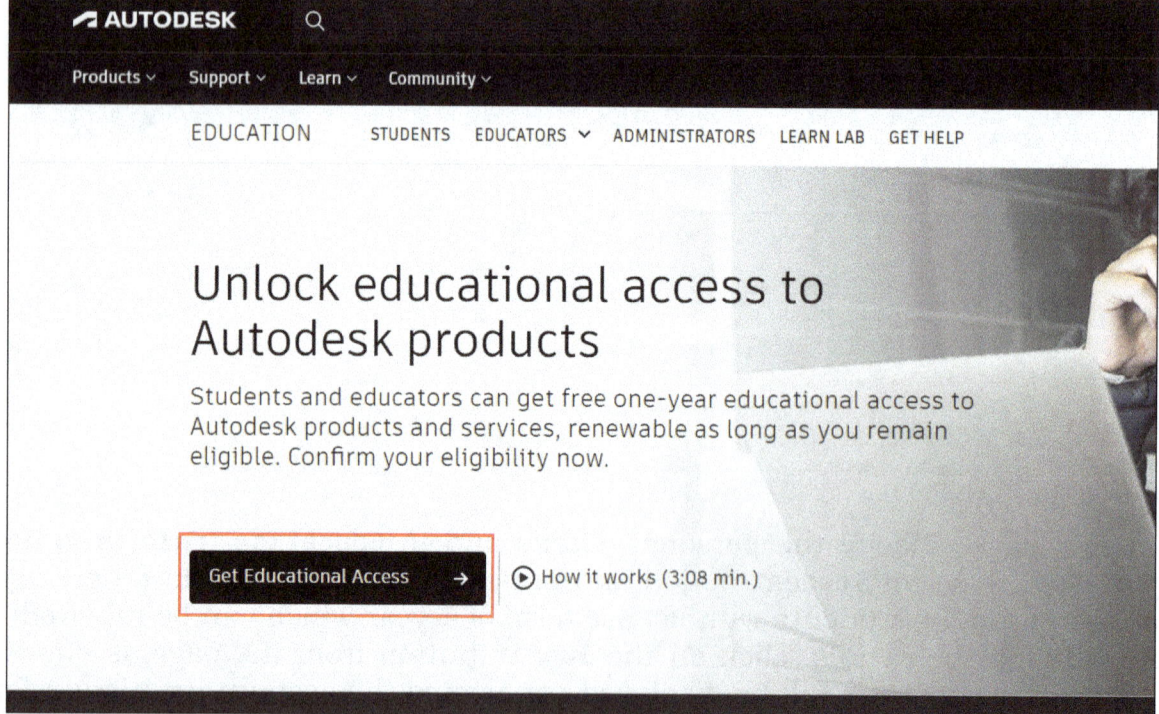

Figure-6. Get Educational Access button

- Provide the details in fields of this page and then click on the **CONFIRM** button. After completion of these processes, you may be asked to upload your documentation like ID card scan or receipt to confirm your student status. Once your account is approved which can take upto 2 days, you need to sign-in to your Autodesk Account and search for the **Autodesk Fusion software**.
- Click on the **Get product** button below the software name. The **Access** button will become active; refer to Figure-7. Click on this button to download the **Autodesk Fusion software**.

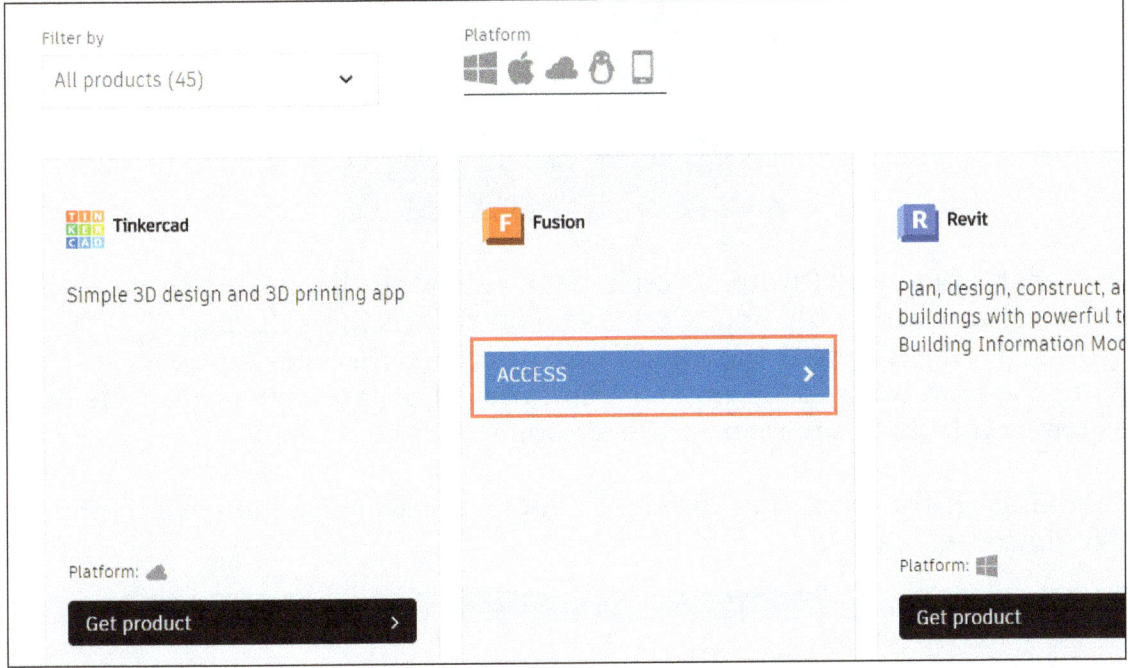

Figure-7. Access button

- Open the download setup file and follow the instructions as per the setup instruction.
- The software will be installed in a couple of minutes.

STARTING AUTODESK FUSION

- To start **Autodesk Fusion** from **Start** menu, click on the **Start** button in the **Taskbar** at the bottom left corner and then select **Autodesk Fusion** option from the **Autodesk** folder. Select the **Autodesk Fusion** icon; refer to Figure-8.

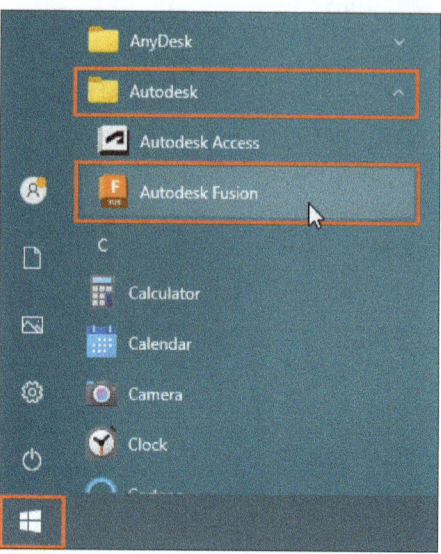

Figure-8. Start menu

- While installing the software, if you have selected the check box to create a desktop icon then you can double-click on that icon to run the software.
- If you have not selected the check box to create the desktop icon and want to create the icon on desktop now in Windows 10 or later, then drag the **Autodesk Fusion** icon from **Start** menu to Desktop.

After clicking on the icon, the Autodesk Fusion software window will be displayed; refer to Figure-9.

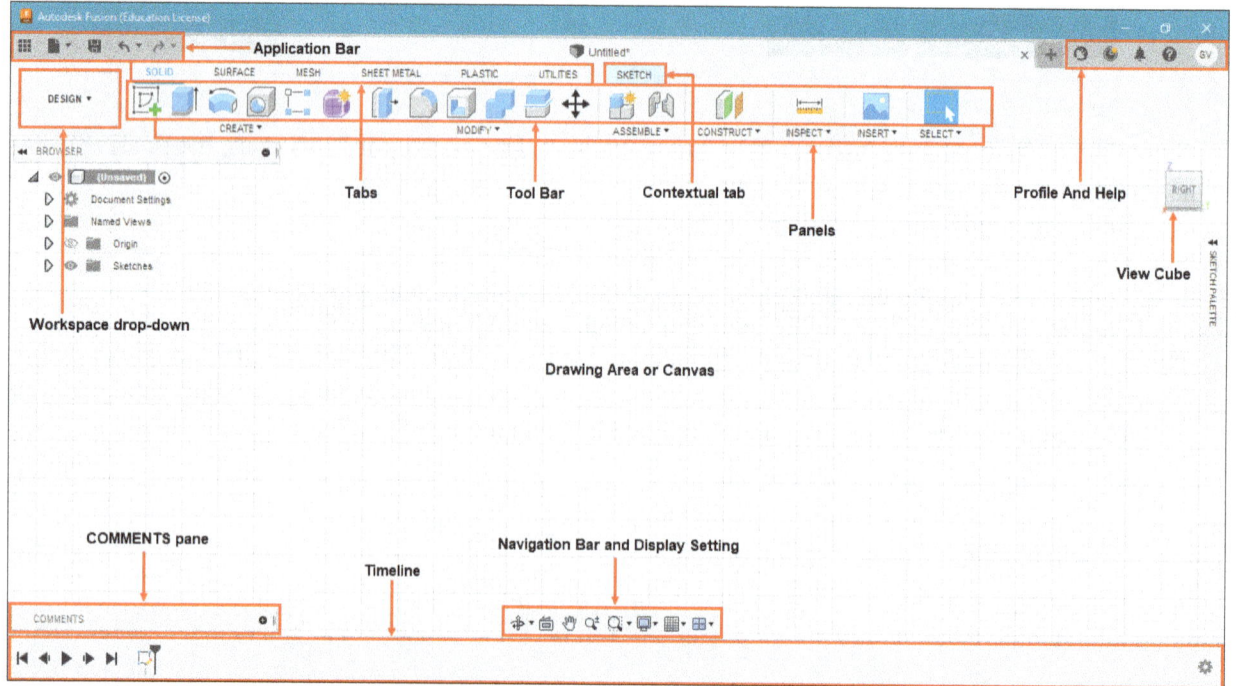

Figure-9. Autodesk Fusion Application Window

STARTING A NEW DOCUMENT

- Click on the **File** drop-down and select the **New Design** tool as shown in Figure-10. A new document will open.

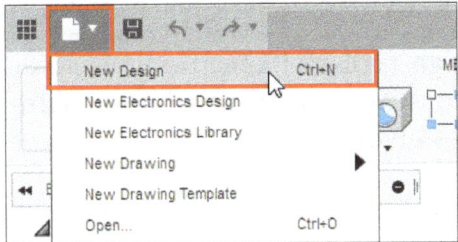

Figure-10. File menu

- Select desired workspace from the **Workspace** drop-down; refer to Figure-11.

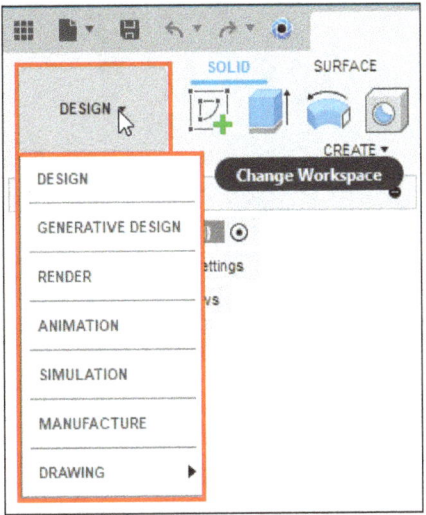

Figure-11. Workspace drop down

- There are six buttons available in **Workspace** drop-down in Educational version; **DESIGN**, **RENDER**, **ANIMATION**, **SIMULATION**, **MANUFACTURE**, and **DRAWING**.

- The **DESIGN** button is used to create solid, surface, and sheet metal designs.

- The **RENDER** button is used to activate workspace for rendering realistic model for presentation.

- The **ANIMATION** button is used to activate workspace for creating automatic or manual exploded views as well as direct control over unique animation of parts and assemblies.

- The **SIMULATION** button is used to perform Engineering Analyses.

- The **MANUFACTURE** button is used to generate G-codes for manufacturing processes like turning, milling, drilling, cutting, probing, and so on.

- The **DRAWING** button is used for generating drawings from model and animation.

If you are working on Professional version of software then another workspace **GENERATIVE DESIGN** is also available using which you can generate different iterations of designs based on specified limits and goals.

You can learn more about these workspaces in our Autodesk Fusion Black Book.

STARTING A NEW ELECTRONICS DESIGN

The **New Electronics Design** tool in **File** menu is used to create PCB design files which include schematics of circuit and manufacturing files for PCB creation. The procedure to use this tool is given next.

- Click on the **New Electronics Design** tool from the **File** menu. A new electronic design file will open; refer to Figure-12.

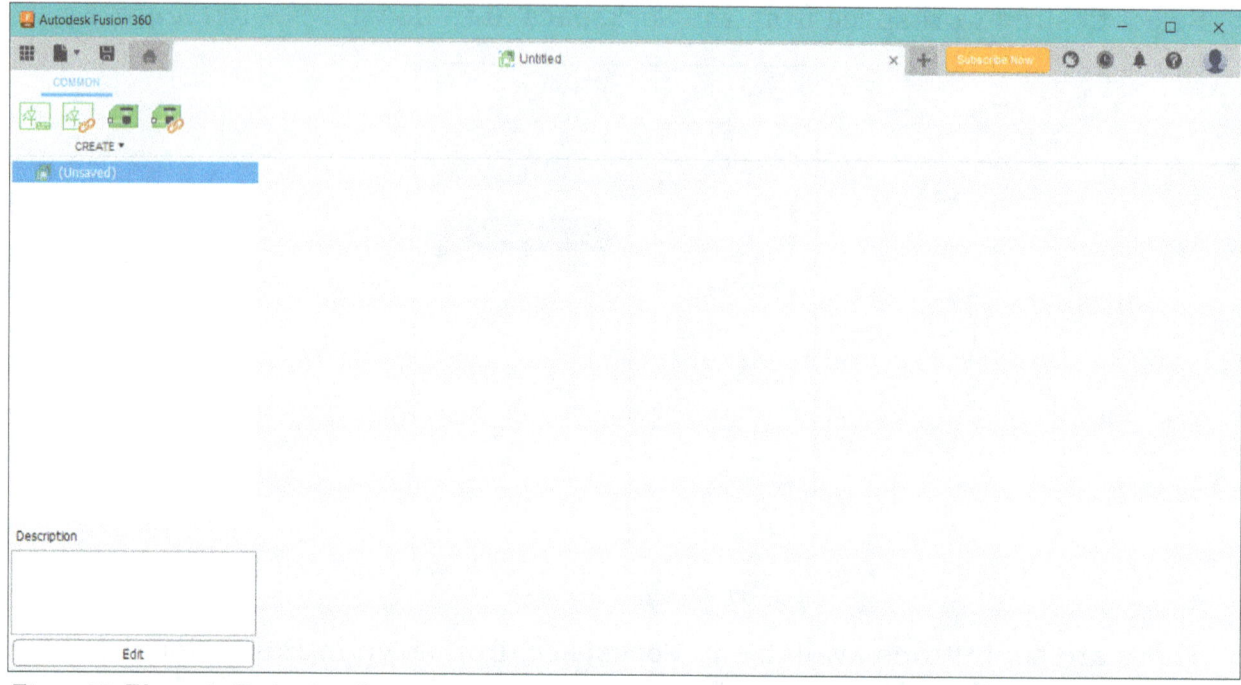

Figure-12. Electronics Design interface

Similarly, you can use the **New Electronics Library** tool to create a new electronics library.

FILE MENU

The options in the **File** menu are used to manage files and related parameters. Various tools of **File** menu are discussed next.

Creating New Drawing

The tools in **New Drawing** cascading menu are used to initialize a new drawing from animation or design; refer to Figure-13.

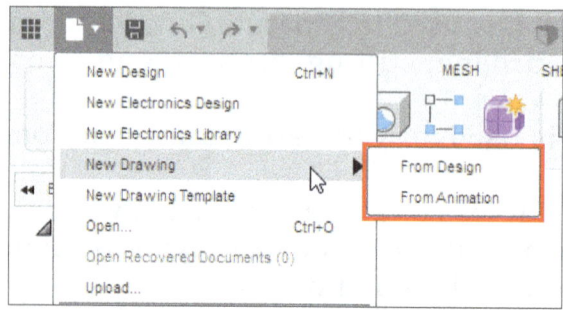

Figure-13. New Drawing cascading menu

Creating New Drawing Template

The **New Drawing Template** tool is used to create a template file for drafting. This template can either be created from scratch or you can use other drawing file as reference for template. This template can later be used for creating engineering drawings of the model.

Opening File

The **Open** tool is used to open files earlier saved in cloud or local drive. You can also use this tool to import supported files of other software. The procedure to use this tool is given next.

- Click on the **Open** tool from the **File** menu or press **CTRL+O** from keyboard. The **Open** dialog box will be displayed; refer to Figure-14. Note that we are working in online mode of Autodesk Fusion.

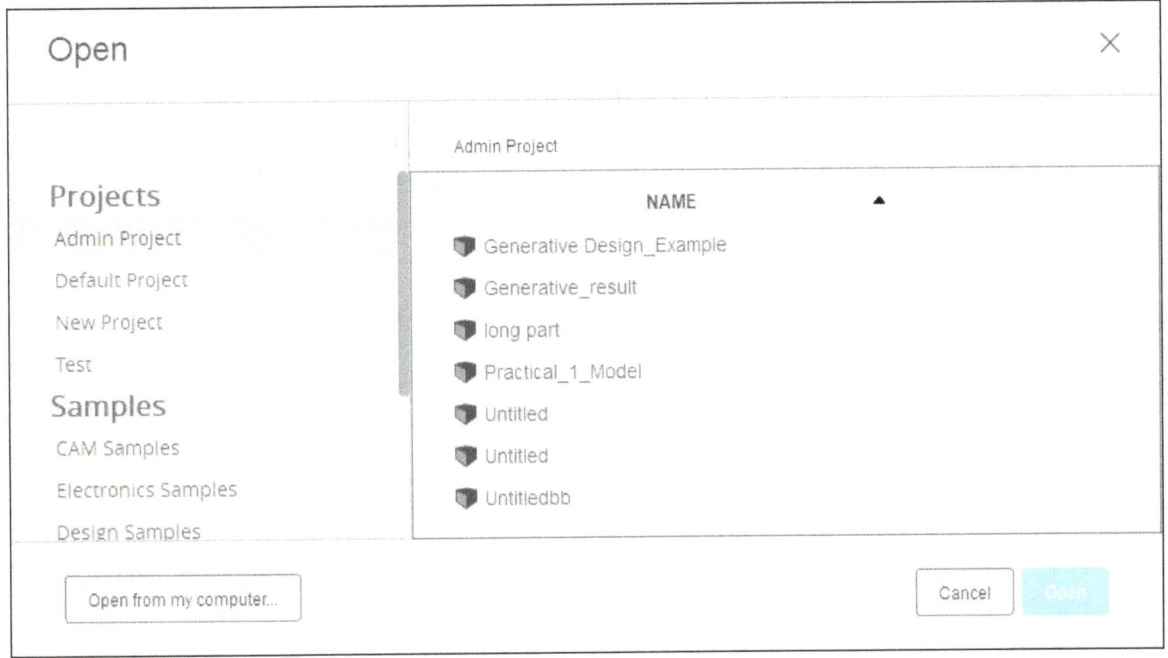

Figure-14. Open dialog box

- By default, files saved on cloud are displayed in the dialog box. Select desired file and click on the **Open** button. The file will open in Autodesk Fusion.
- If you want to open a file stored in local drive then click on the **Open from my computer** button. The **Open** dialog box will be displayed; refer to Figure-15.

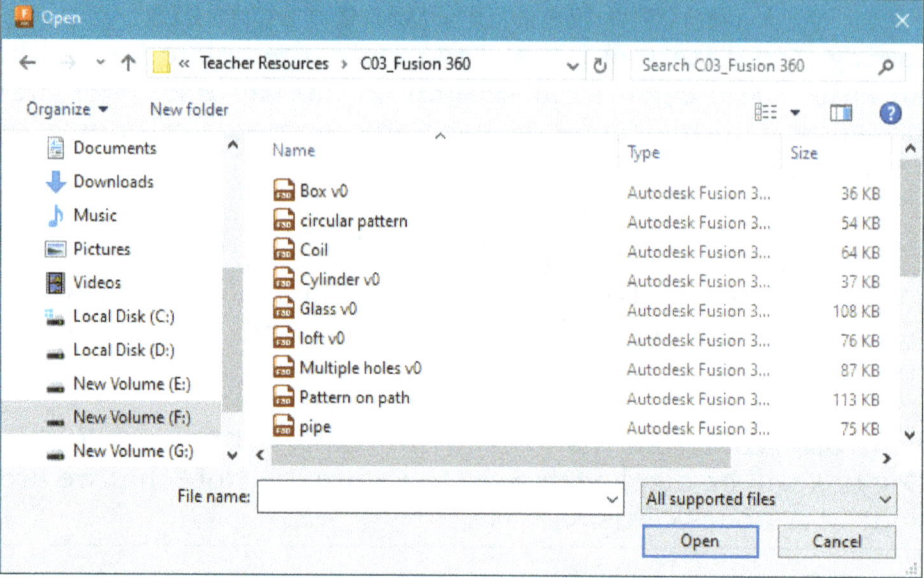

Figure-15. Open dialog box

- Select desired file and click on the **Open** button. The file will be displayed in the application window. If you have selected the file of different format from native format then the `Job Status` dialog box will be displayed notifying you the status of file import; refer to Figure-16.

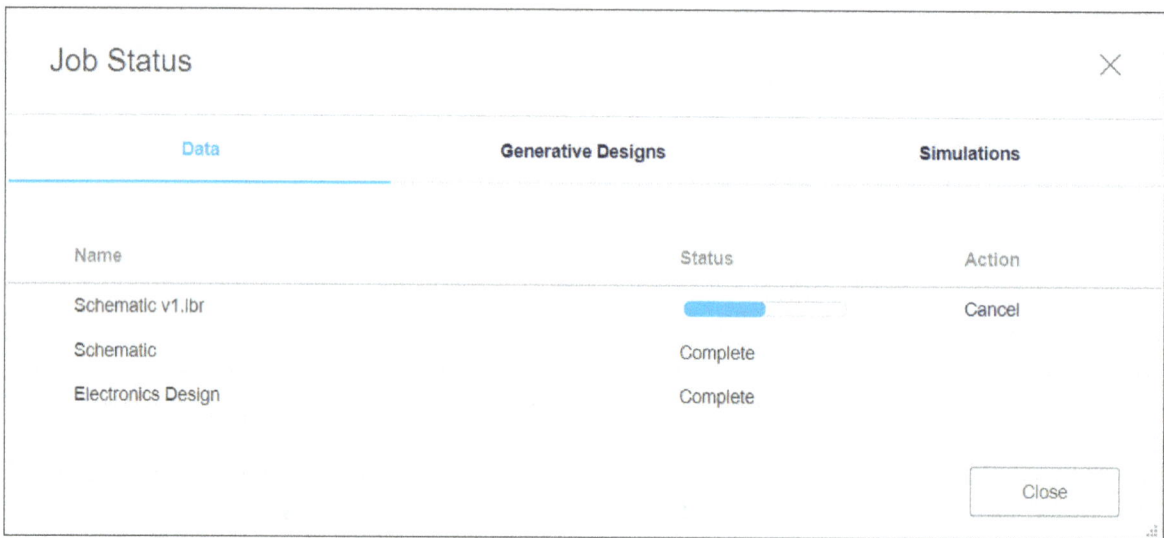

Figure-16. Job Status dialog box

- Once the status of import is complete then click on the **Open** button from the **Action** column in the dialog box. The model will be displayed; refer to Figure-17 (Creo Parametric file imported in Fusion).

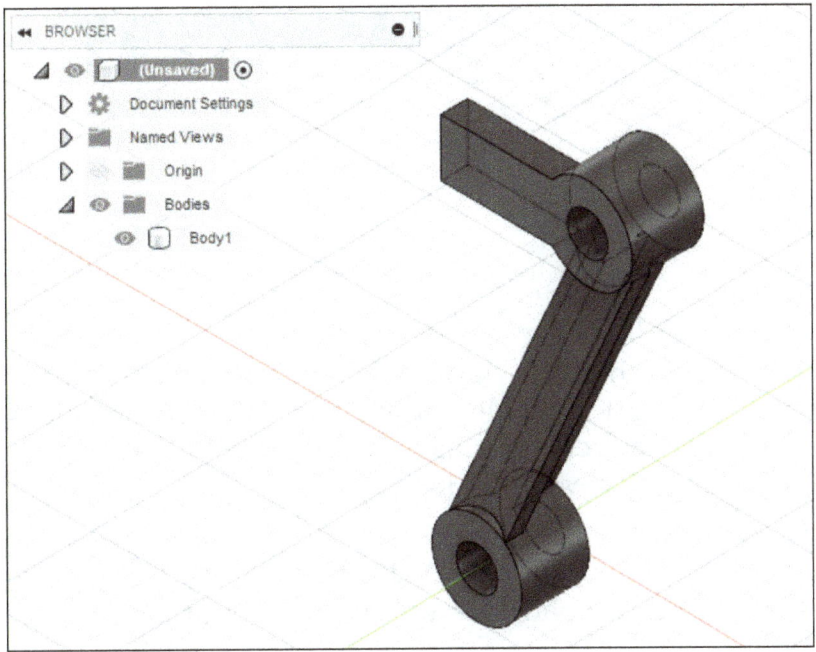

Figure-17. Model imported from Creo Parametric file

Recovering Documents

The **Recover Documents** tool is used to recover unsaved file versions which are created when software closes unexpectedly; refer to Figure-18. Note that there is an auto save feature in Autodesk Fusion which saves versions of file at a specific time intervals. If software has created the auto save version of your file then only you will be able to recover the file. By default, this time interval is 5 minutes and it can be changed in **Preferences** dialog box which will be discussed later. The procedure to use this tool is given next.

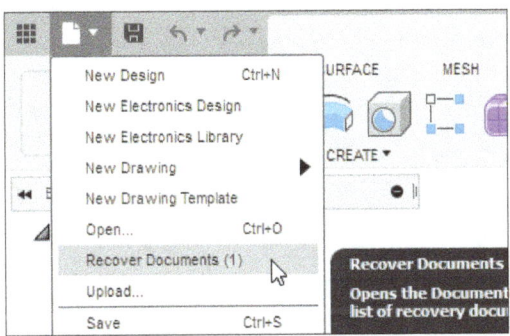

Figure-18. Recover Documents tool

- Click on the **Recover Documents** tool from the **File** menu. The **File Recovery** dialog box will be displayed where you need to select auto save version of your file; refer to Figure-19.

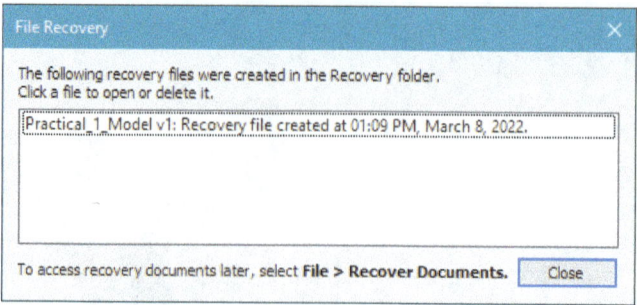

Figure-19. File Recovery dialog box

- Click on the file version that you want to be recovered and click on the **Open** button from the displayed menu. The file version will open in `Autodesk Fusion`. If the opened version is as desired then save it otherwise close it and open the other version from the `File Recovery` dialog box.

Upload

The **Upload** tool is used to upload files on cloud. The procedure to use this tool is discussed next.

- Click on **Upload** button from the `File` menu. The **Upload** dialog box will be displayed; refer to Figure-20. Click on the `Select Files` button or drag & drop the files to be uploaded in the `Drag and Drop Here` area of the dialog box.

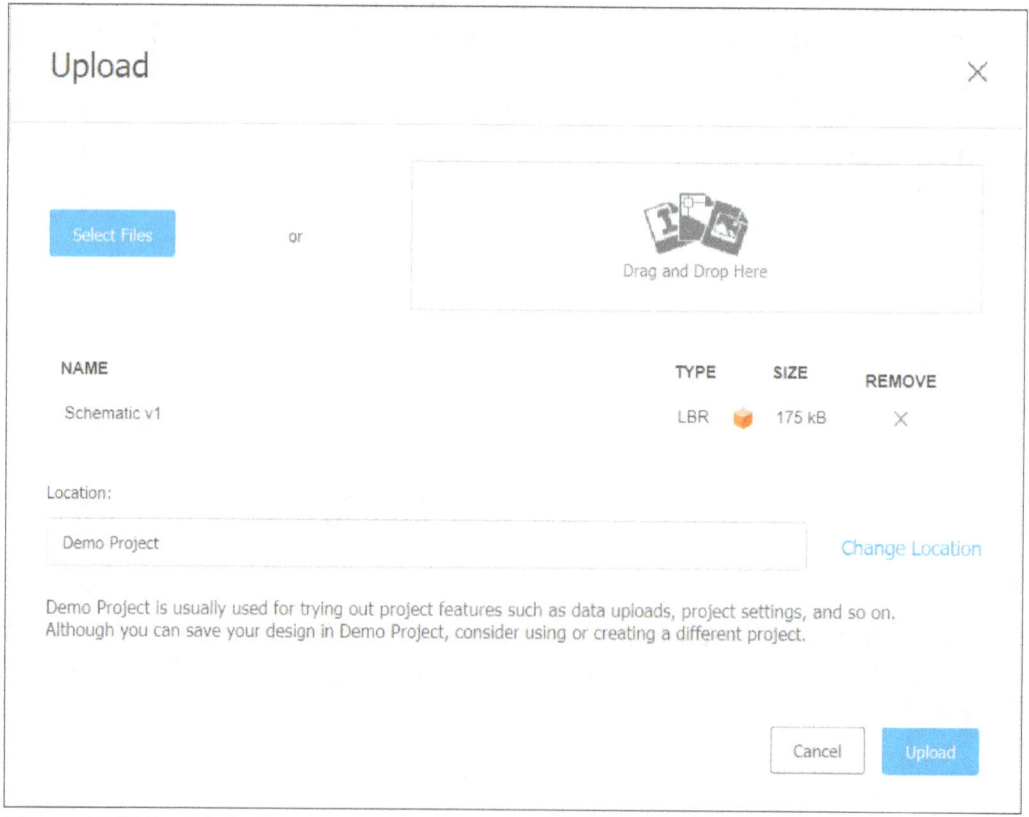

Figure-20. Upload dialog box

- If you want to change the location where files will be uploaded then click on the `Change Location` button and select desired location from the `Change Location` dialog box; refer to Figure-21.

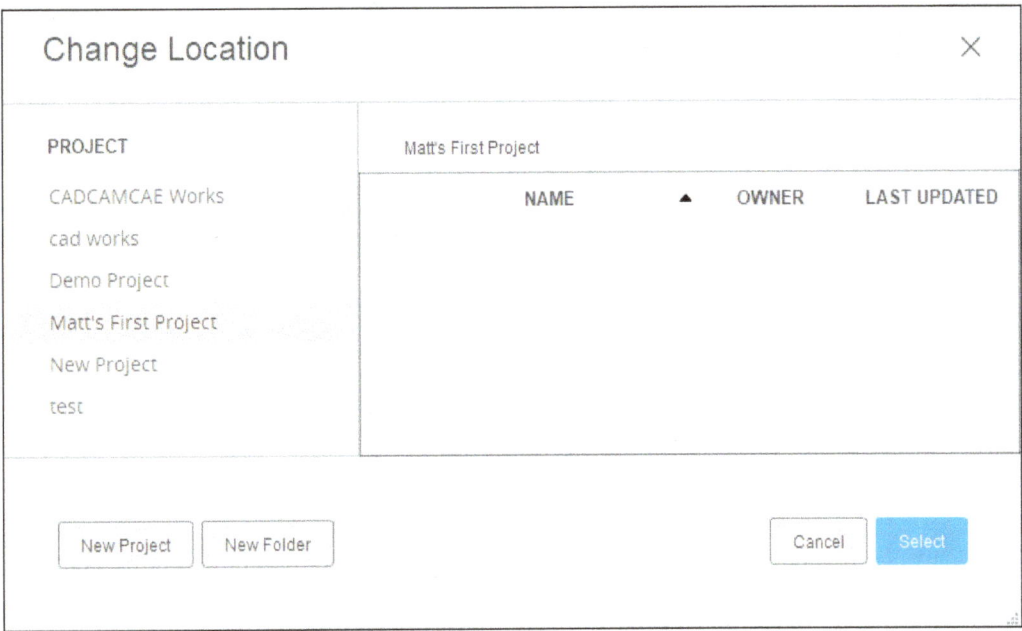

Figure-21. Change Location dialog box

- After selecting desired location, click on the **Select** button. You will return to **Upload** dialog box with updated location. Note that you can also create new folders and new locations by using the options in **Change Location** dialog box which will be discussed later in this book.
- Click on the **Select Files** button to upload files and select multiple files while holding the **CTRL** key from the **Open** dialog box displayed; refer to Figure-22.

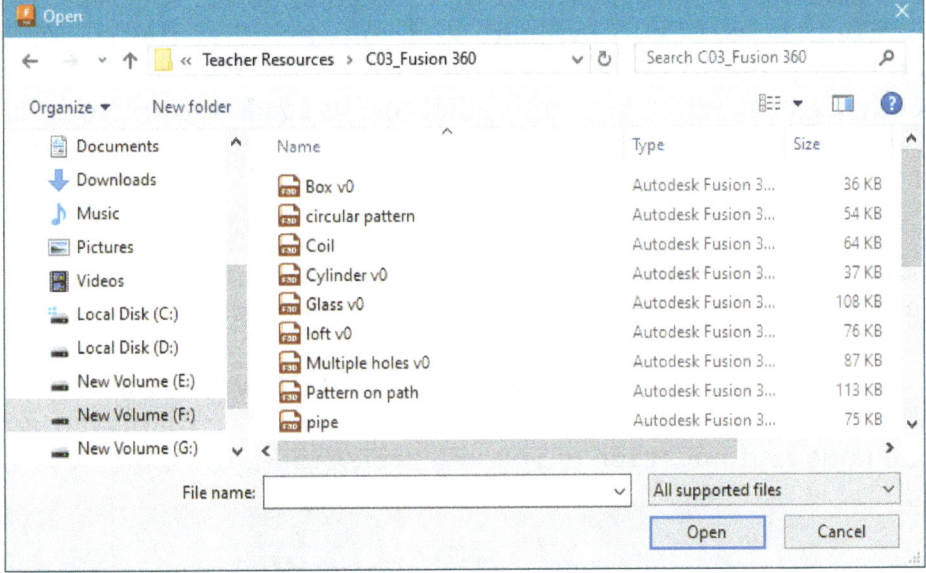

Figure-22. Open dialog box for file upload

- Click on the **Open** button from the dialog box to upload files. Or, drag the files in the dialog box and drop them. The options in the **Upload** dialog box will be updated; refer to Figure-23.

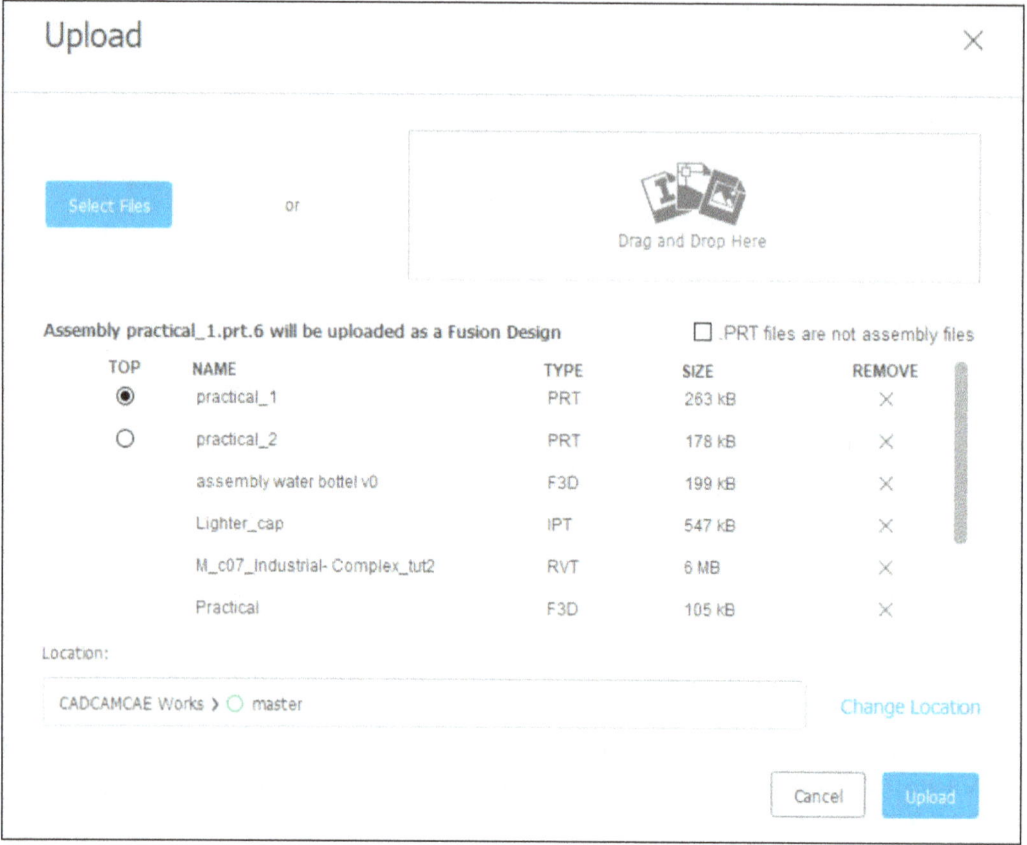

Figure-23. Updated Upload dialog box

- If you have selected a **.PRT** file of NX or Creo then select the **.PRT files are not assembly files** check box to explicitly tell software that these PRT files are not connected to any assembly.
- After specifying desired parameters, click on **Upload** button from **Upload** dialog box. Files will be uploaded and status will be displayed in **Job Status** dialog box.

Save

The **Save** tool is used to save the current file on cloud. You can press **CTRL+S** key from keyboard to save file. The procedure to save file is given next.

- Click on the **Save** tool from **File** menu or select the **Save** button from **Application bar** as shown in Figure-24. The **Save** dialog box will be displayed asking you to specify the **Name** and **Location** of your file; refer to Figure-25.

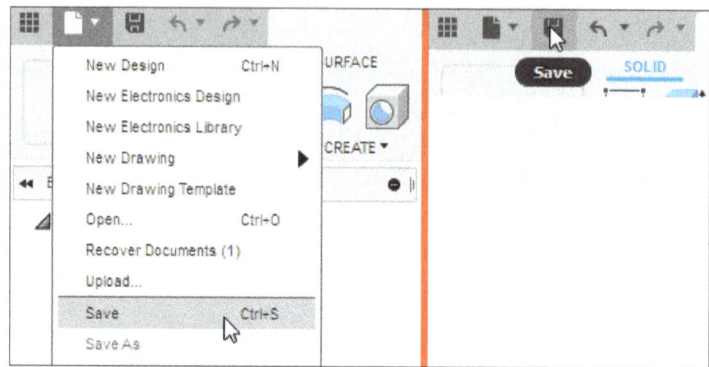

Figure-24. Save tool

Autodesk Fusion 360 PCB Black Book

1-15

Figure-25. Save dialog box

- Specify desired name in the **Name** edit box.
- Click on the down arrow next to **Location** edit box in the dialog box if you want to select a different location on cloud. A list of locations created for/by you in Autodesk Cloud will be displayed; refer to Figure-26.

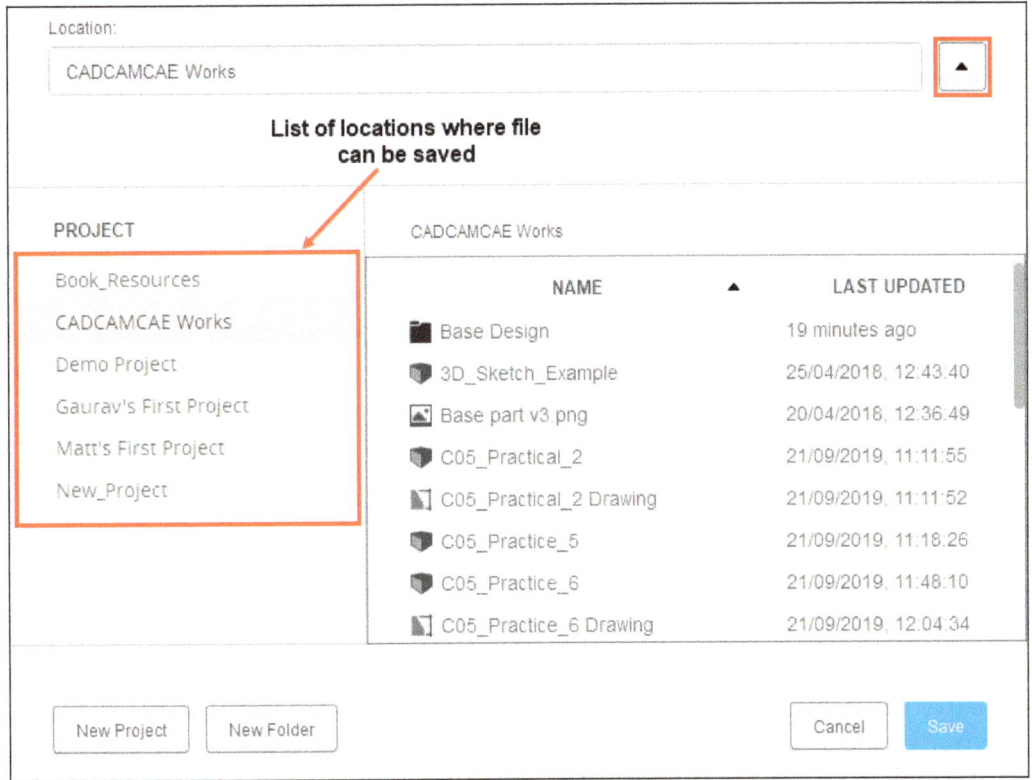

Figure-26. Locations to save files

- By default, Demo Project is added in the Project list. To add a new project, click on the **New Project** button at the bottom left corner of the dialog box. You will be asked to specify name of the new project; refer to Figure-27.
- Specify desired name for the project in displayed edit box and click anywhere in blank area of the dialog box. The new location will be created. Select this new location to save the file; refer to Figure-28.

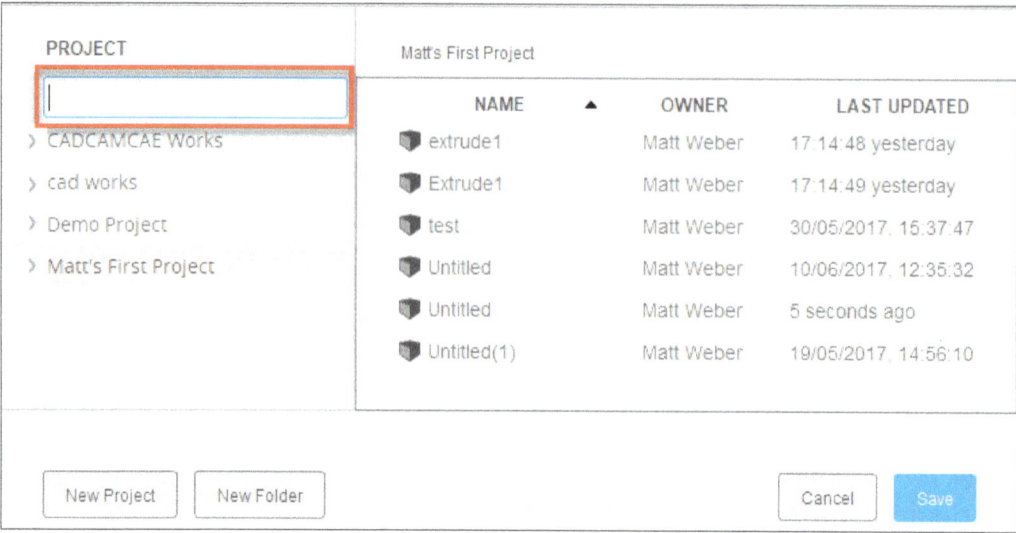

Figure-27. Specifying project name

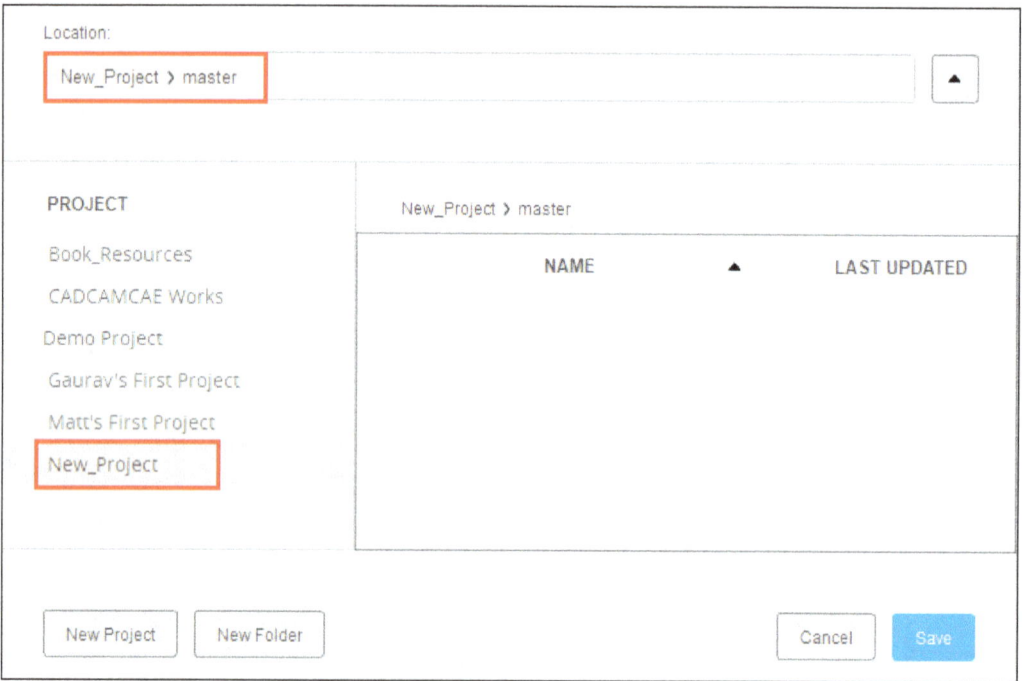

Figure-28. Location selected for saving file

- If you want to create folder in the project with desired name then click on the **New Folder** button from the dialog box. You will be asked to specify the name of the folder. Specify the name of folder and click in blank area of the dialog box.
- To save your file in the folder, double-click on folder name to enter the folder and then click on the **Save** button. The file will be saved at specified location.

Note that when next time you will use **Save** tool after making changes then the **Save** dialog box will be displayed asking you to specify description for this new version of same file; refer to Figure-29. Type a short description about what has changed in this version. Select the **Milestone** check box, type desired milestone name in the edit box, and click on the **OK** button. You will notice that file name has changed from xxxxx v0 to xxxxx v1 which describes that this is the version 1 of same file. These versions can be accessed by **Data Panel** which will be discussed later.

Autodesk Fusion 360 PCB Black Book   1-17

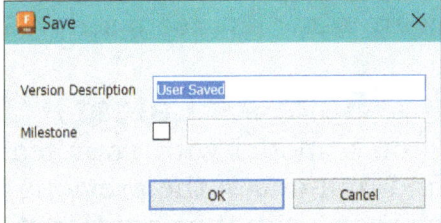

Figure-29. Save dialog box

Save As

Using the **Save As** tool, you can save the file with different name at different location. The procedure to use this tool is discussed next.

- Click on **Save As** tool from **File** menu; refer to Figure-30. The **Save As** dialog box will be displayed as shown in Figure-31.

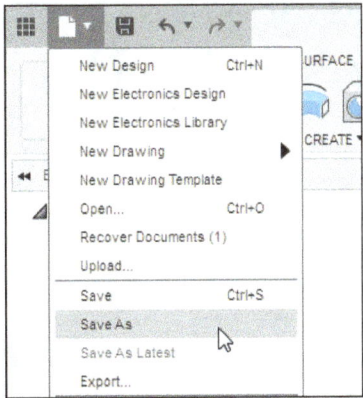

Figure-30. Save As tool

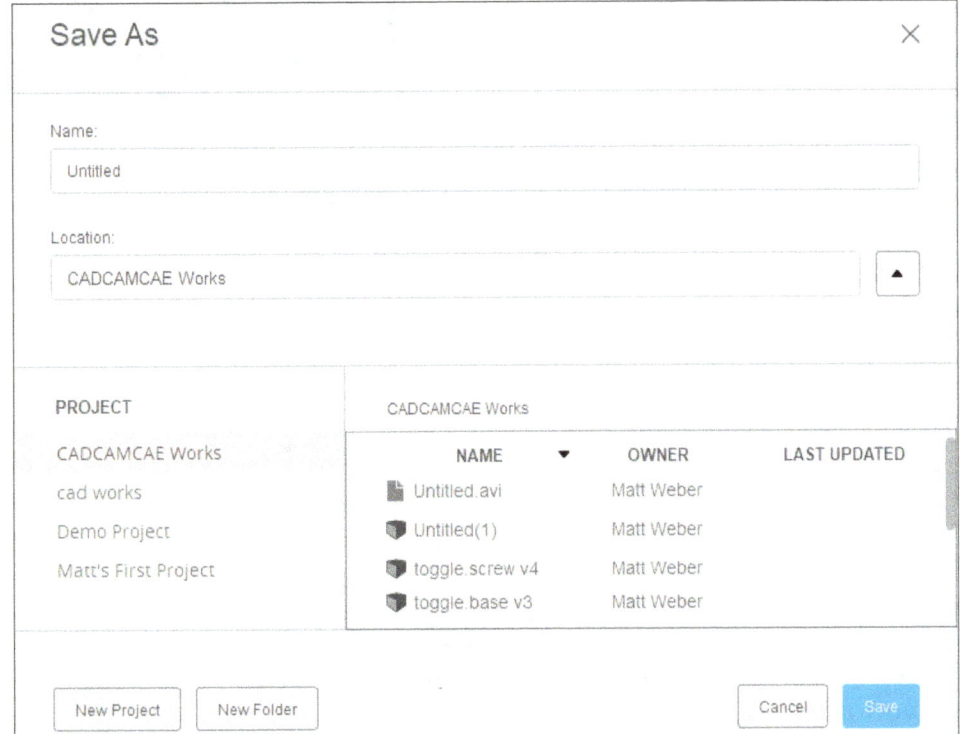

Figure-31. Save As dialog box

- Specify desired name and location for the file, and click on the **Save** button. The options in this dialog box are same as discussed in **Save** dialog box.

Save As Latest

When you are collaborating on a model with your team and after creating many versions of the model, you find that one of the previous versions is to be considered as final model for use. In such case, open the previous version file of Autodesk Fusion using **Open** tool or **Data Panel** and the **Save As Latest** tool will be active in **File** menu to save this version as latest; refer to Figure-32. On clicking the **Save As Latest** tool, an information box will be displayed showing how versions will be renumbered by making current version latest. Click on the **Continue** button from the information box. The **Save As Latest** dialog box will be displayed; refer to Figure-33. Specify desired description text about this version in **Version Description** edit box. Select the **Milestone** check box if you want to mark it as achievement of one of your design goals and click on the **OK** button from dialog box. The file will be saved on cloud.

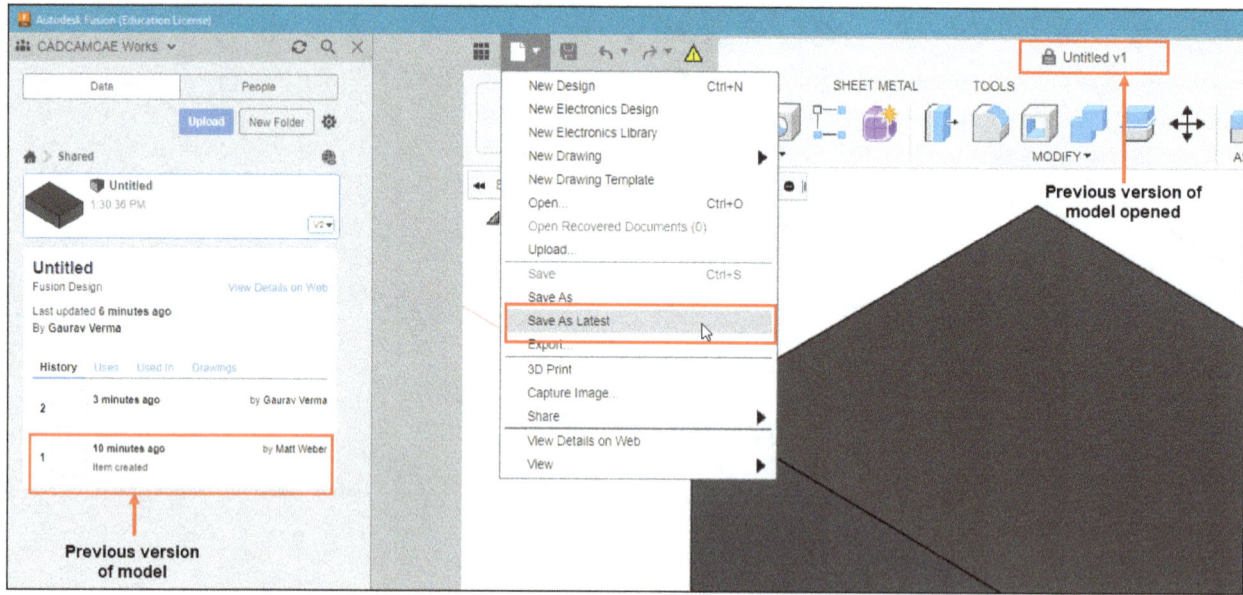

Figure-32. Save As Latest tool

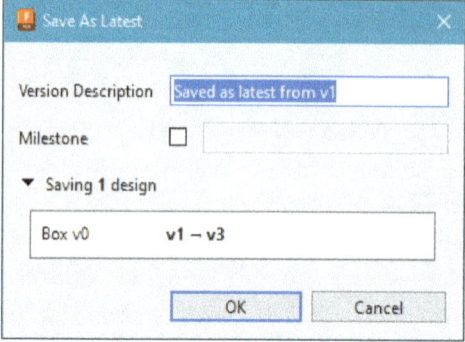

Figure-33. Save As Latest dialog box

Export

The **Export** tool is used to export the file in different formats. The procedure to use this tool is given next.

- Click on the **Export** tool from the **File** menu; refer to Figure-34. The **Export** dialog box will be displayed; refer to Figure-35.

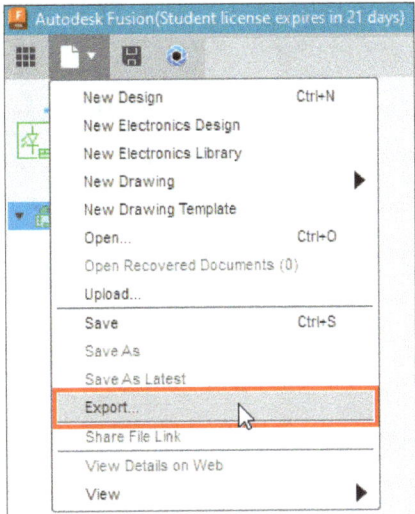

Figure-34. Export tool

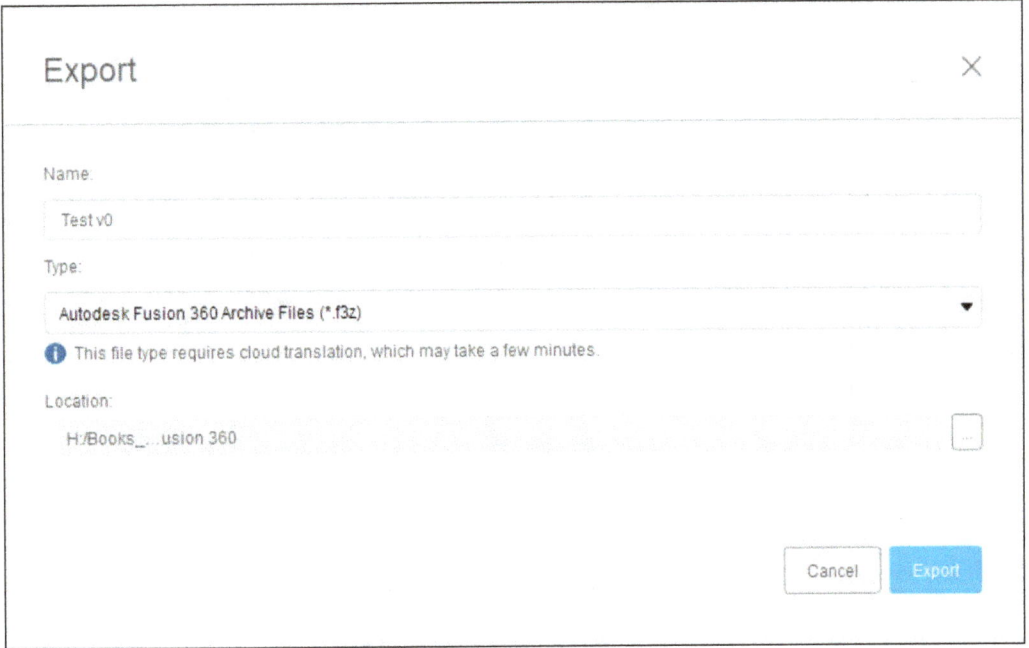

Figure-35. Export dialog box

- In the **Type** drop-down, there is currently only one format ***.f3z** in which you can export the electronic design files. Note that a cloud conversion will be required for exporting file and it may take some time. You will be able to see progress of conversion in **Job Status** dialog box.
- Specify desired name and location for the file. Note that you can save the files only in the local drive by using this dialog box.
- Click on the **Export** button from the dialog box to save the file.

Capture Image

The **Capture Image** tool is used to capture the image of model in current state. The procedure is discussed next.

- Click on the **Capture Image** tool from **File** menu; refer to Figure-36. The **Image Options** dialog box will be displayed; refer to Figure-37.

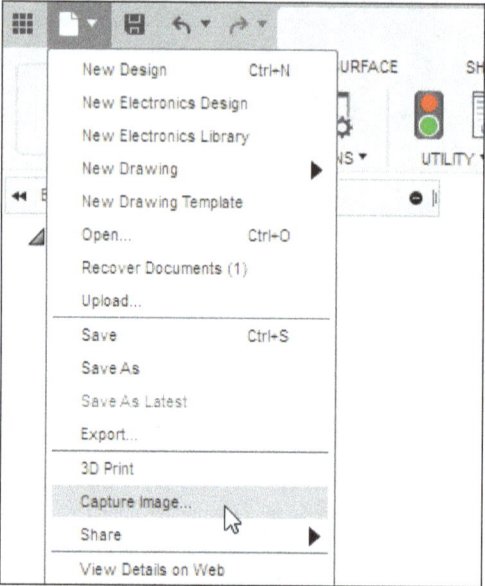

Figure-36. Capture Image tool

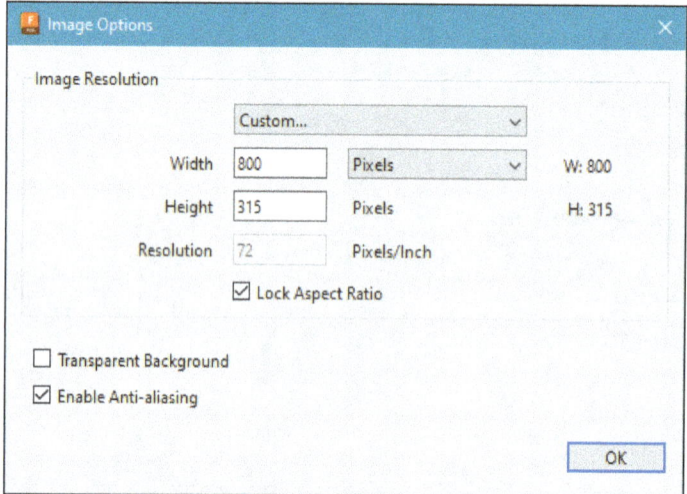

Figure-37. Image Options dialog box

- Select desired option from the drop-down in **Image Resolution** area to define size of image. All the other parameters of **Image Resolution** area will be defined automatically.
- If you want to set image resolution other than the standard one then select the **Custom** option from the drop-down and define the parameters. By default, you are asked to specify size in pixels but you can also define the image size in inches by selecting the **Inches** option from the drop-down next to **Width** edit box. Now, you will be able to define pixel density in the **Resolution** edit box manually. Clear the **Lock Aspect Ratio** check box to specify values of height and width individually.
- Select **Transparent Background** check box for setting the background of the image to be transparent.
- Select the **Enable Anti-aliasing** check box to smooth jagged lines or textures by blending the color of an edge with the color of pixels around it.
- After setting desired parameters, click on the **OK** button from **Image Options** dialog box. The **Save As** dialog box will be displayed; refer to Figure-38.

Autodesk Fusion 360 PCB Black Book   1-21

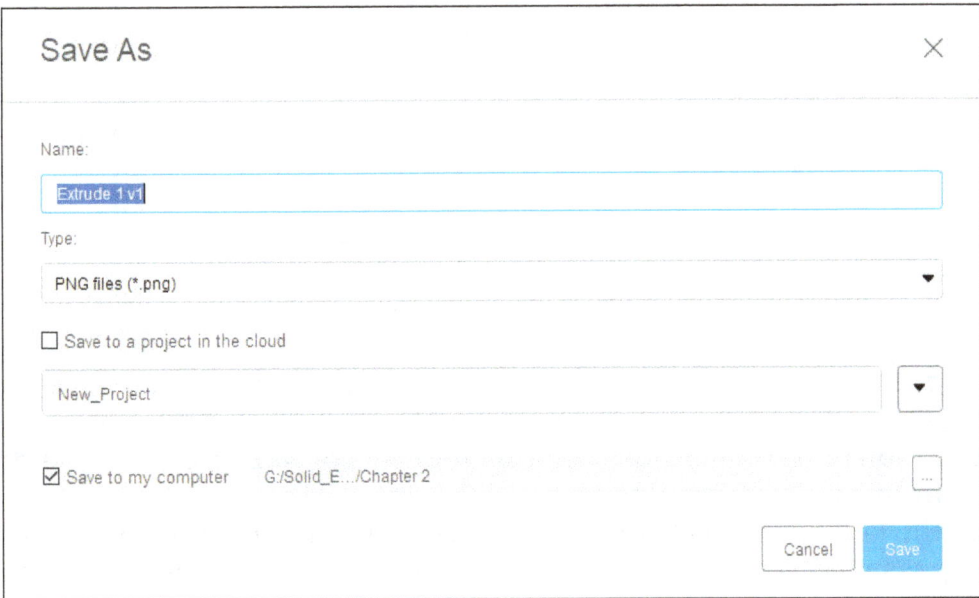

Figure-38. Save As dialog box for image capturing

- Select desired format from the **Type** drop-down for image. There are three formats available: PNG, JPG, and TIFF.
- Set the other parameters as discussed earlier and click on the **Save** button to save the image.

Sharing File Link

The **Share File Link** tool is used to share the file or project using a link. The procedure to use this tool is given next. Click on the **Share File Link** tool from **File** menu; refer to Figure-39.

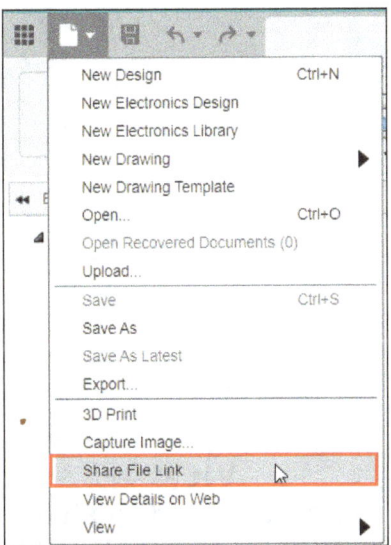

Figure-39. Share File Link option

- On selecting this tool, the **Share** dialog box will be displayed, refer to Figure-40.

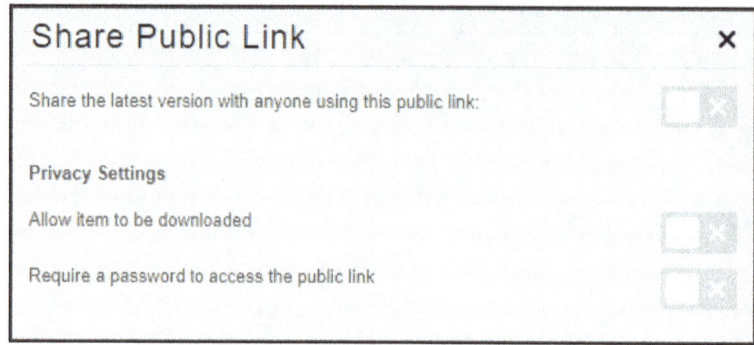

Figure-40. Share Public Link dialog box

- Click on the toggle buttons to activate or de-activate respective functions.
- After activating the sharing function, copy the link from **Copy** box and share the link via E-mail or other method with anyone. Set the **Allow item to be downloaded** toggle button to allow downloading of file.
- Activate the **Require a password to access the public link** option and specify desired password in the edit box displayed below it to secure your link with a password.

View details On Web

The **View Details On Web** tool is used to display the details of current file on Autodesk cloud in web browser; refer to Figure-41. You can use this option only after you have saved your file on Autodesk cloud.

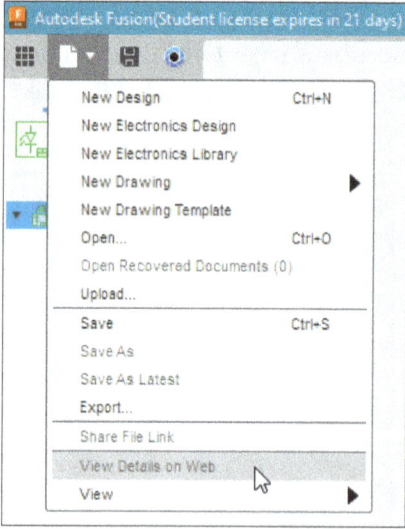

Figure-41. View Details on Web tool

View

The tools in **View** cascading menu are used to show and hide various elements of Fusion application window. The list of tools displayed in this cascading menu depend on the design file you are working with.

- Click on the **View** cascading menu from **File** menu. The tools will be displayed as per the opened file; refer to Figure-42.

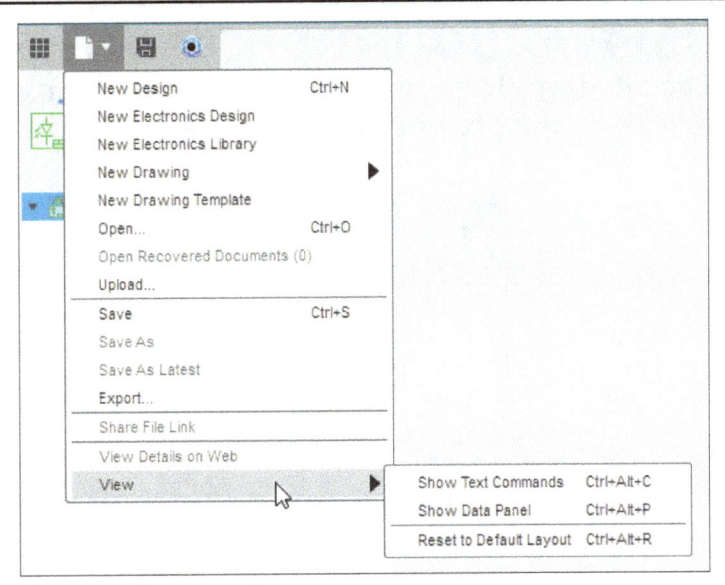

Figure-42. View cascading menu

Various tools in this cascading menu are given next.

Show/Hide Text Commands

The **Show/Hide Text Commands** tool is used to display or hide the command bar for entering various commands. You can also press **CTRL+ALT+C** key to display/hide command bar.

Show Data Panel

The **Show Data Panel** tool is used to show the **Data Panel**. If the **Data Panel** is already showing then **Hide Data Panel** tool will be displayed in place of it. This tool can also be activated by pressing **CTRL+ALT+P** together.

Reset To Default Layout

The **Reset To Default Layout** tool is used to reset user interface to default settings. This tool can also be used by pressing **CTRL+ALT+R** together.

UNDO AND REDO BUTTON

The **Undo** tool is used to revert back to condition before performing most recent action. The procedure to use this tool is discussed next.

- Click on **Undo** tool in **Application bar** or press **CTRL+Z** keys; refer to Figure-43.
- You can't undo some actions, like clicking commands on the **File** tab or saving a file.

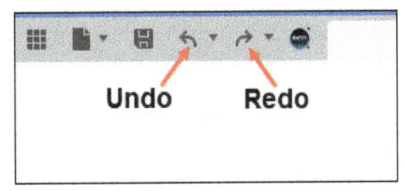

Figure-43. Undo And Redo Button

- The **Redo** tool is used to redo an action reverted by **Undo** tool. To use this tool, select the **Redo** button in **Application bar** or press **CTRL+Y** keys.
- The **Redo** button activates only after you have undone an action.

USER ACCOUNT DROP-DOWN

The tool in **User Account** drop-down are used to manage user account details and preferences for Autodesk Fusion; refer to Figure-44. The tools in this drop-down are discussed next.

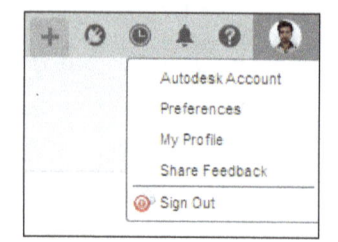

Figure-44. User Account drop down

Preferences

The **Preferences** tool is used to set the preferences for various functions of the software like you can set units, material libraries, display options, and so on. The procedure to use this tool is given next.

- Click on **Preferences** tool from **User Account** drop-down; refer to Figure-44. The **Preferences** dialog box will be displayed where you can specify various parameters for application; refer to Figure-45.

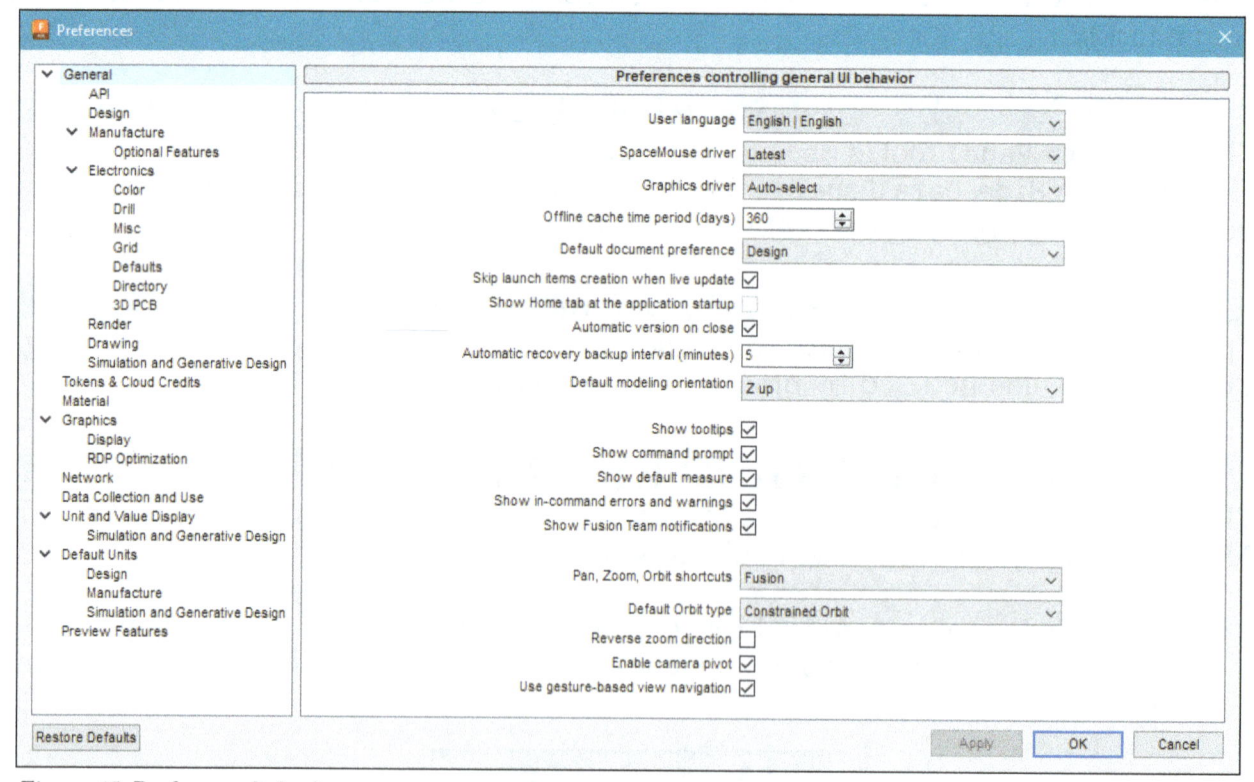

Figure-45. Preferences dialog box

General

In **General** node, you can specify the preferences for language, graphics, mouse functioning, default workspace, and so on. Click on the **General** node at the left in the dialog box to display general options. Various important options are discussed next.

- Click on the **User language** drop-down and select desired language for Autodesk Fusion interface.

Autodesk Fusion 360 PCB Black Book   1-25

- Select desired version of the SpaceMouse SDK used by the Fusion from **SpaceMouse driver** drop-down.
- Click on the **Graphics driver** drop-down to select the graphics driver to be used by Autodesk Fusion.
- **Offline cache time period (days)** edit box/spinner is used to specify the number of days up to which your documents can be in the cache memory of Autodesk Fusion before you go back to online mode of Autodesk Fusion. If you specify high value then more local memory will be used by Autodesk Fusion to save temporary copy of your documents. If you specify a low value here then system will prompt you soon to go back online. The maximum number of days specified here can be 360 and minimum number of days can be 7.
- Whenever Autodesk Fusion is updated, launch icons are created automatically in various locations. Select the **Skip launch items creation when live update** check box to skip creating new launch icons.
- Select the **Automatic version on close** check box to automatically save the newer version of file when you close Autodesk Fusion.
- Specify desired time in minute after which a recovery copy of your model will be created automatically in the **Recovery time interval (min)** edit box.
- Click on the **Default modeling orientation** drop-down and select desired direction option to define default orientation of model. In most of the CAD software, Z axis upward is the default orientation of model. To set the common orientation, select the **Z up** option from the drop-down.
- Select the **Show tooltips**, **Show command prompt**, **Show default measure**, **Show in-command errors and warnings**, and **Show Fusion Team notifications** check boxes to display respective interface elements.
- Select desired software style from the **Pan, Zoom, Orbit shortcuts** drop-down to define which software style for shortcuts of pan, zoom, and orbit should be used. If you are switching from Alias, Inventor, Solidworks, or Tinkercad to Autodesk Fusion then you can select respective option from the drop-down to use familiar shortcuts.
- Select desired option from the **Default Orbit type** drop-down to define the default orbit type for rotating model view. There are two options available for orbit; **Constrained Orbit** and **Free Orbit**.
- Select the **Reverse zoom direction** check box to reverse the zoom direction of mouse scroll and other zoom methods.
- Select the **Enable camera pivot** check box to display camera pivot point used as center for orbiting.
- Select the **Use gesture-based view navigation** check box to use touch gestures on trackpad/touchpad for performing navigation operations.

Note that by default the pivot point is placed on center of mass point of model but if you want to change the position of pivot point then right-click in blank area of canvas and select the **Set Orbit Center** option. The green point will get attached to cursor and you will be asked to define the location of orbit point. Click at desired location to place the orbit center.

API Options

- Click on the **API** option under **General** node in the left of the dialog box and set desired parameters to define default location and language of scripts/Add-Ins.

Electronics Options

- Click on the **Electronics** option from the left area in the dialog box to define parameters related to PCB design. The options will be displayed as shown in Figure-46.

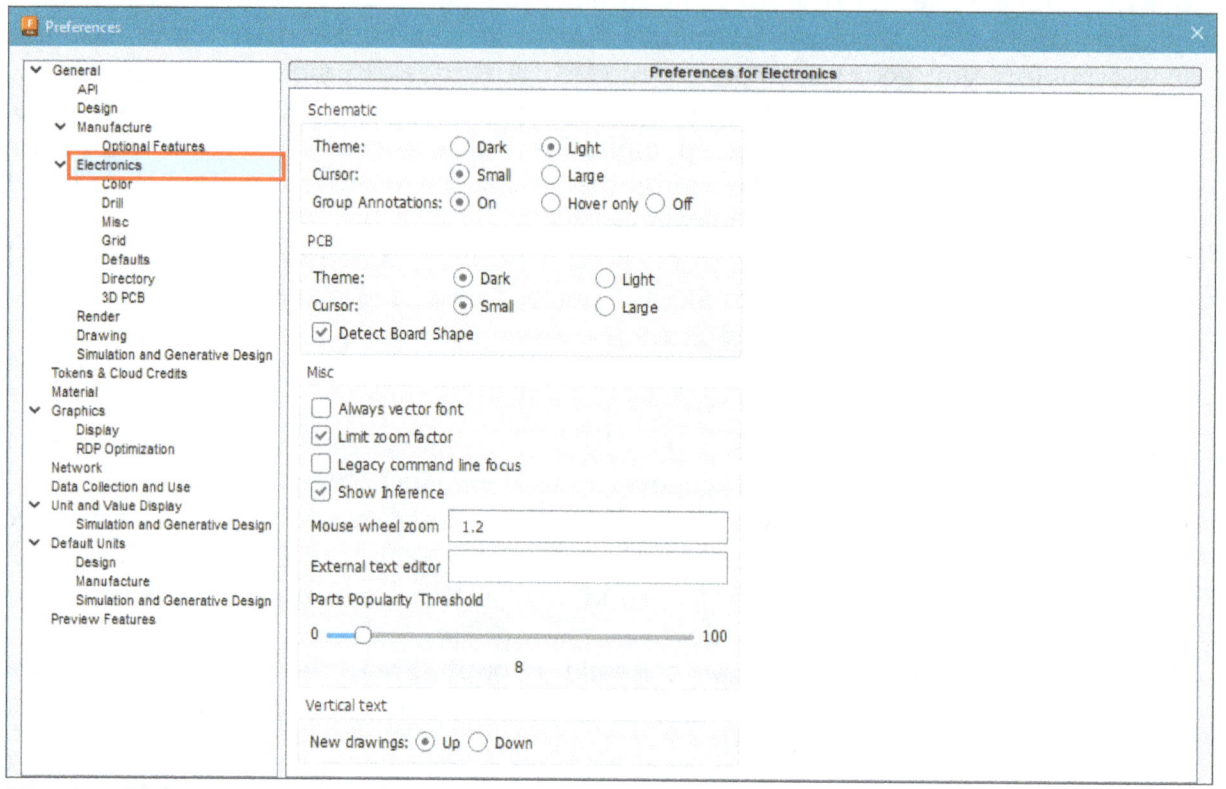

Figure-46. Electronics parameters

- Select the **Dark** option from **Theme** section if you want to display schematic design environment in dark theme or select the **Light** option to display the environment in light colored theme.
- Select the **Small** or **Large** option from **Cursor** section to define the size of cursor.
- Select the **On** option from the **Group Annotations** section to group all annotations of a component together. Select the **Hover only** option if you want to group them only when you hover the cursor on the component. Select the **Off** option if you do not want to group the annotations of component together.
- Similarly, specify the theme and cursor options for PCB environment in the **PCB** section.
- Select the **Detect Board Shape** check box to automatically detect shape of board from schematic in PCB environment.
- Select the **Always vector font** check box if you do not want to use rendered fonts in creating PCB design.
- Select the **Limit zoom factor** check box to limit the amount of zoom performed in one step.
- Select the **Legacy command line focus** check box to use Eagle command line in Fusion.
- Select the **Show Inference** check box to display soft guidelines for alignment near vertices and edges.
- Specify desired value in the **Mouse wheel zoom** edit box to define zoom ratio with respect to mouse wheel rotation.

- Specify the path of alternative application to be used for editing text in the **External text editor** edit box.
- Set desired value in **Parts Popularity Threshold** slider to define the minimum number of parts to existing the public library before it is considered display in part filtering.
- Select **Up** or **Down** radio button from **Vertical Text** section to display vertically aligned text right hand side or left hand side, respectively.
- Select the **Color** option from the left area and set colors for various elements of design in schematic and PCB designs.
- Select the **Drill** option from the left area to define list of drill sizes that can be created on PCB for assembly of components; refer to Figure-47.

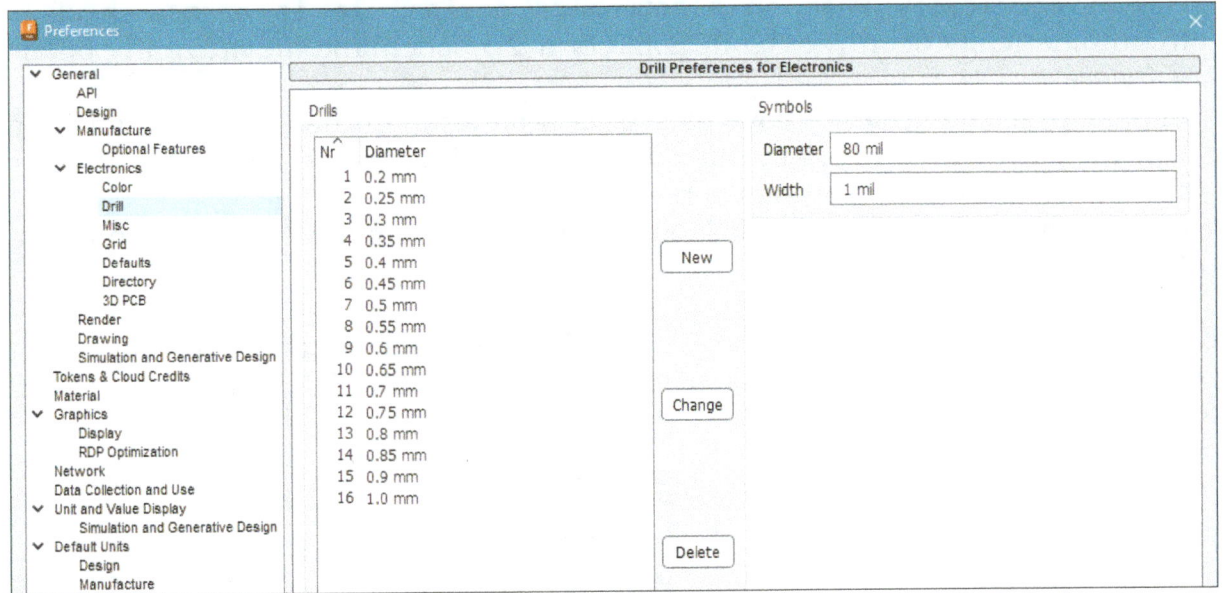

Figure-47. Drill options

- Select the **Misc** option from the left area of dialog box to activate/deactivate various parameters related to schematic or 3D PCB design file.
- Select the **Grid** option from the left area to define size of default grid for snapping. The options will be displayed as shown in Figure-48.

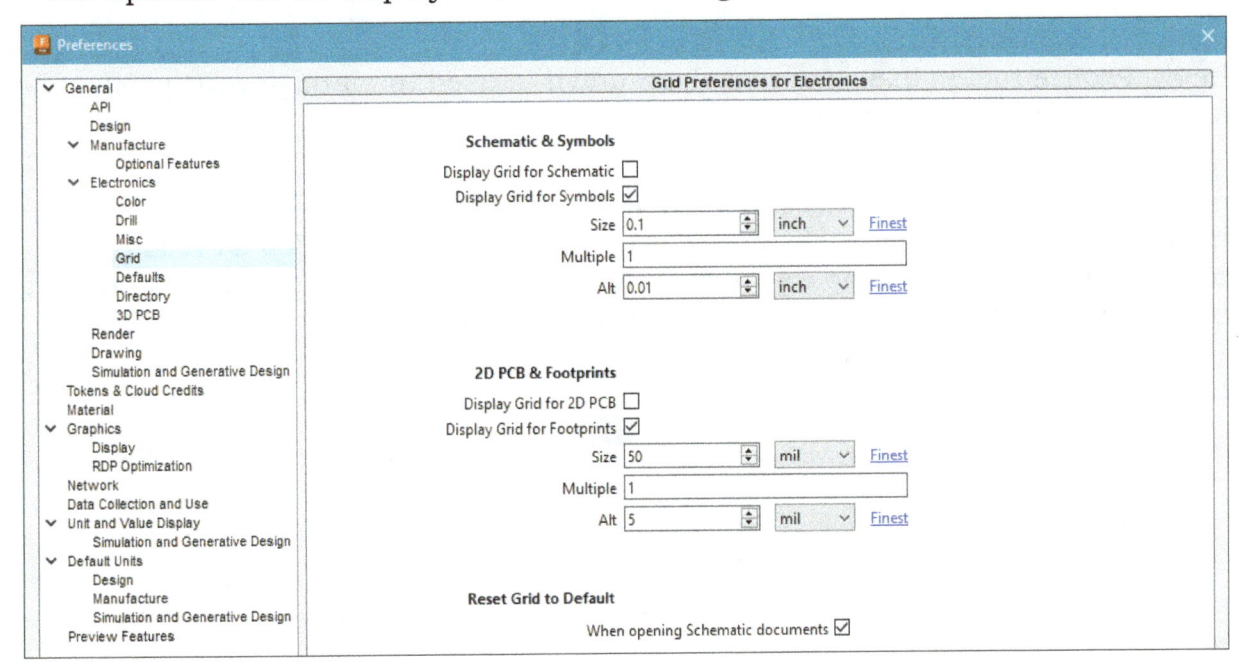

Figure-48. Grid options

- Select the **Display Grid for Schematic** check box to display in the grid lines in schematic drawing environment. Similarly, select the **Display Grid for Symbols** check box to display grid lines when editing/creating symbols.
- Set the size of each grid box in the **Size** edit box with corresponding unit system.
- Specify desired value in the **Multiple** edit box to define multiplication factor for grid sizing.
- Specify desired size value in the **Alt** edit box to define size of alternative grid displayed on pressing and holding the **ALT** key.
- Select the **Defaults** option from the left area to define the default sizes of various elements in the design.
- Select the **3D PCB** option from the left area in dialog box and select the **Open 3D PCB automatically when an electronics design file is opened** check box to automatically open the 3D PCB file of a 2D schematic file currently open in software.
- Similarly, set the other options as required on different nodes of the dialog box. We will work with options in this dialog box during rest of the book whenever a change in application settings is required.
- After setting the parameters, click on the **OK** button from the dialog box and restart Autodesk Fusion if needed.

You can learn more about all the Electronics preferences parameters by visiting the link: https://help.autodesk.com/view/fusion360/ENU/?guid=ECD-SET-PREFERENCES

Autodesk Account

The **Autodesk Account** tool in **User Account** drop-down is used to manage your Autodesk account profile in default web browser.

My Profile

The **My Profile** tool in **User Account** drop-down is used to manage your project data stored on cloud via web browser.

Work Offline/Online

The **Work Offline/Online** toggle button is displayed on clicking **Job Status** button at the left of **User Account** drop-down; refer to Figure-49. Click on this button to toggle between offline and online mode of Autodesk Fusion.

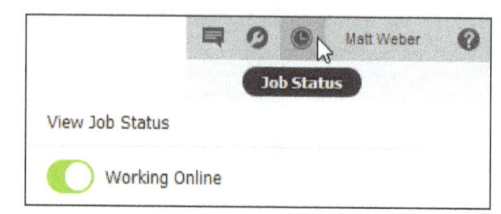

Figure-49. Work Offline Online

Extensions

Click on the **Extensions** button to display the list of current available extensions in the **Extension Manager** dialog box for adding more functionality to Autodesk Fusion; refer to Figure-50. If you want to use these extensions then select the extension to be used and click on **Purchase** button or activate button if available.

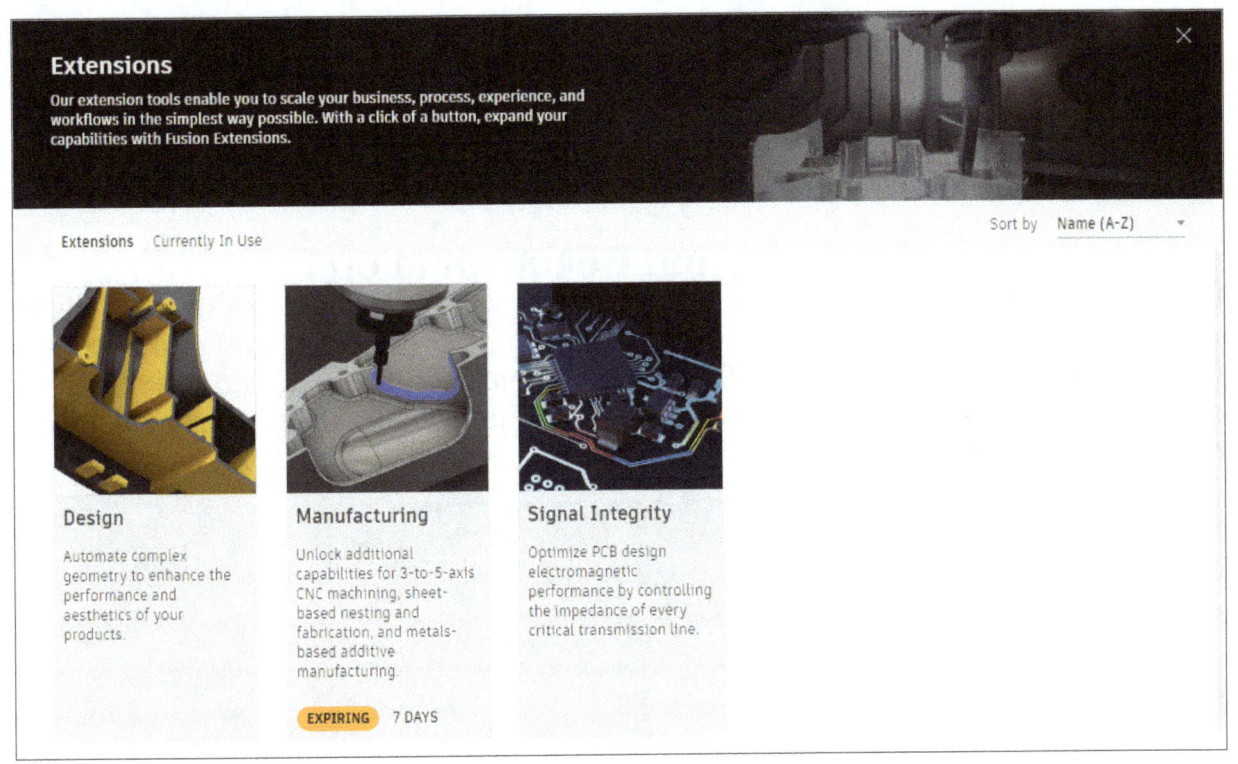

Figure-50. Extension Manager dialog box

HELP DROP-DOWN

When you face any problem working with Autodesk Fusion, the tools in **Help** drop-down can be useful; refer to Figure-51. Tools in the drop-down are given next.

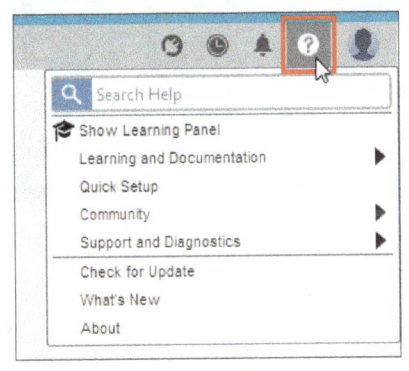

Figure-51. Help drop down

Search Help Box

When you are facing any problem regarding tools and terms in the software, then you need to type keywords for your problem in **Search Box** and press **ENTER** to find a suitable solution to your problems; refer to Figure-52.

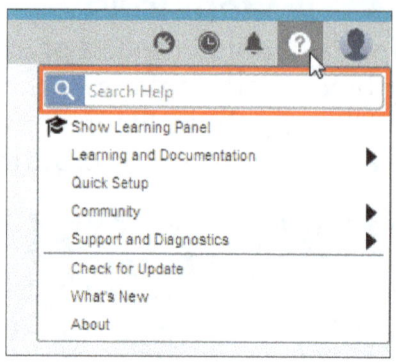

Figure-52. Search box

Learning and Documentation

The tools in the **Learning and Documentation** cascading menu are used to display video tutorials, help files, and Fusion api related content in web browser; refer to Figure-53. On clicking the **Self-Paced Learning** button in the **Learning and Documentation** cascading menu of **Help** menu, the website of Autodesk Fusion will be displayed; refer to Figure-54.

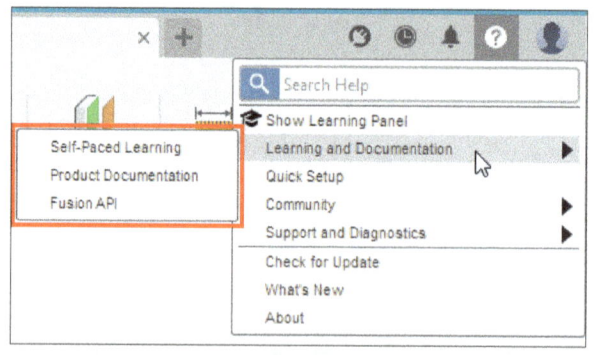

Figure-53. Learning and Documentation

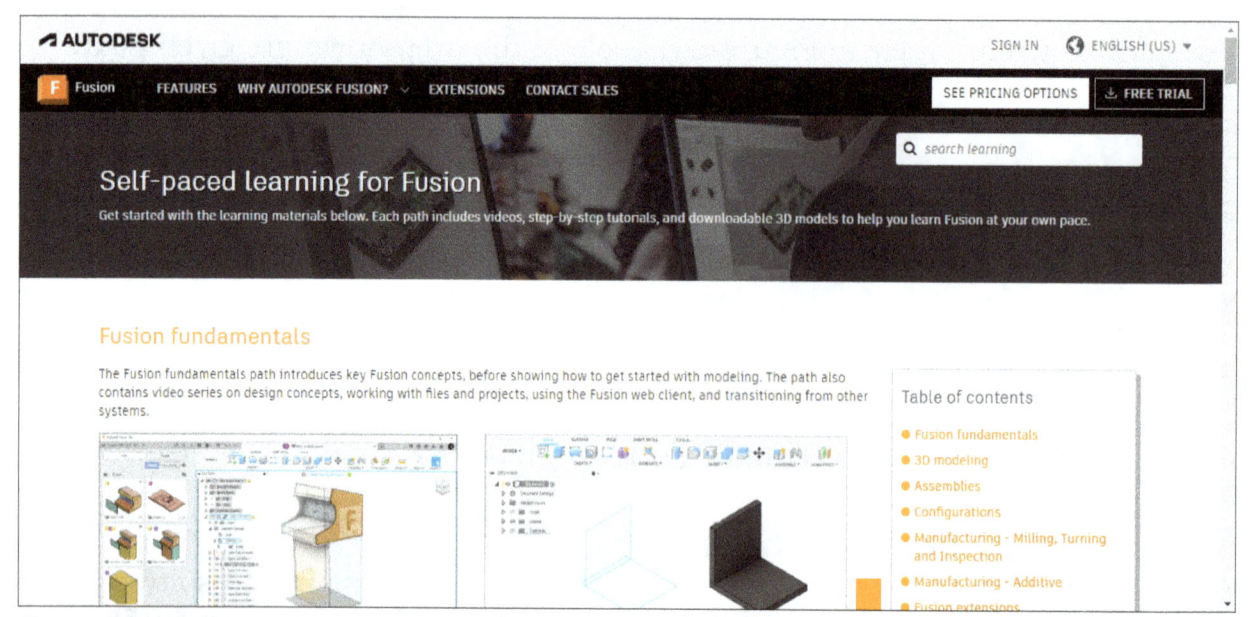

Figure-54. Fusion 360 Learning Website

Similarly, you can use **Product Documentation** and **Fusion API** tools to learn about Fusion interface, design methodology, and APIs.

Quick Setup

When you are new to Autodesk Fusion then this tool will help you to understand this software terminology and basic settings.

- Click on **Quick Setup** tool from **Help** menu. The **QUICK SETUP** dialog box will be displayed; refer to Figure-55.

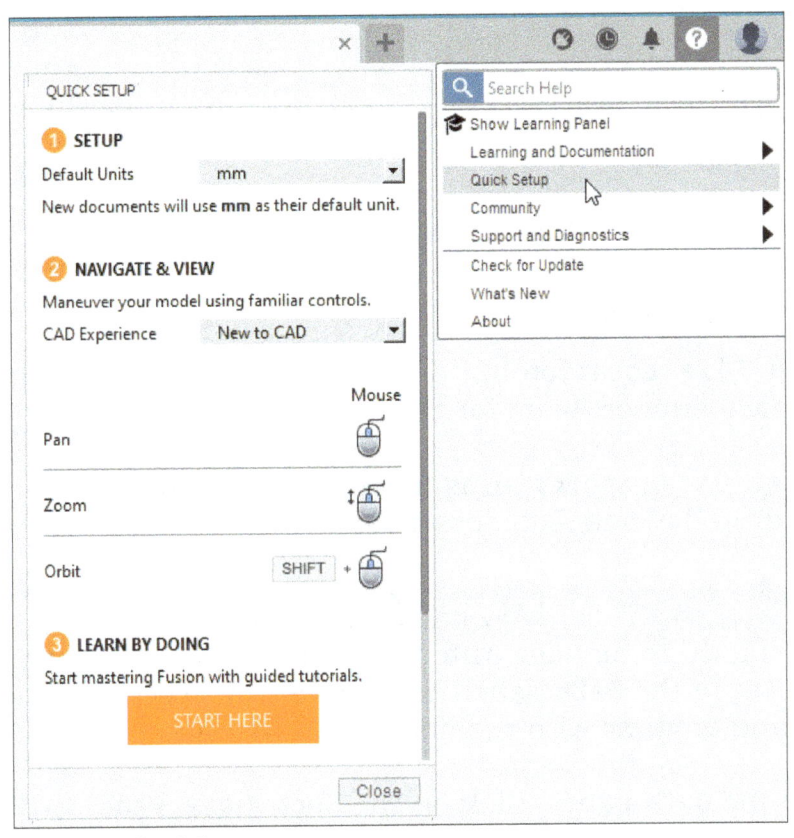

Figure-55. Quick Setup

- Set desired unit in the **Default Units** drop-down.
- The navigation functions of mouse can be set as per Autodesk Inventor, SolidWorks or Alias software by selecting respective option from the **CAD Experience** drop-down.
- If you are new to CAD then you can select the **New to CAD** option in **Cad Experience** drop-down. The mouse will function as per Autodesk Fusion.

Similarly, you can use **Community Forum, Feedback Hub, Gallery, Roadmap, Blog** tool to get help from Autodesk Fusion users or provide feedback.

Support and Diagnostics

When you are having a problem in Fusion and want **Technical support** team to look at software problem data, then you can create the log files and send it to the support team of Autodesk.

- To create log files, select **Diagnostic Log Files** tool under the **Support and Diagnostics** cascading menu in **Help** drop-down. The **Diagnostic Log Files** dialog box will be displayed; refer to Figure-56.

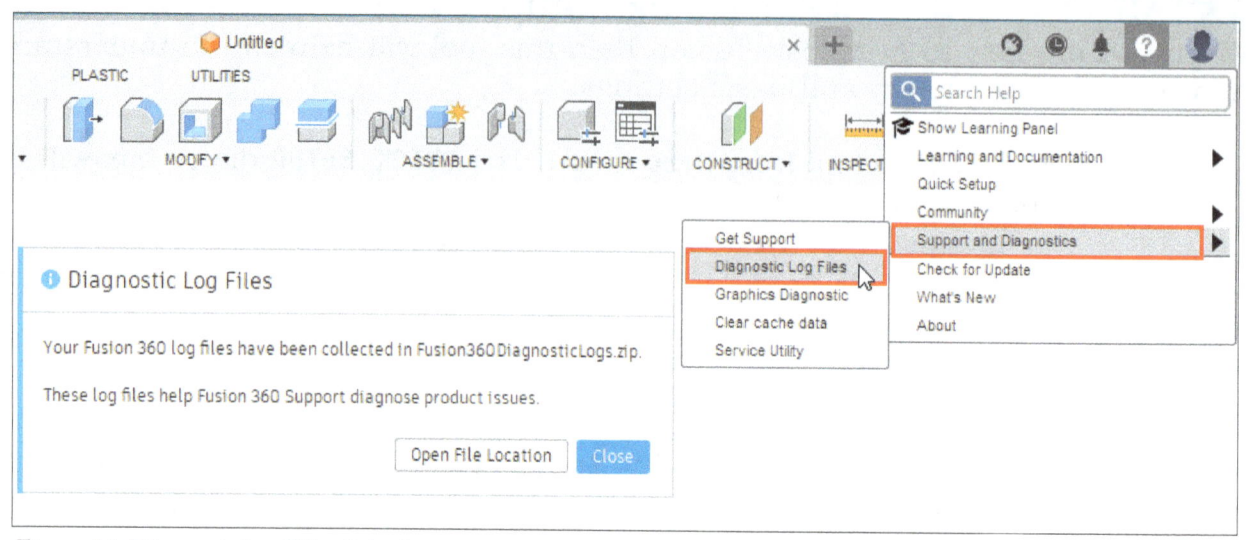

Figure-56. Diagnostic Log Files dialog box

- Click on **Open File Location** button and select the log file which you want to send to technical support team for the solution of problem.

Similarly, you can use **Graphics Diagnostic**, **Clear user cache data**, and other tools from **Help** drop-down in Autodesk Fusion.

DATA PANEL

The **Data** panel is used to manage Autodesk Fusion projects. The projects you save in cloud are shown in the **Data** panel. You can easily access the saved project files from anywhere and anytime with the help of internet access.

To show or hide the **Data** panel, click on the **Show Data Panel** or **Hide Data Panel** button at upper left corner of the Fusion window; refer to Figure-57. A box will be displayed with files of current project. In this box, there are two sections; **Data** and **People**.

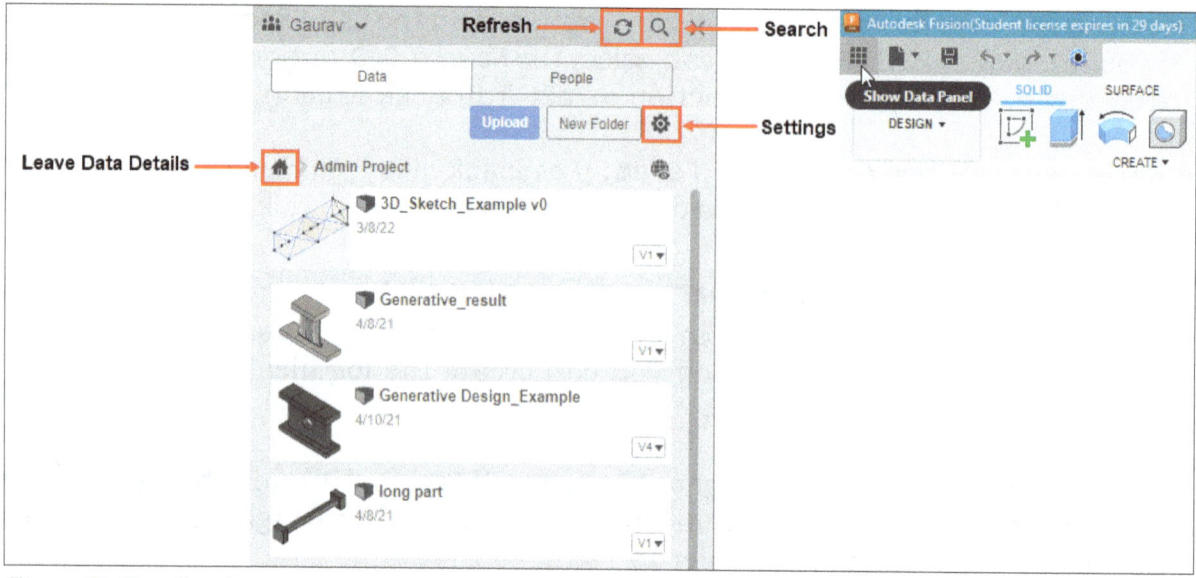

Figure-57. Data Panel

Autodesk Fusion 360 PCB Black Book 1-33

In **Data** section, the files which you have saved earlier in current project will be displayed. You can open any file by double-clicking on it.

In **People** section, the information of users will be displayed who have access to the files of current project; refer to Figure-58.

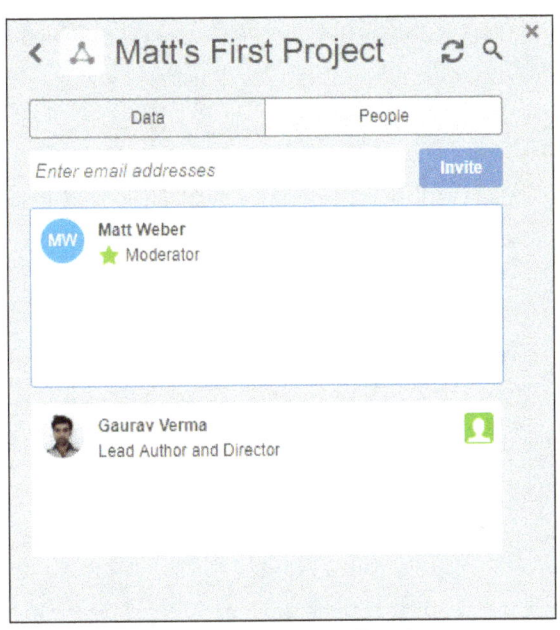

Figure-58. People Section In Data Panel

Working in a Team

Using the tools in **User** drop-down of **Data Panel**, you can create a new team or work in other's team to create designs in collaboration with others; refer to Figure-59. Various tools of this drop-down are discussed next.

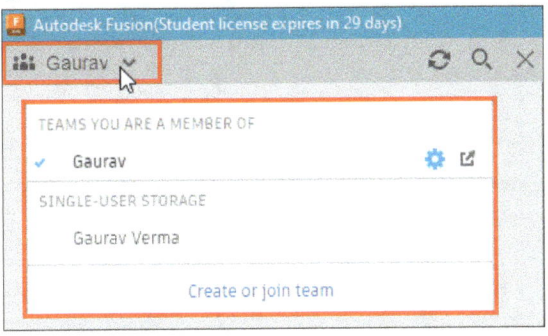

Figure-59. User drop-down

Creating a Team

- To create your own team for a given project, click on the **Create or join team** tool from the drop-down. An information box will be displayed about Fusion Team; refer to Figure-60. Note that this information box will be displayed for only one time when you are creating team using educational version of software as Educational version can create only 1 team for students & educators.

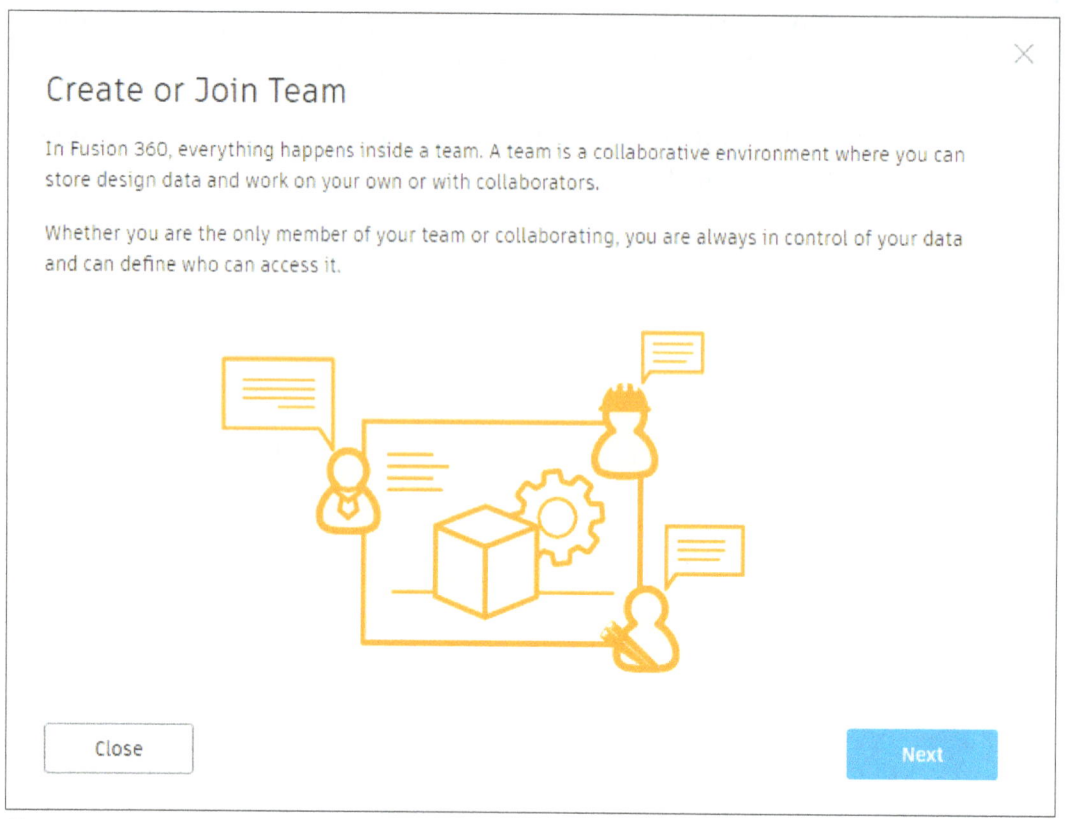

Figure-60. Create or join team

- Click on the **Next** button from the information box. The options to create or join teams will be displayed in a dialog box; refer to Figure-61.

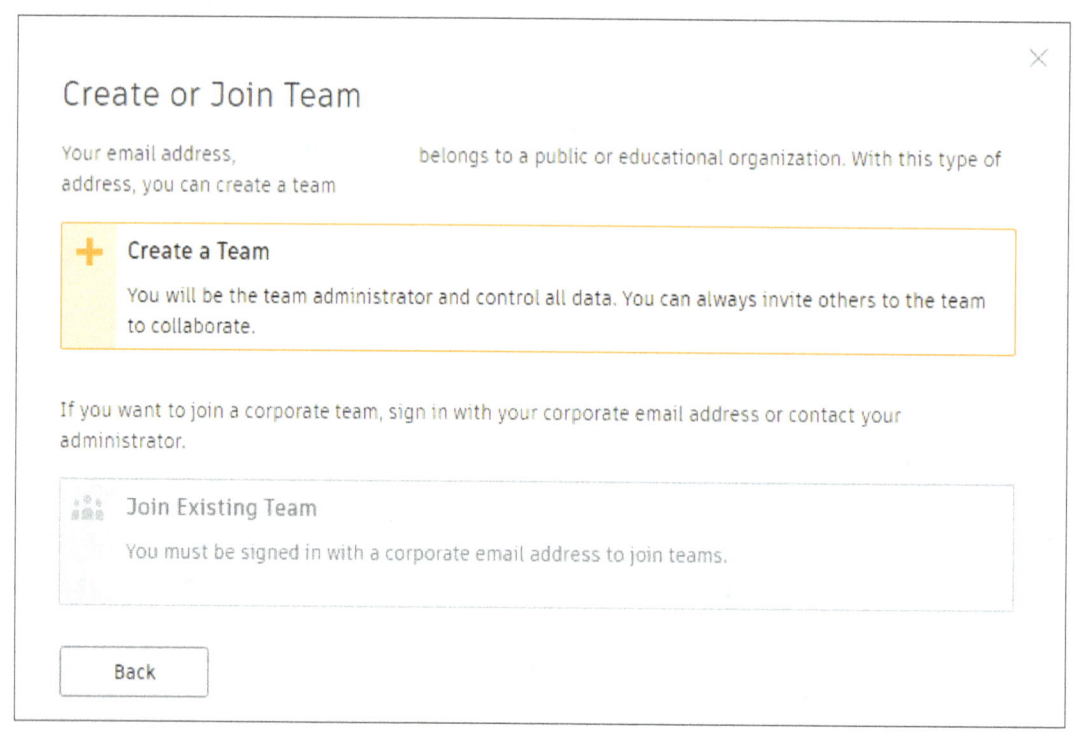

Figure-61. Create a team or join existing team options

- Click on the **Create a Team** option from the dialog box displayed. You will be asked to specify name of the team in next page of dialog box; refer to Figure-62.

Autodesk Fusion 360 PCB Black Book

Figure-62. Specifying name of team

- Specify desired name for the team to be created and click on the **Next** button. The next page in dialog box will ask you whether to make the team discoverable on cloud server or make it private.
- Select desired radio button from the dialog box and click on the **Create** button. A message box will be displayed tell you that a team with your specified name is created. Click on the **Go to team** button from the information box. Name of the team will be added in the **Data Panel**; refer to Figure-63.

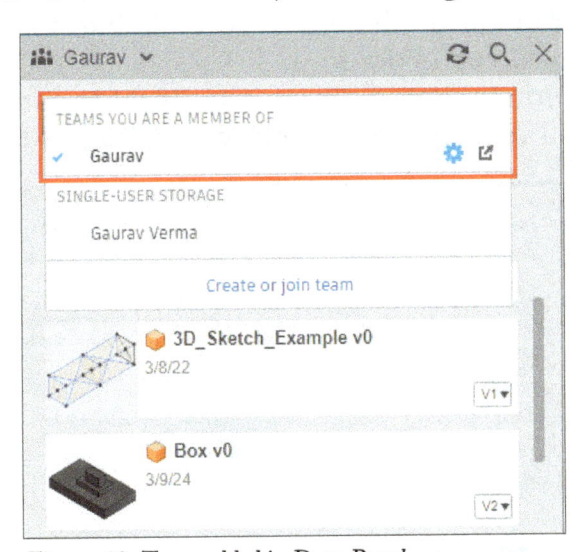

Figure-63. Team added in Data Panel

- Click on the **Open Administrator Console on the Web** button (similar to common settings button icon) displayed next to name of the team. The administration page for team will be displayed in the default web browser; refer to Figure-64.

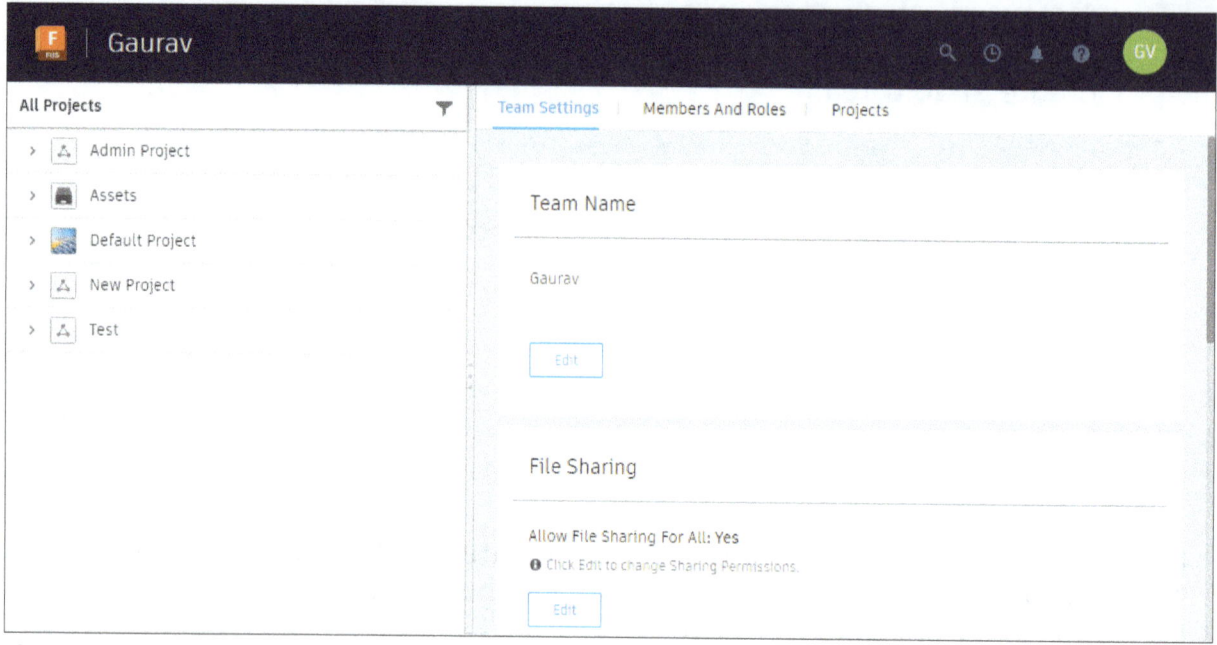

Figure-64. Administration page for team

- Specify desired parameters in this page to define settings related to file sharing, access, and collaboration. Note that you can set auto-approval settings for specified domains by using the **Add domain** button on this page. You will be asked to specify the text in email ID displayed after @ like for matt@cadcamcaework.com the `cadcamcaework.com` is domain.
- To add members in your team, click on the **Members and Roles** tab in the web page. The options will be displayed as shown in Figure-65.

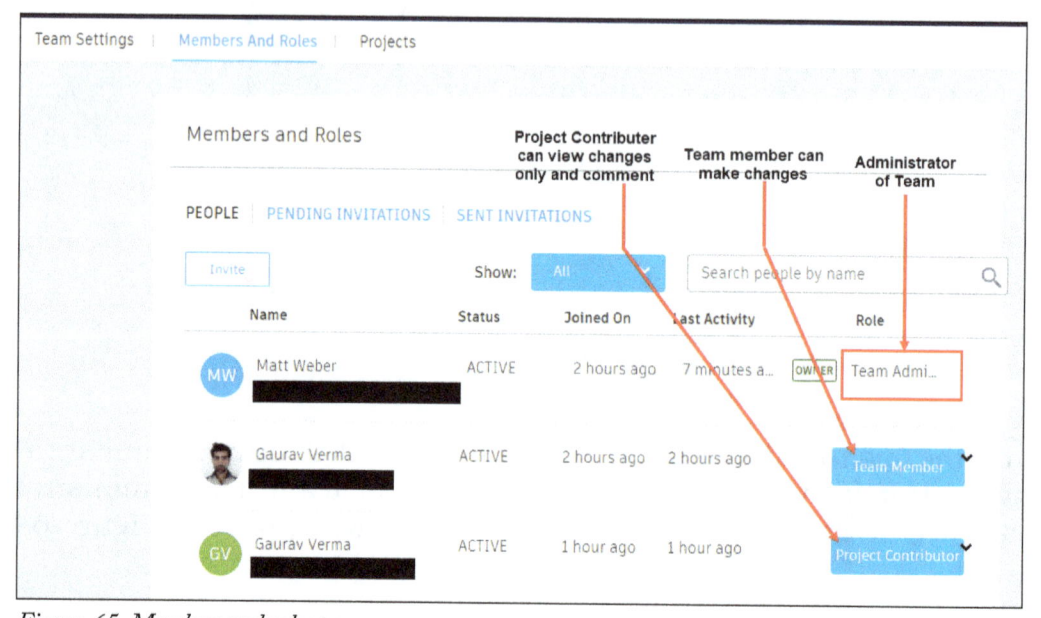

Figure-65. Members and roles page

- To invite a new member, click on the **Invite** button on this page. The **Invite Team Members** dialog box will be displayed; refer to Figure-66.

Figure-66. Invite Team Members dialog box

- Specify the email id's of team members separated by comma (,) in the `Enter email addresses` edit box and click on the `Invite` button. Once they accept invitation on their email ids, they will become members of the team. You can later assign the roles to members as desired using the `Administrator` page. You can now logout and close the browser to exit the admin panel of Fusion team.

After creating a team, make sure you are log-in by team name in the `Data Panel` and create a new project using `New Project` button in the `Data Panel`. This new project will be shared with all the team members for modifications based on roles assigned to them; refer to Figure-67.

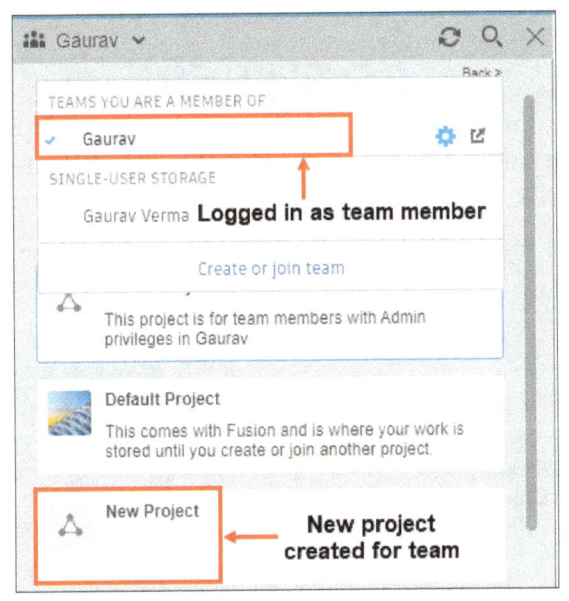

Figure-67. Sharing project with team

Cloud Account of User

To open the cloud account of user, you need to click on the `Open on the Web` button next to the name of user in `Data Panel`; refer to Figure-68. On selecting this button, details of user's cloud storage will be displayed in web browser. You can check and modify the files as desired. Note that you can delete any file available in Cloud account using the tools in web browser which is not possible in `Data Panel`; refer to Figure-69.

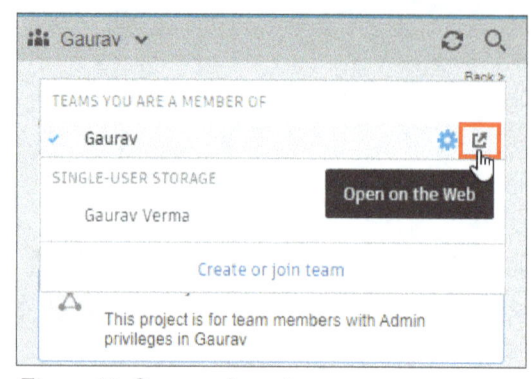

Figure-68. Open on the web

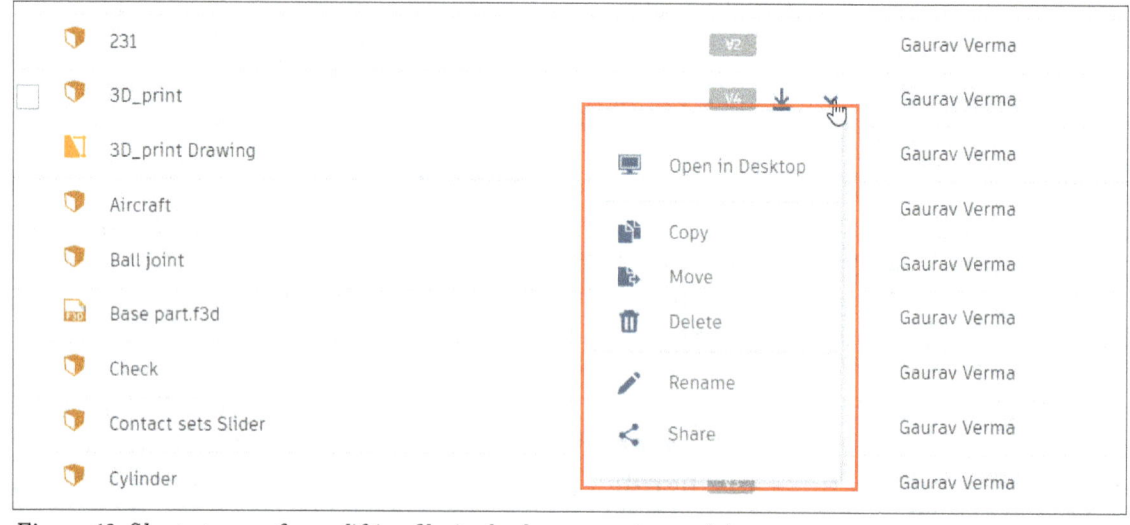

Figure-69. Shortcut menu for modifying files in cloud account using web browser

BROWSER

The **BROWSER** presents an organized view of your design steps in a tree like structure on the left of the Fusion screen; refer to Figure-70. When you select a feature or component in the **BROWSER**, it is also highlighted in the graphics window.

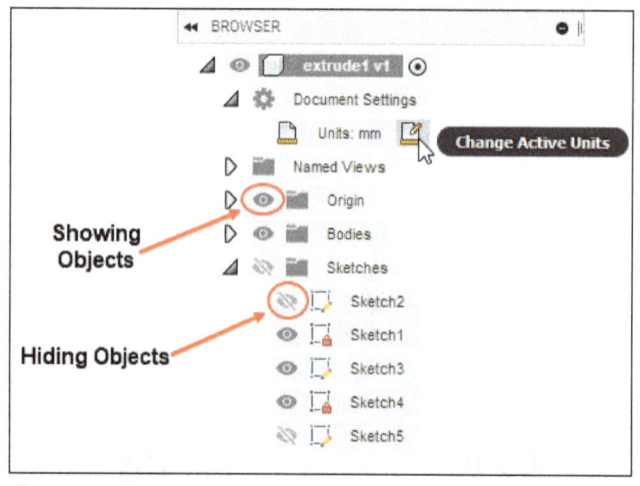

Figure-70. Browser

- Click on the **Eye** button to toggle the visibility of respective object.
- If you select any object in **BROWSER** then it is highlighted in blue color which indicates that the respective object is active for various operations.

Autodesk Fusion 360 PCB Black Book 1-39

- To change the units of the design or sketch, select the **Change Active Units** button as highlighted in the above figure. The **CHANGE ACTIVE UNITS** dialog box will be displayed. Select desired unit and click on the **OK** button.

CUSTOMIZING TOOLBAR AND MARKING MENU

In Autodesk Fusion, you can add or remove any tool in the displaying panels of **Toolbar**. You can also add or remove tool from right-click marking menu. These operations are discussed next.

Adding Tools to Panel

- Click on the **Three dot** button next to tool to be added in the panel from the drop-down of **Toolbar**. The shortcut menu to customize toolbar will be displayed; refer to Figure-71.

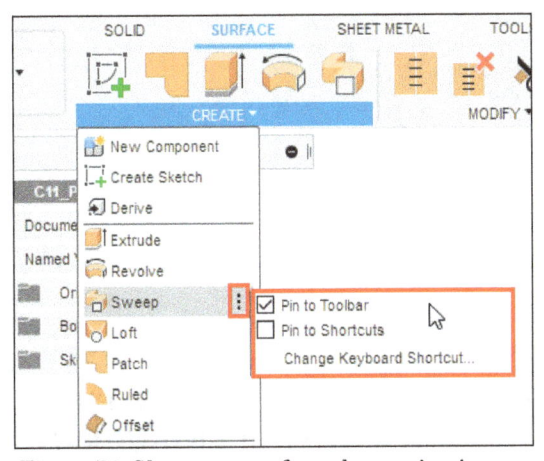

Figure-71. Shortcut menu for tool customization

- Select the **Pin to Toolbar** check box from the shortcut menu. The tool will be added in the panel.

Assigning Shortcut Key

- Click on the **Change Keyboard Shortcut** option from the shortcut menu as shown in Figure-71. The **Change Keyboard Shortcut** dialog box will be displayed; refer to Figure-72.

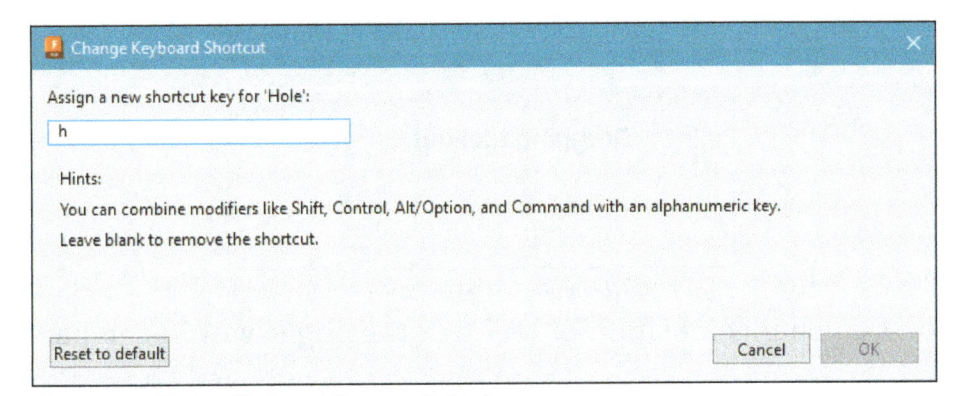

Figure-72. Change Keyboard Shortcut dialog box

- Use desired combine modifier like **SHIFT**, **CTRL**, or **ALT** and then press desired alphanumeric key to create a shortcut button.

- If you want to remove shortcut key then use **Backspace** to delete shortcut key from edit box. Click on the **OK** button from the dialog box to apply the settings.

Adding or Removing tool from Shortcut menu/Marking menu

- Select the **Pin to Shortcuts** check box from the shortcut menu for tool that you want to add in right-click shortcut menu/marking menu; refer to Figure-73.

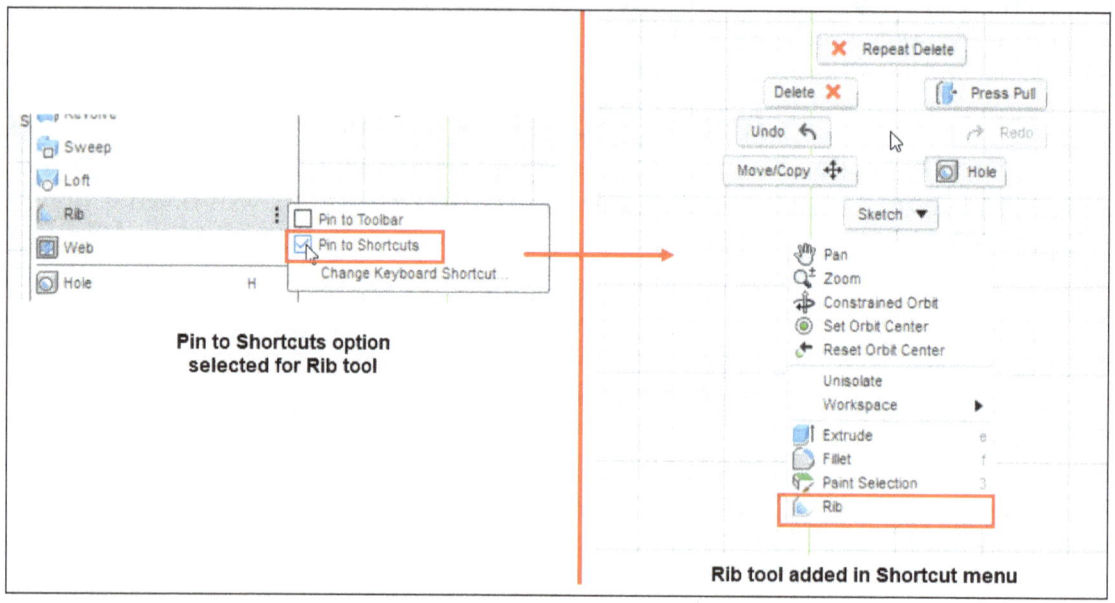

Figure-73. Adding tool to shortcut menu

- Clear the **Pin to Shortcuts** check box from the shortcut menu for tool that you want to remove from right-click shortcut menu/marking menu.

Note that the shortcut menu is workspace specific so what you find in shortcut menu for **DESIGN** workspace will be different from **MANUFACTURE** workspace, and so on.

Removing Tool from Panel in Toolbar

To remove a tool from panel in **Toolbar**, select the tool and drag it to canvas; refer to Figure-74.

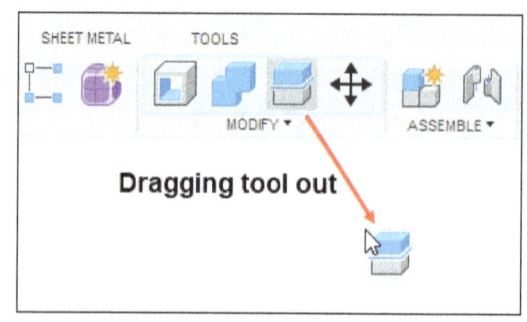

Figure-74. Removing tool from panel

Similarly, you can change position of tool by dragging it within the panel.

Resetting Panel and Toolbar Customization

Right-click on any of the tool in **Toolbar**. A shortcut menu will be displayed as shown in Figure-75.

Autodesk Fusion 360 PCB Black Book   1-41

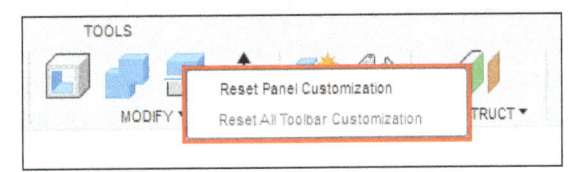

Figure-75. Shortcut menu for Resetting Toolbar

Click on the **Reset Panel Customization** option to reset current panel to original state. Note that this option will be available only when you have added or removed a tool from the panel.

Click on the **Reset All Toolbar Customization** option from the shortcut menu if you want to reset the complete toolbar including all the panels to its original state.

SELF ASSESSMENT

Q1. .... is used to manage the Fusion project.

Q2. Which tool is used to share the files to GrabCAD?

Q3. Which of the following workspace is not available in Autodesk Fusion application?

a. DESIGN Workspace
b. ANIMATION Workspace
c. Assemble Workspace
d. Drawing Workspace

Q4. Discuss the use of **Save** tool with example.

Q5. Discuss the procedure of starting a new electronics design file with example.

Q6. How to reset the default layout?

Q7. Explain the process of creating **Script** with example.

Q8. Mark the following statements as True or False:

a. Autodesk Fusion software can be used to perform CAD operations.
b. Autodesk Fusion software can be used to perform CAM operations.
c. Autodesk Fusion software can be used to perform CAE operations.
d. Autodesk Fusion software can be used to perform PLM operations.
e. Autodesk Fusion software can be used to perform direct 3D printing on 3D Printer.

Q9. Which of the following tool is used to create a new part model in Autodesk Fusion?

a. New Design
b. New Drawing->From Animation
c. New Drawing->From Design
d. New Drawing->New Drawing Template

Q10. Which of the following workspace is not available in educational version of Autodesk Fusion?

a. Animation
b. Manufacturing
c. Generative Design
d. Simulation

Q11. Which of the following tool is used to save the file in local drive in Autodesk Fusion?

a. Save
b. Save As
c. Export
d. Upload

Q12. Which of the following is the shortcut key for displaying/hiding Data Panel?

a. CTRL+ALT+P
b. CTRL+ALT+C
c. CTRL+ALT+D
d. CTRL+ALT+R

Chapter 2

Introduction to Electronics Schematic

Topics Covered

The major topics covered in this chapter are:

- *Overview of Electronics*
- *Starting a new Schematic Drawing*
- *Schematic Design User Interface*
- *Workflow of Electronics Schematic Design*
- *Placing Components And Blocks*

OVERVIEW OF ELECTRONICS

Electronics is the branch of electrical engineering and physics which deals with behavior of electrons in electronic devices. Electronic devices actively perform amplification and rectification of electron flow in the circuit. In electronic devices, components are generally soldered to a PCB which has the conducting wire lines printed on it. As you know there are two types of currents which are Alternating Current (AC) and Direct Current(DC). Most of the electronic circuits use DC current to perform their operations.

Components used in Electronics

The components used in electronics can be divided into three categories: active components, passive components, and electromechanical components. These components are discussed next.

Active Components

Active components are those which rely on DC power source. They are used to inject power in circuit and amplify the signals. Some examples of active electronic components are transistors, diodes, integrated circuit(IC), programmable logic devices (PLCs), optical switches, and so on.

Transistors

Transistor is one of the biggest invention in electronics sector of 20th century. They are semiconductor devices used to amplify and switch digital signals as well as electrical power sources. There are various type of transistors like MOSFET (Metal Oxide Semiconductor Field Effect Transistors), PMOS (P-type MOS), NMOS (N-type MOS), CMOS (Complementary MOS), JFET (Junction FET), Bipolar Junction Transistors, Thyristors, and so on; refer to Figure-1.

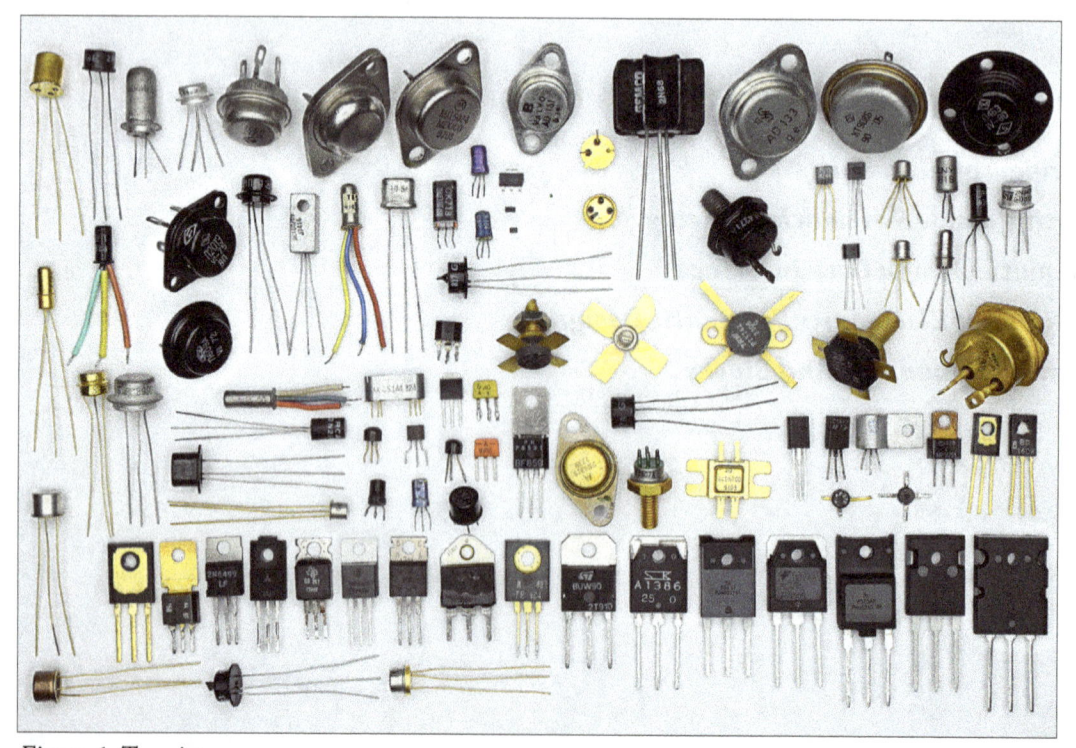

Figure-1. Transistors

Diodes

Diodes are electronic devices whose main purpose is to allow current to pass in one direction and restrict the current in other direction. There are various other functions performed by different types of diodes like Light Emitting Diodes (LEDs) are used for lighting. Some common types of diodes are Rectifier diodes, Zener diodes, LEDs, Thermal diodes, Schottky diodes, Laser diodes, DIAC, Tunnel diodes, and so on; refer to Figure-2.

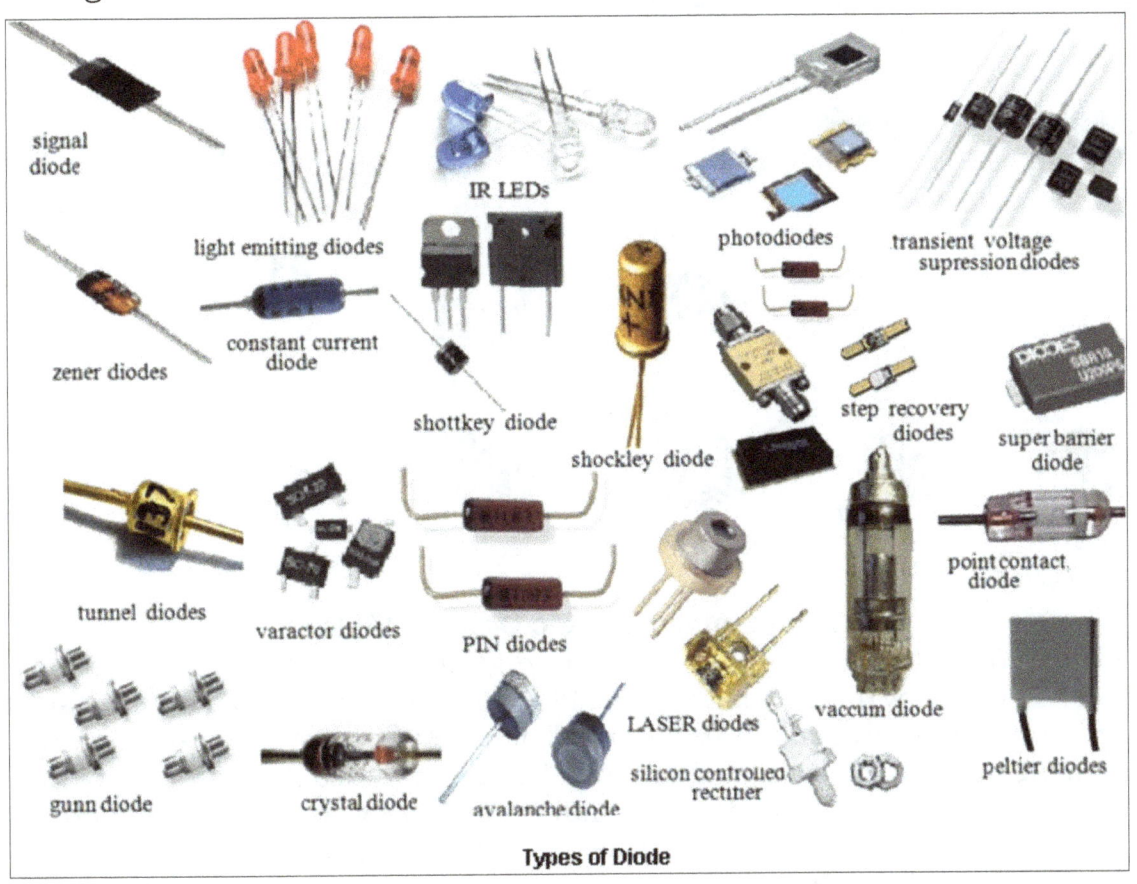

Figure-2. Diodes

Integrated Circuits (ICs)

Integrated Circuit as the name suggests is integration of tiny circuits consisting of many components like transistors, resistors, diodes, and so on in a small encapsulation. There are various operations performed by ICs like timer, digital to analog signal converters, logical operators, signal amplifiers, and so on. There are various types of ICs like Digital ICs (microprocessors, micro controllers, memory chips, level shifter, power management ICs, and programmable devices), Analog ICs (RF circuits and Linear integrated circuits), Mixed signal ICs (A/D converters, D/A converters, Clock/Timing ICs, RF CMOS, and Switched Capacitor ICs), and 3D ICs. Different types of ICs based on packaging are shown in Figure-3.

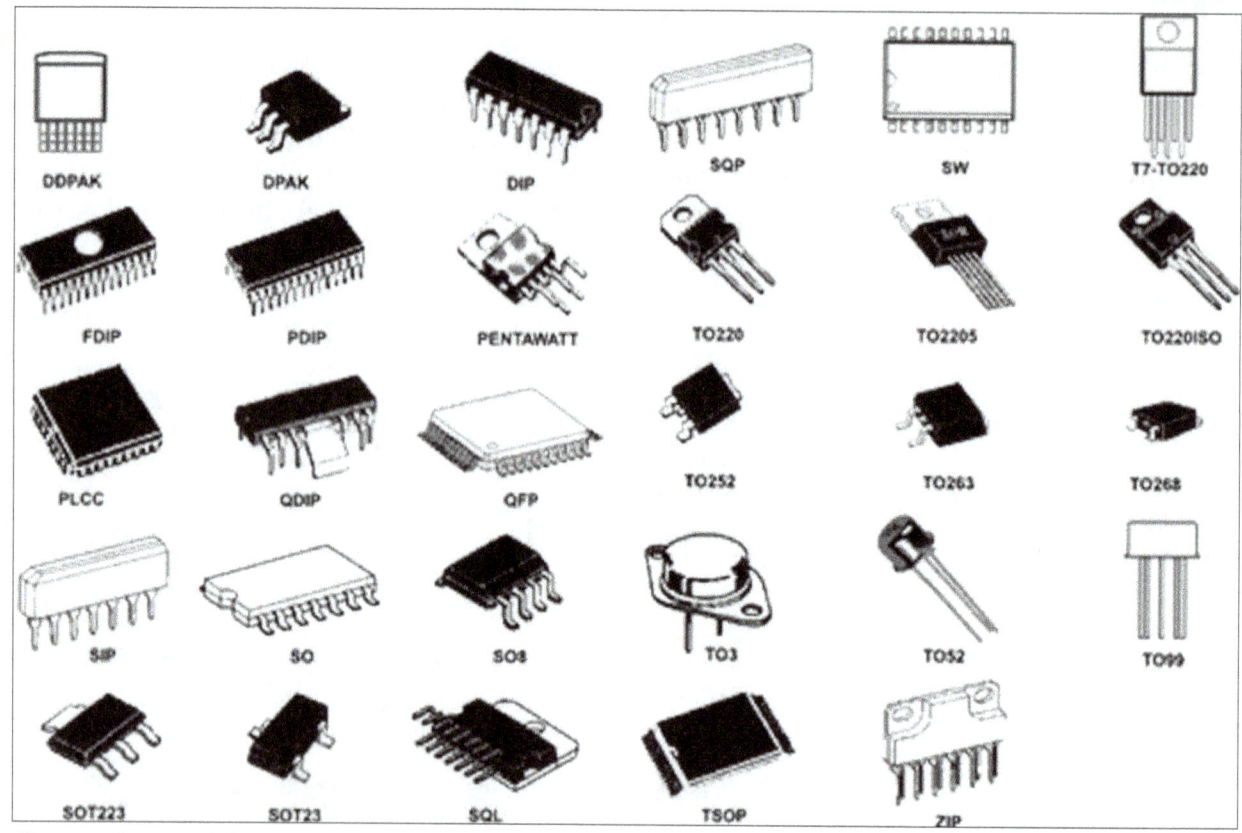

Figure-3. Integrated Circuits

Programmable Logic Devices (PLDs)

The PLD is used to create re-configurable digital circuits. As compared to digital logics constructed by discrete logic gates, PLDs (Programmable Logic Devices) are not manufactured with predefined functions. These devices are generally programmed before using them in their application area. Based on increasing complexity the PLDs can be categorized as Simple Programmable Logic Devices (SPLDs) which include programmable array logic, programmable logic array, and generic array logic; Complex Programmable Logic Devices (CPLDs), and Field-Programmable Gate Arrays (FPGAs); refer to Figure-4 and Figure-5.

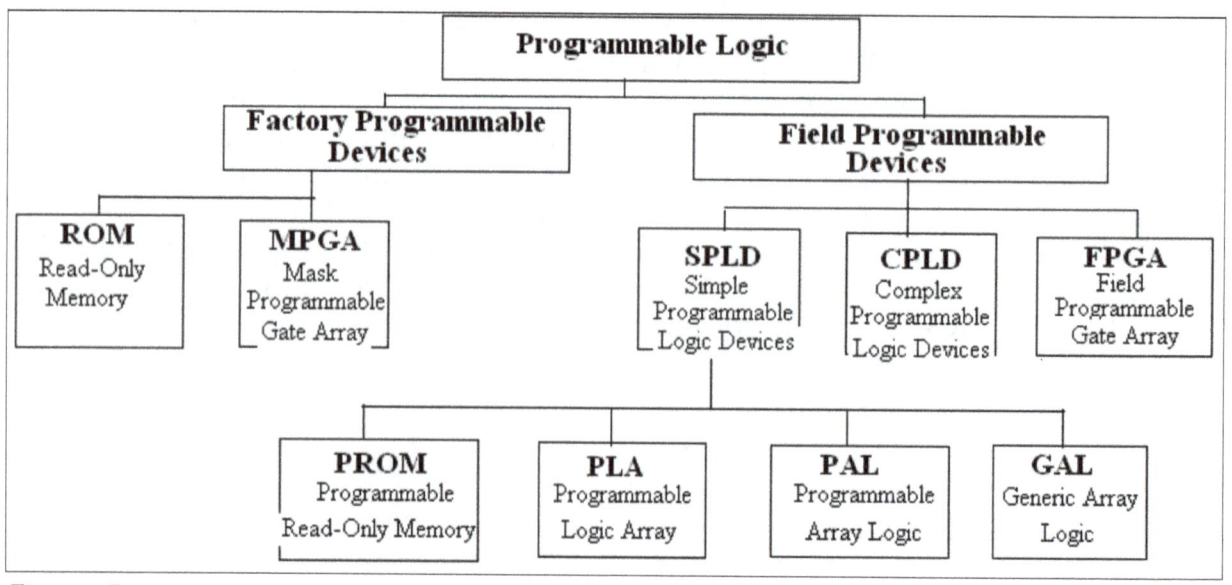

Figure-4. Programmable logic devices

Figure-5. PLD

Optoelectronic (Optronic) Devices

Optoelectronic devices are those which perform optical to electrical or electrical to optical energy conversions. There are various optoelectronic devices used in electronics like LEDs, photodiodes (Solar Cells), photo-transistors, photo-resistors, and so on; refer to Figure-6.

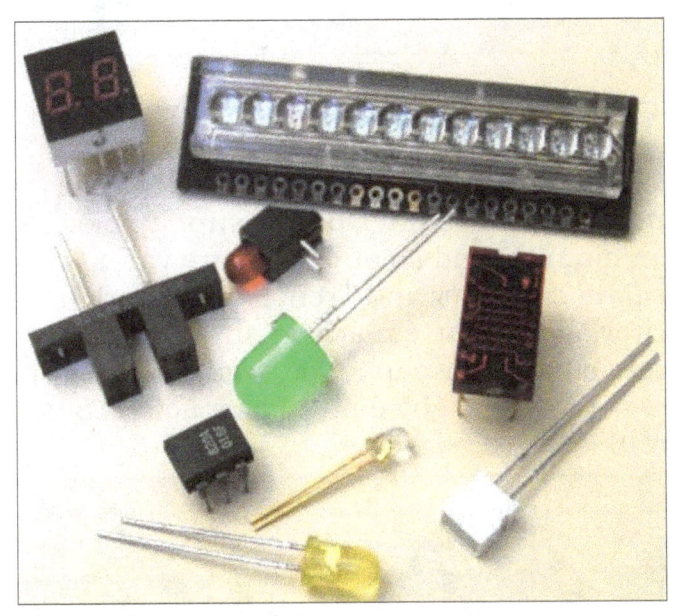

Figure-6. Optoelectronic devices

Vacuum Tube (Valve)

The vacuum tubes are the components that work by passing current through vacuum tubes to amplify or rectify current. They can also be used in oscillators.

Passive Components

Passive components are those which cannot be controlled by electrical signal. Some examples of passive components are resistors, capacitors, inductors, transformers, and so on.

Resistors

The resistors are used to restrict the amount of current passing through them based on Ohm's law. Resistors are broadly classified into two categories: fixed value resistor and Variable resistor. There are various types of resistors used in electronics like power resistor and SIP/DIP resistor network which are fixed value resistors and variable resistors like rheostat, potentiometer, trim pot, thermistor, and so on; refer to Figure-7.

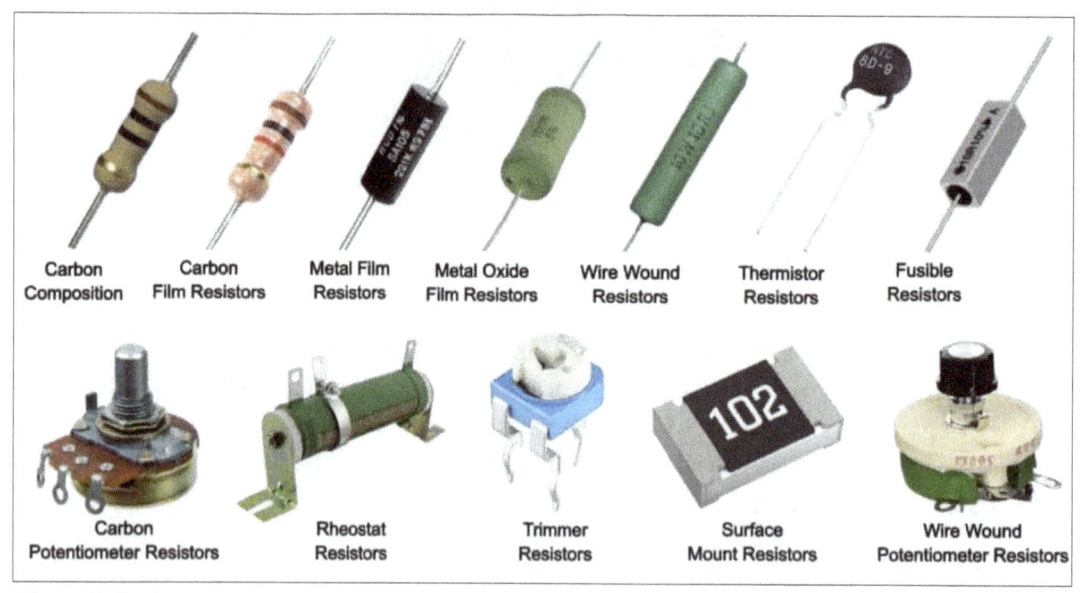

Figure-7. Resistor types

Capacitors

The capacitors are used to store electrical charge and release it when required in the circuit. Capacitors are also used to filter power supply, block DC voltages, pass AC signals, tune resonant circuits, and so on. There are various types of capacitors like integrated capacitors (MIS capacitor and Trench capacitor), fixed capacitors (ceramic capacitor, film capacitor, electrolyte capacitor, super capacitor, and so on), variable capacitors (tuning capacitor, trim capacitor, and so on), and special application capacitors (like power capacitor, Light-emitting capacitor, photoflash capacitor, and so on); refer to Figure-8.

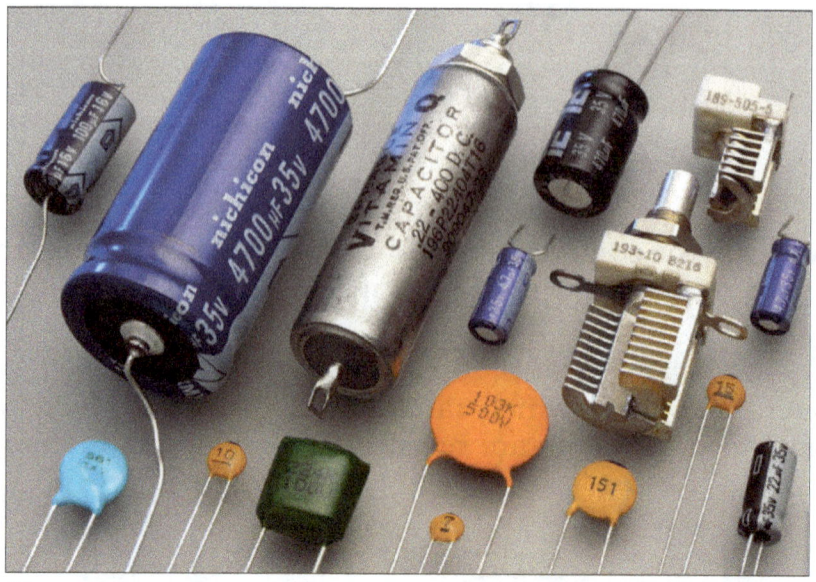

Figure-8. Capacitor types

Electro-magnetic Devices

Magnetic (inductive) devices use magnetic field to store energy and release them as electrical discharge when required. Various components that fall under this category are inductor coils, variable inductor, transformer, solenoid, loudspeaker, microphone, and so on; refer to Figure-9.

Figure-9. Magnectic devices

Transducers and Sensors/Detectors

Transducers are used to generate physical effects when current passes through them.

Sensors/Detectors are devices which convert environmental conditions in to electric signals.

Some common components that fall under this category are motion sensor, flow meter, accelerometer, thermistor, hygrometer, magnetometer, photo resistor, and so on.

Switches

The switches are used to pass current when closed and stop flow of current when the position is open. There are various types of switches available in electronic and electrical circuits like keypad, manual switches, knife switch, relay, contactors, thermostat, and so on.

Circuits in Electronics

A circuit is an arrangement of conducting wires that provide current to different electronic components for performing actions. There are two types of circuits viz. open circuit and close circuit. In open circuit, the power and ground points of the circuit are not connected and there is a break in the circuit where as a close circuit has its terminals connected to power and ground. Based on connection types, there are two types of circuits: series circuits and parallel circuits. In Series circuits, all the components are connected one after other in series. In Parallel circuits, all the components are connected in parallel branches of wires. Refer to Figure-10. Scan the QR code given next to learn about basics of Electronics if you have no prior experience with electronic circuits.

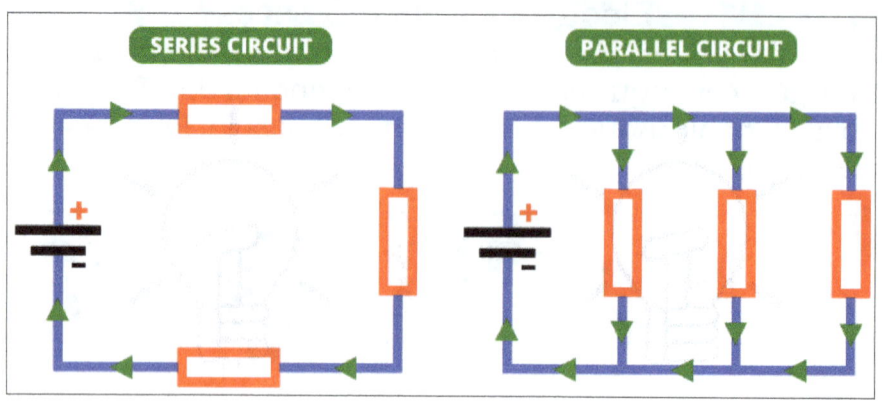

Figure-10. Circuit types

Electronic Components Symbols

Symbols for various commonly used electronic component are shown in Figure-11.

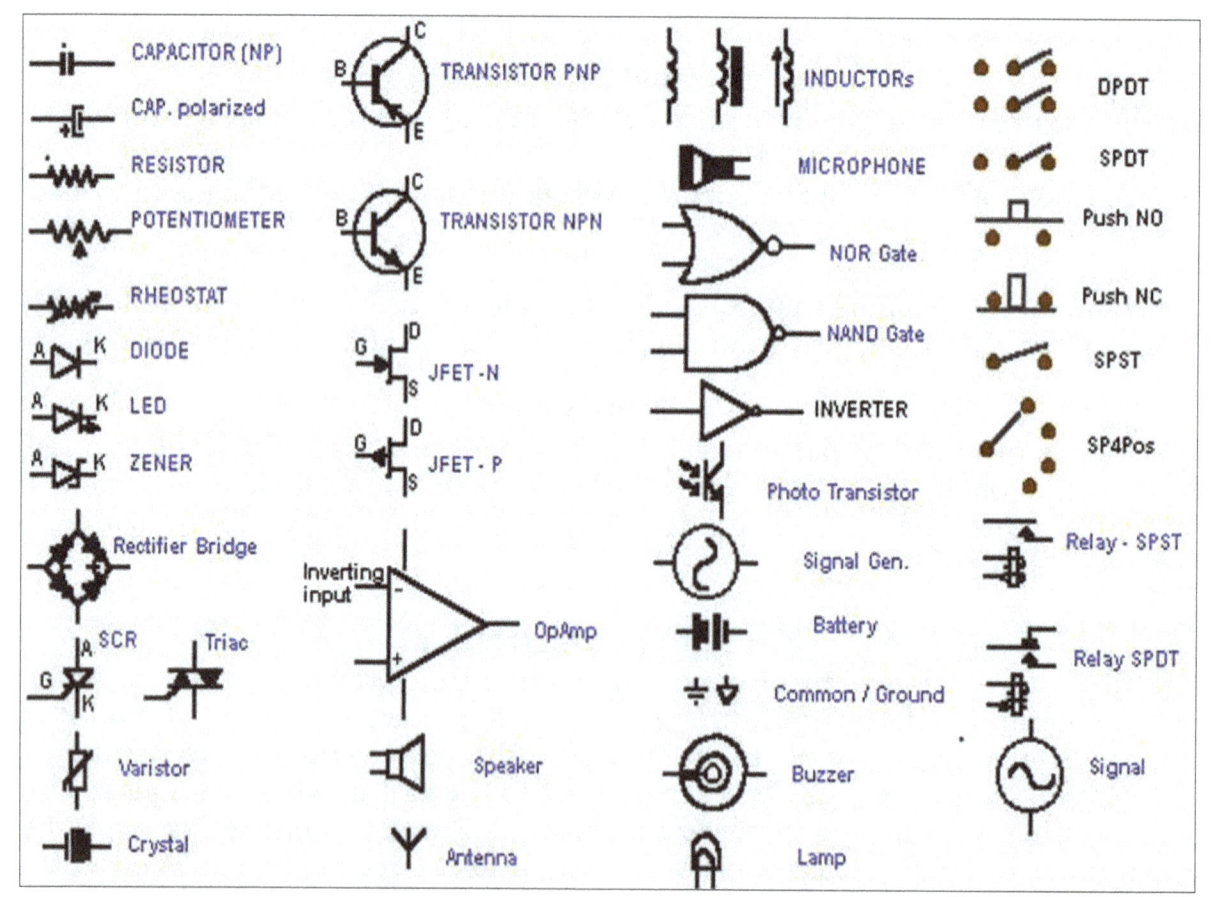

Figure-11. Common electronic symbols

STARTING NEW SCHEMATIC DRAWING

Schematic and PCB are two interlinked parts of an electronic design in Autodesk Fusion 360. A schematic is circuit diagram of electronic device represented by various symbols and lines to act as electronic components and wires, respectively. A schematic diagram does not show terminal connections of components in circuit. To show the connections between various components of design, you will need a PCB diagram. The procedure to start a new schematic drawing is given next. You will learn about starting a new PCB file later in the book.

- Click on the **New Electronics Design** tool from the **File** menu. A new electronic design file will open and options will be displayed as shown in Figure-12.

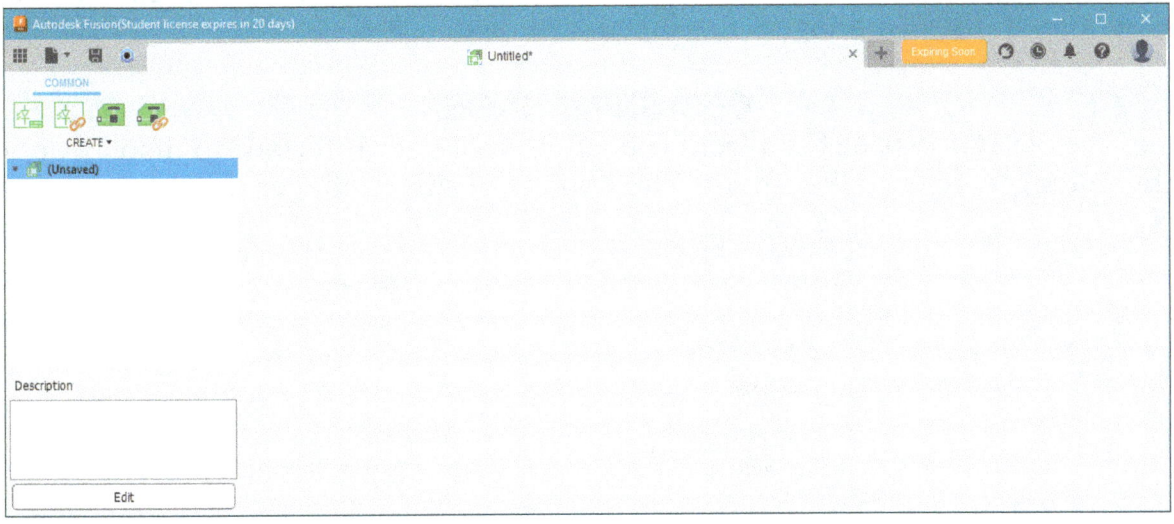

Figure-12. New electronics design file created

- Click on the **New Schematic** tool from the **COMMON** tab in the **Ribbon**. The interface to create new schematic design will be displayed; refer to Figure-13.

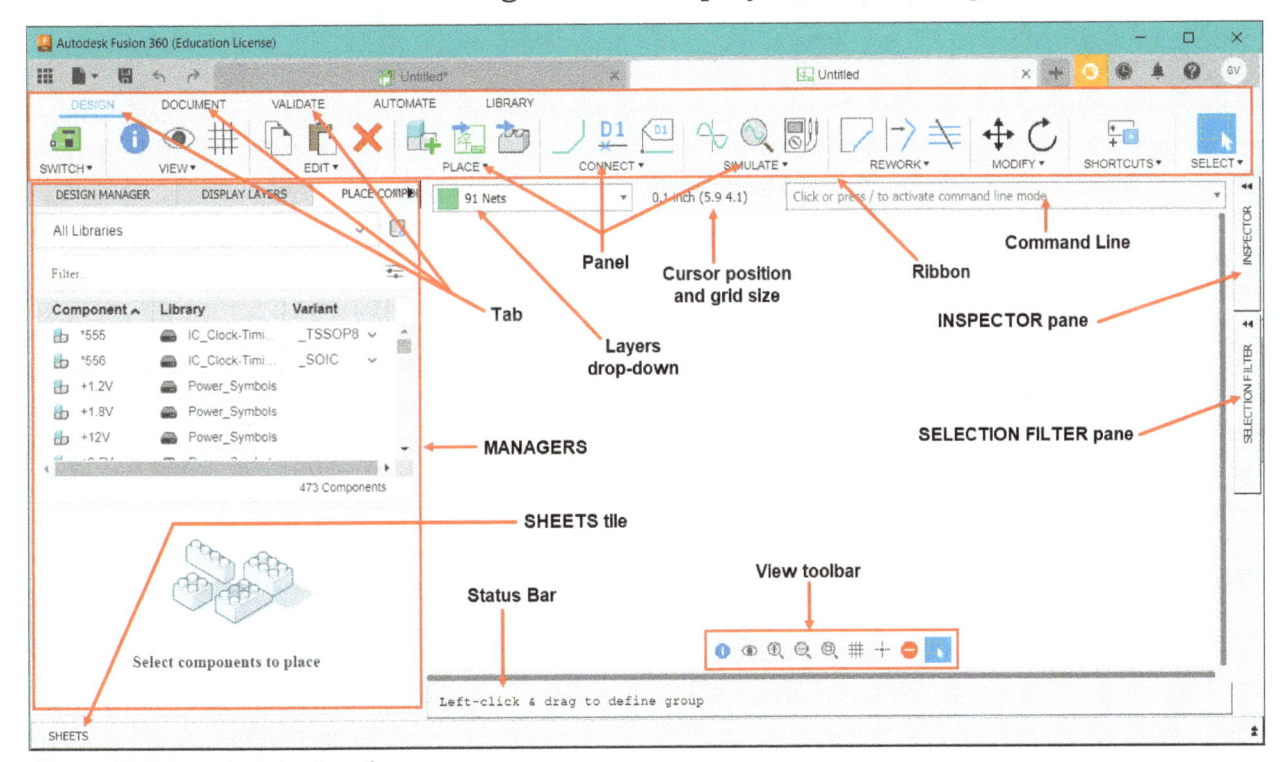

Figure-13. Schematic design interface

SCHEMATIC DESIGN USER INTERFACE

Various interface elements of the schematic design are discussed next.

Toolbar, Tabs, and Panels

The **Toolbar** is component of user interface of Autodesk Fusion where you can find all the tools arranged in the form of **Tabs** which are further categorized as **Panels** based on their similarity in functioning.

Managers

There are three managers available in Autodesk Fusion 360 Schematic Design interface; **DESIGN MANAGER**, **DISPLAY LAYERS MANAGER**, and **PLACE COMPONENTS MANAGER**. The options in these managers are discussed next.

DESIGN MANAGER

Select the **DESIGN MANAGER** tab from the left area of user interface to display options of the manager; refer to Figure-14. The options of this manager are discussed next.

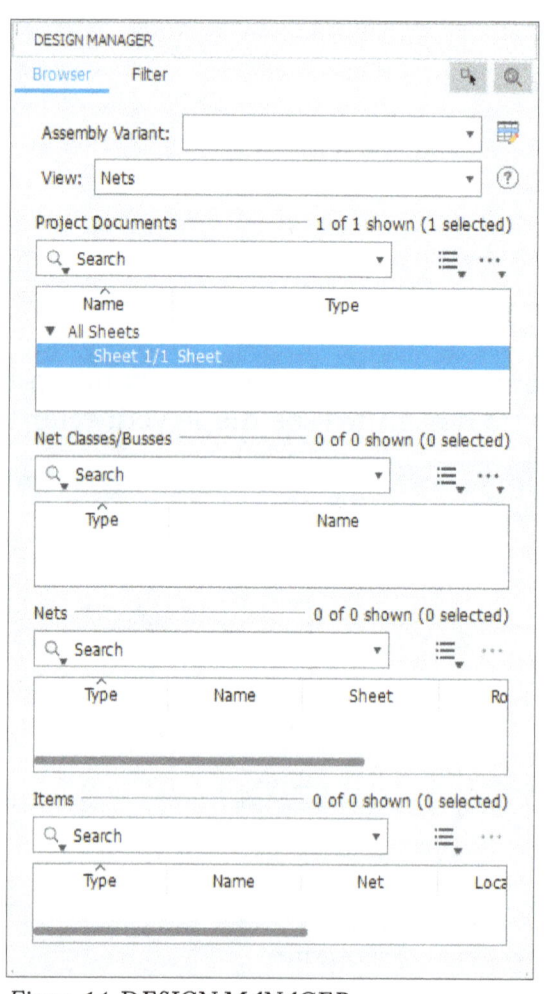

Figure-14. DESIGN MANAGER

- Select desired option from the **Assembly Variant** drop-down to define the design variation for which you want to check the design. For example, you have a circuit in which 10k resistor is used and you want to check the response of same circuit with 8k resistor then you can create two assembly variants and compare their output. You can create and edit the assembly variants of design by using the **Edit assembly variants** button. On selecting this button, the **Main assembly variants** dialog box will be displayed; refer to Figure-15. Select the components and click on the **OK** or **Delete** button that you want to include or remove from the assembly variant. To create a new variant of assembly, click on the **New** button. The **New assembly variant** dialog box will be displayed; refer to Figure-16. Specify desired name of variant in the **Name** edit box and click on the **OK** button from **New assembly variant** dialog box to create the variant. Using the **Rename** button, you can rename recently created variant. Click on the **OK** button from the dialog box to apply changes in assembly variants.

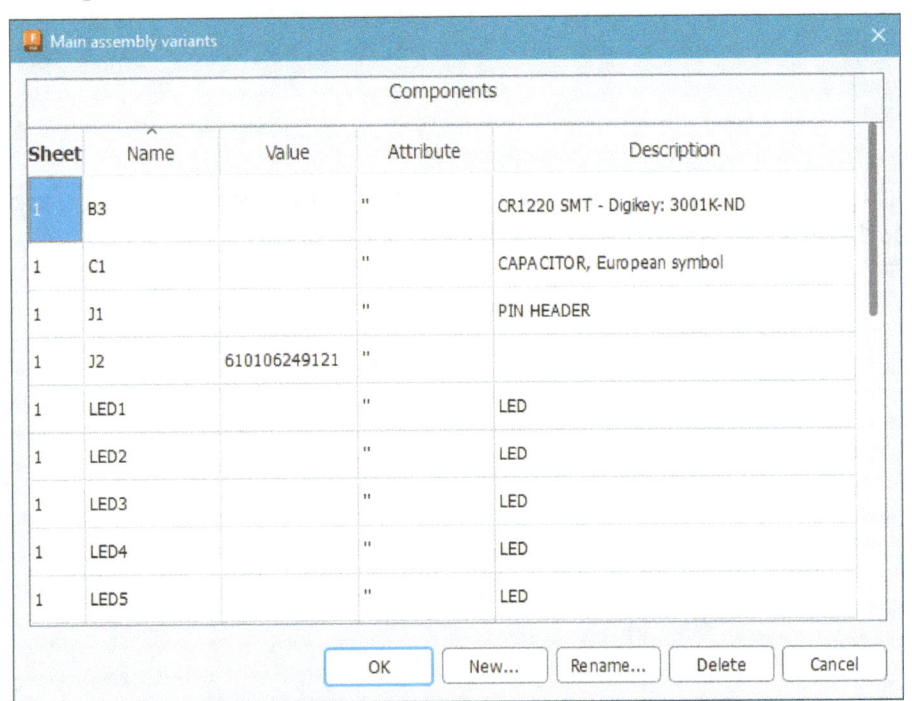

Figure-15. Main assembly variants dialog box

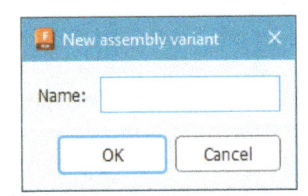

Figure-16. New assembly variant dialog box

- Select desired option from the **View** drop-down to define which types of entities you want to check in the drawing area. Select the **Components** option from the **View** drop-down to check electronic components in the drawing. Select the **Nets** option from the drop-down to check sections of wire net in PCB schematic. Select the **Groups** option from the drop-down to check various grouped and un-grouped components.
- Select the other parameters from other sections of the **DESIGN MANAGER** to check respective object of design; refer to Figure-17.

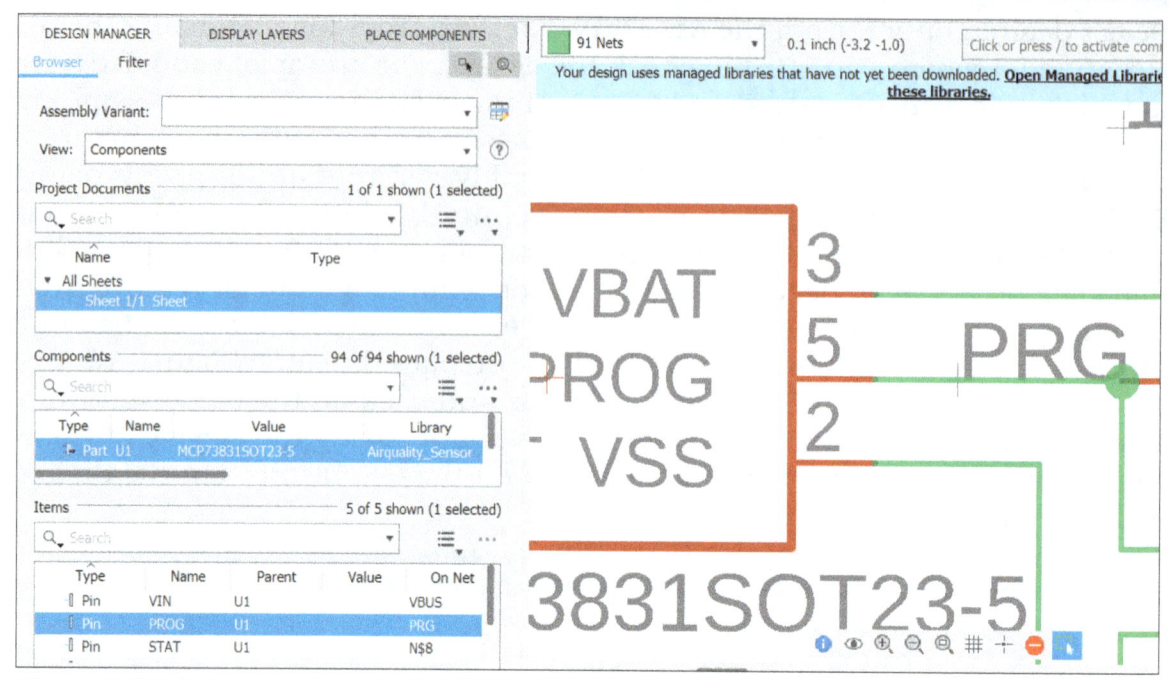

Figure-17. Displaying component using DESIGN MANAGER

- You can also use **Filter** tab from the **DESIGN MANAGER** to select component based on specified filter parameters; refer to Figure-18.

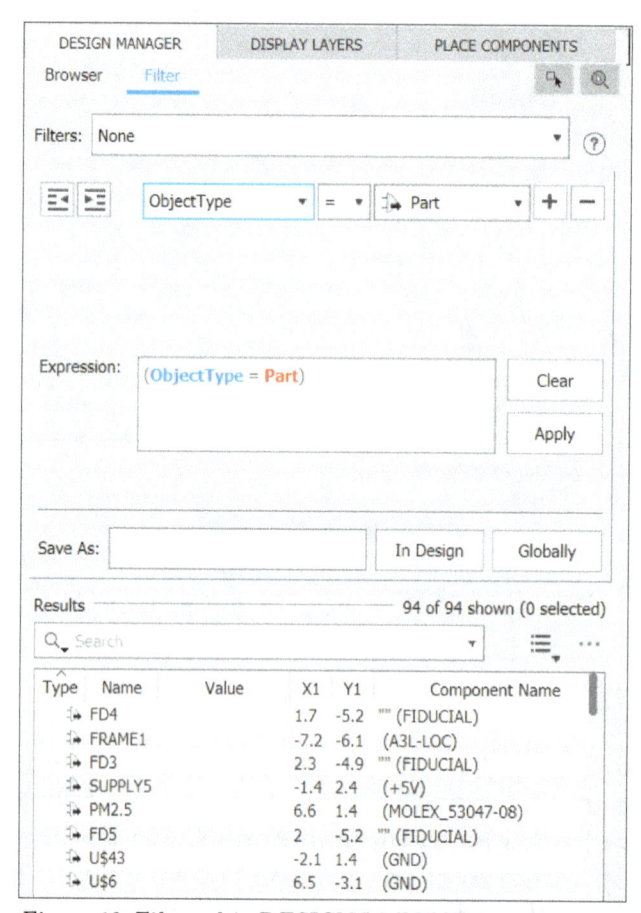

Figure-18. Filter tab in DESIGN MANAGER

DISPLAY LAYERS MANAGER

Select the **DISPLAY LAYERS** tab from the left area of user interface to display options of the manager; refer to Figure-19.

Figure-19. DISPLAY LAYERS tab

- Click on the **Show/Hide** button before desired layer to display/hide its content from drawing area.
- Select the radio button next to desired layer in **DISPLAY LAYERS MANAGER** to make it current layer for placing schematic design component.
- To modify color and pattern for a layer, select respective button from **Appearance** section at the bottom of manager; refer to Figure-20.

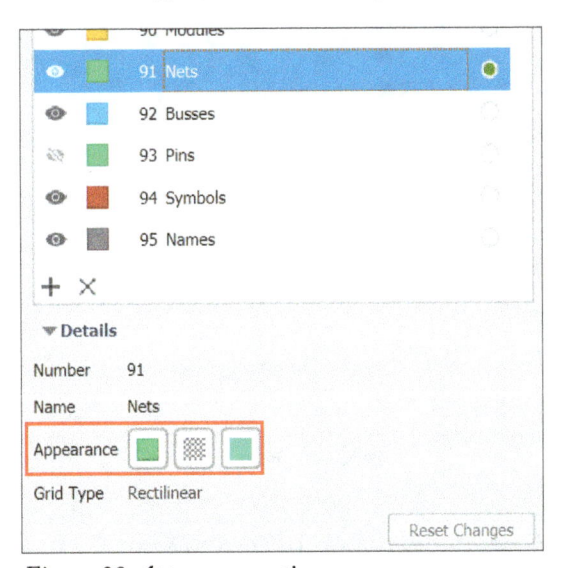

Figure-20. Appearance section

- To create a new layer, select any custom layer from the bottom in the list and specify desired name in the **Name** edit box of **Details** section; refer to Figure-21. Click on the **Create a new layer** button from the bottom of layer list in the **Manager** to create a new layer in schematic design.

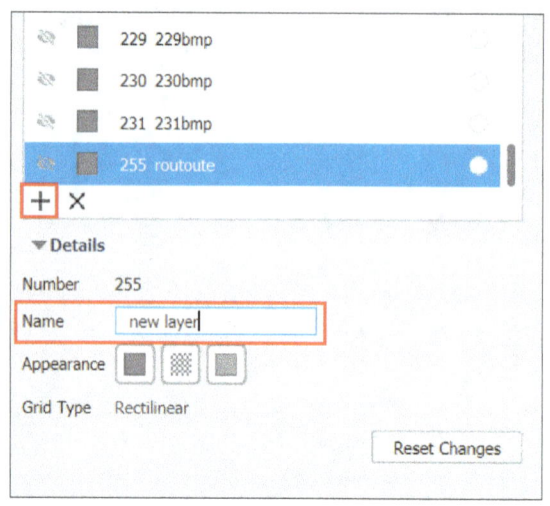

Figure-21. Specifying name of layer

PLACE COMPONENTS MANAGER

Select the **PLACE COMPONENTS** tab from the left area of user interface to display options of the manager; refer to Figure-22.

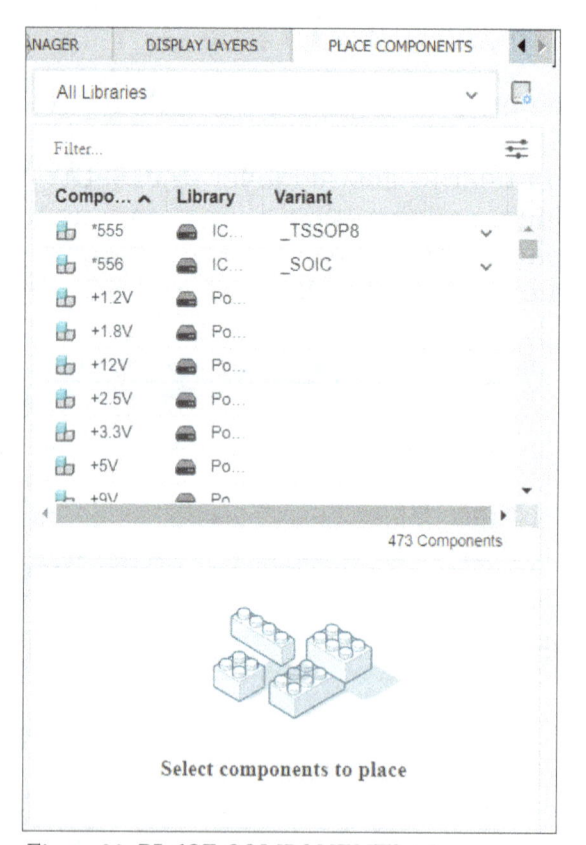

Figure-22. PLACE COMPONENTS tab

- Select desired option from the **Libraries** drop-down at the top in the **PLACE COMPONENTS MANAGER**; refer to Figure-23. Related components of selected library will be displayed.
- Select desired variant of the component using drop-down in **Variant** column from the list; refer to Figure-24.

Autodesk Fusion 360 PCB Black Book   2-15

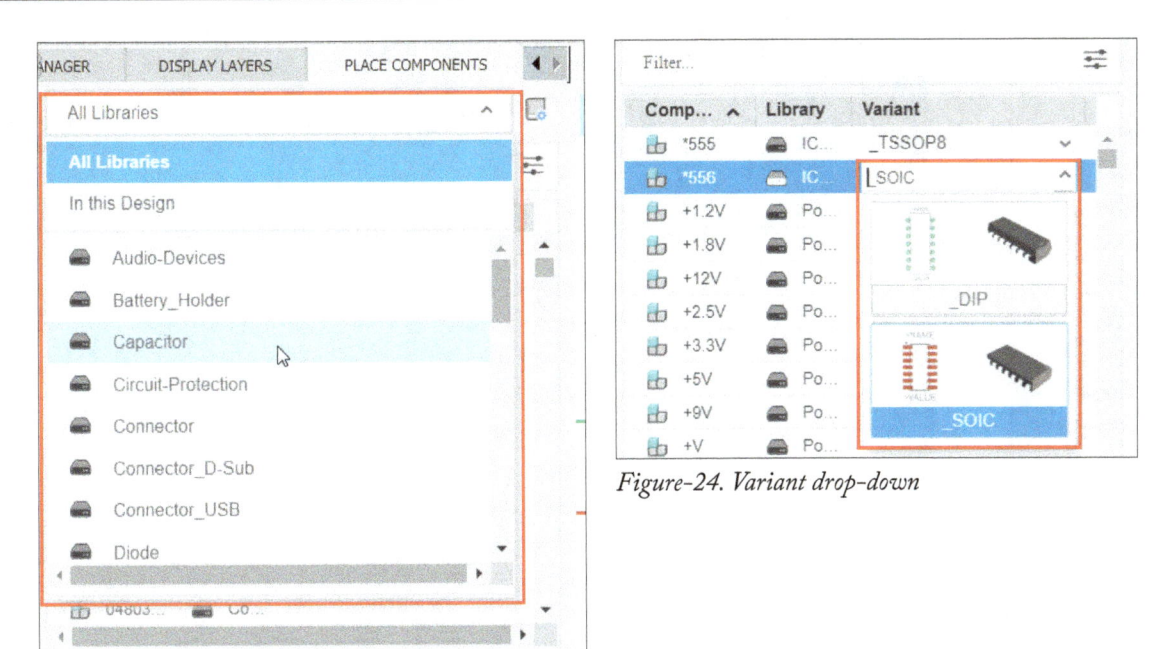

Figure-23. Libraries drop-down

Figure-24. Variant drop-down

- Scroll down in the **Details** section at the bottom in the **Manager** and click on the **Datasheet** link button to check details regarding the selected component like operating voltage, temperature range, applications, and so on; refer to Figure-25. Similarly, you can check the attributes of component from the **Attributes** tab at the bottom in Manager.

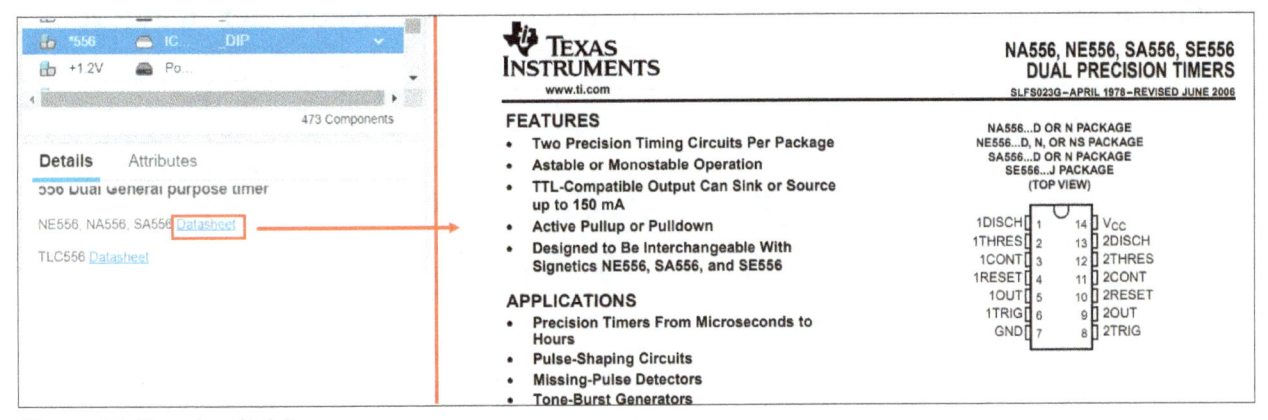

Figure-25. Datasheet link button

- Double-click on the component after selecting variant. The component will get attached to cursor and tools to rotate/mirror component will be displayed; refer to Figure-26.

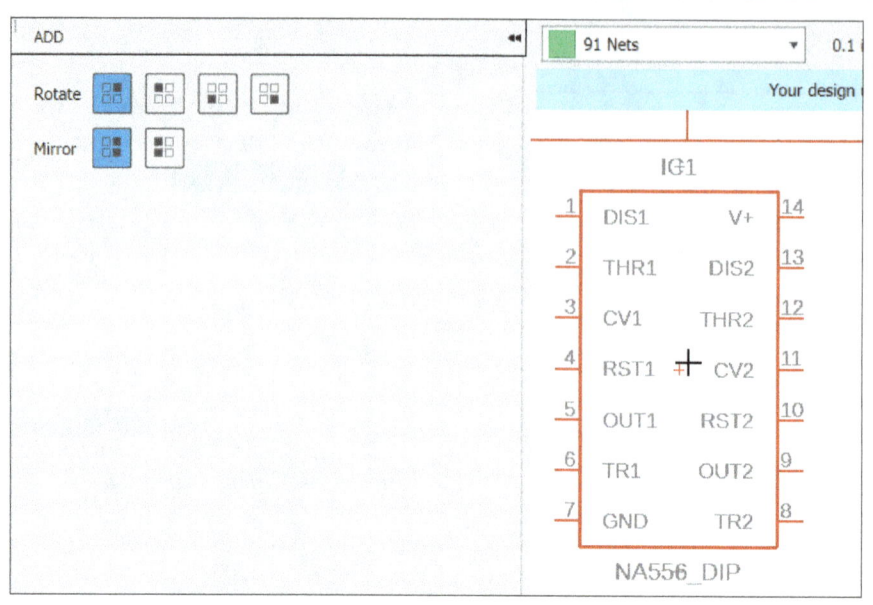

Figure-26. Component attached to cursor

- Select desired buttons from the **Rotate** and **Mirror** sections in the **Manager** to change the orientation of component. After setting the orientation, click at desired location to place the component. Press **ESC** to exit the tool.

INSPECTOR Pane

The options in **INSPECTOR** pane are used to check and modify properties of selected component(s). Click on the expand button for the **INSPECTOR** pane from the right in the application window to display the pane and select desired components from drawing area. The list of selected components will be displayed; refer to Figure-27.

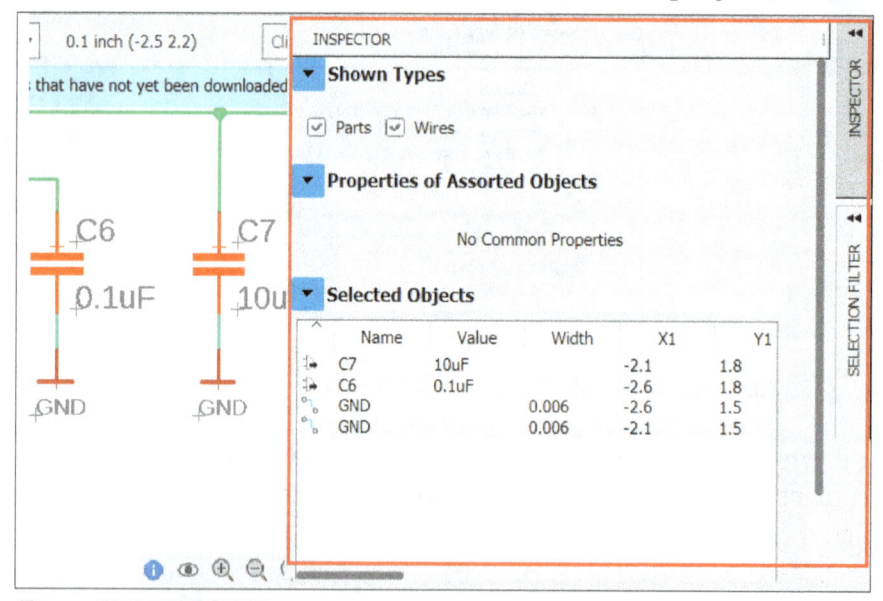

Figure-27. INSPECTOR pane

Select desired component from the list to check its properties; refer to Figure-28. Set desired values in the **Properties of Part** section to modify the component. Click in the empty area of drawing to exit selection. Click on the **<<** button in **INSPECTOR** tile from right area of application window to keep it open.

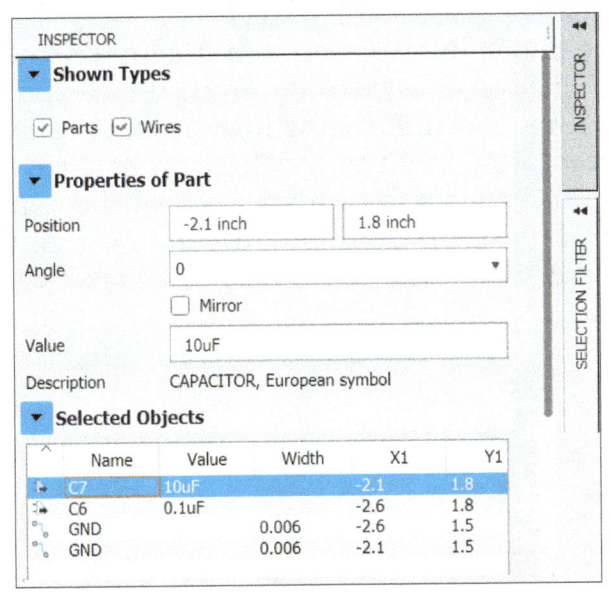

Figure-28. Properties of selected component

SELECTION FILTER Pane

The options in **SELECTION FILTER** pane are used to preset the type of objects that can be selected by cursor from drawing area; refer to Figure-29. By default, all the objects are selected in the selection filter list. Click on desired object types to include them in selection while holding the **CTRL** key. Click on the **Reset** button at the bottom to revert to default selection filters.

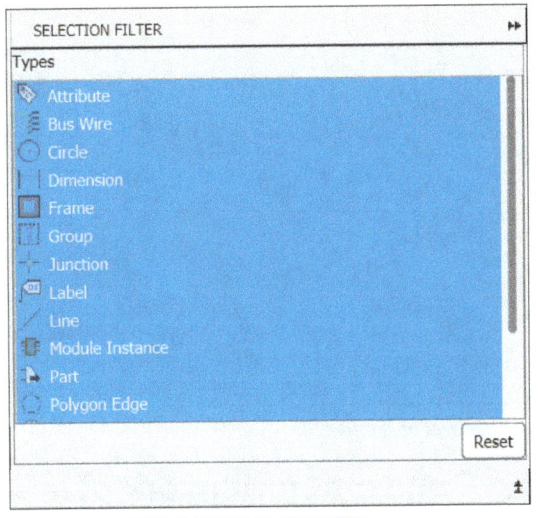

Figure-29. SELECTION FILTER pane

View Toolbar

The tools in **View** toolbar are used to change view orientations and check properties of components; refer to Figure-30. The toolbar is available at the bottom of drawing area. Various tools of this toolbar are discussed next.

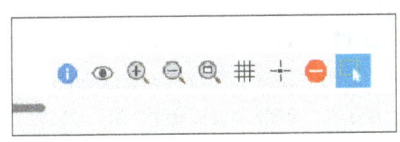

Figure-30. View toolbar

Info Tool

The **Info** tool is used to check information about component selected from drawing area. To use this tool, click on the **Info** tool from the toolbar. You will be asked to select the component from drawing area whose properties are to be checked. Click on the + sign near desired component to check its properties; refer to Figure-31. Note that some of the fields in **Properties** dialog box will be available for modification whereas manufacturer data of the component will be fixed. Click on the **OK** button to apply changes and exit the dialog box. Press **ESC** to exit the tool.

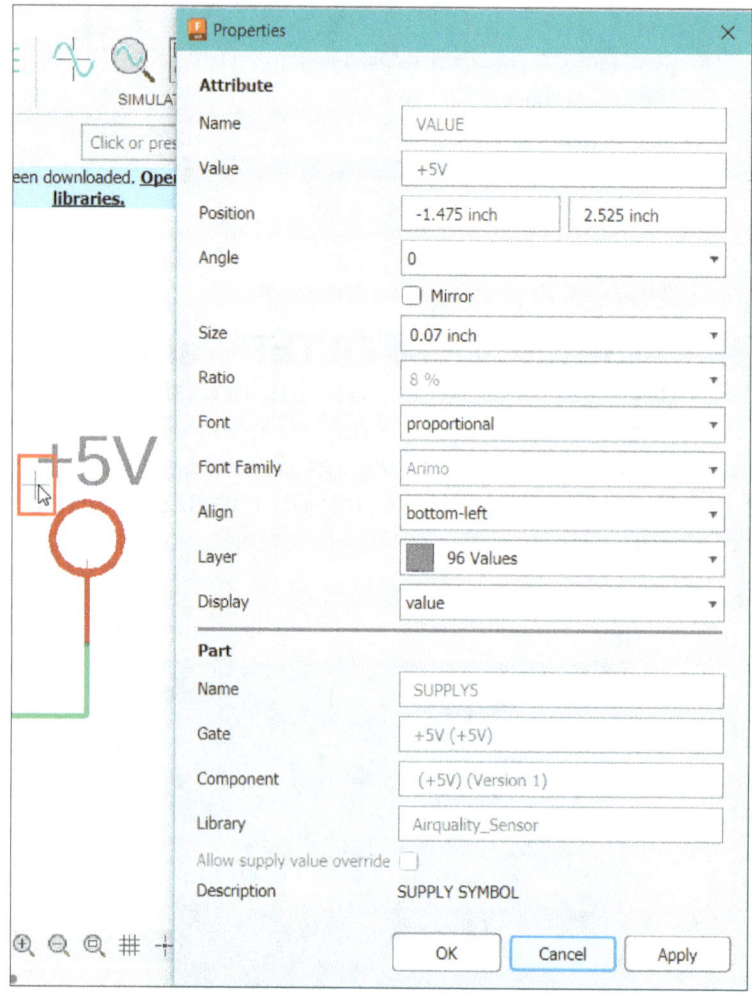

Figure-31. Properties of component

Highlighting Connection Network of a Component

The **Show** tool in **View** toolbar is used to check the network connected to selected component. Click on the **Show** tool from toolbar and click on net/component to get info about them while highlighting the chain. (Note that you need to click on + node of component to select it). Press **ESC** to exit the tool.

Zoom In/Zoom Out and Pan

The **Zoom In** tool is used to enlarge the view of current objects using the center of screen as reference. The **Zoom Out** tool is used to diminish the view of current objects using the center of screen as reference. You can scroll up mouse wheel for zoom in and scroll down mouse wheel for zoom out. Press the middle mouse button and drag to pan the screen.

Zoom to Fit

The `Zoom to Fit` tool is used to fit all the objects of drawing in current view boundaries.

Grid Settings

The `Grid Settings` tool is used to modify the size of grid. On clicking this tool, the `Grid` dialog box will be displayed; refer to Figure-32. Specify the parameters in dialog box as discussed earlier in `Preferences` section. After setting the parameters, click on the `OK` button from the dialog box.

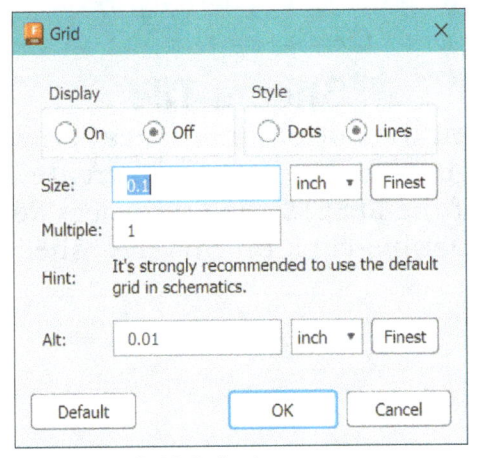

Figure-32. Grid dialog box

Marking Local Origin

The `Mark Local Origin` tool is used to change the position of local origin of drawing with respect to which other entities of drawing are measured. After clicking this tool, click at desired location in drawing to define the origin location. You can check the position of cursor in drawing from the `Cursor Position and Grid Size` section at the top of drawing area; refer to Figure-33.

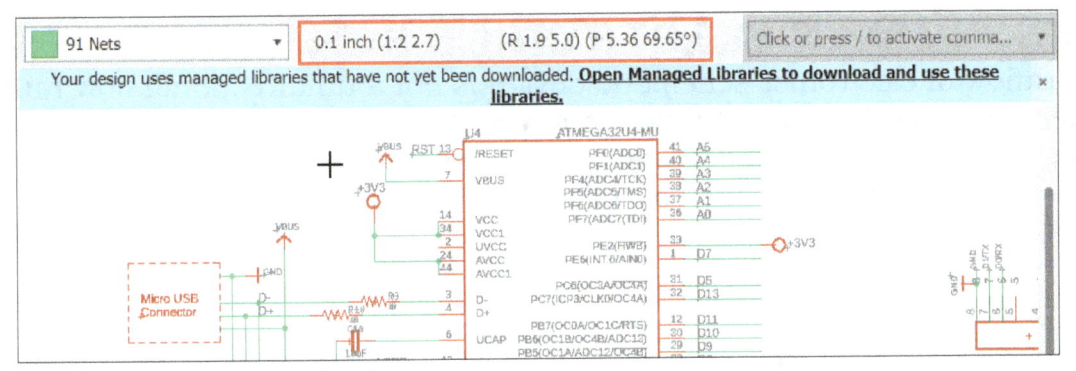

Figure-33. Cursor position and grid size

Stop Command

The `Stop Command` tool is used to exit from the current active commands.

By default, the `Group` tool is active in the `View` toolbar if no other tool is active. Note that you can also access these tools from the `VIEW` drop-down in the `DESIGN` tab of `Ribbon`.

Command Input Box

The **Command Input** box is used to active various electronics design command; refer to Figure-34. Type the command in input box and press **ENTER** to activate it. Most of the Eagle commands can be accessed from **Command Input** box.

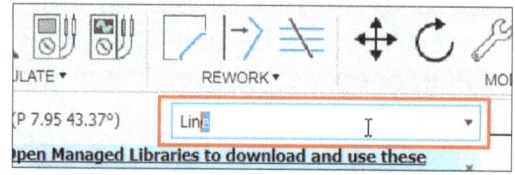

Figure-34. Command input box

Sheets Tile

The **Sheets** tile is available at the bottom in application window. Click on the **SHEETS** tile to expand it; refer to Figure-35. If you want to create a new schematic sheet then right-click in empty area of the **SHEETS** tile and select **New** option from the shortcut menu; refer to Figure-36. Double-click on this new sheet from the tile to open it.

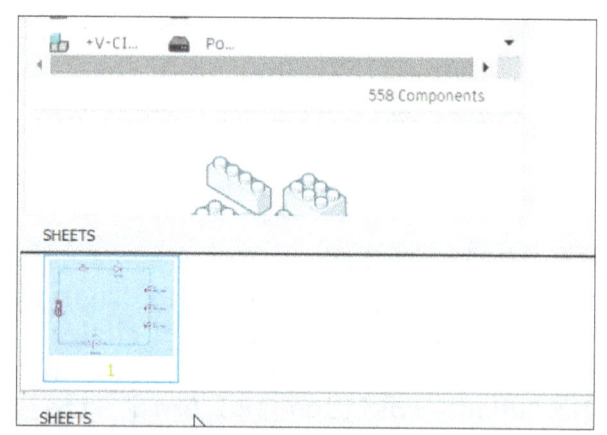

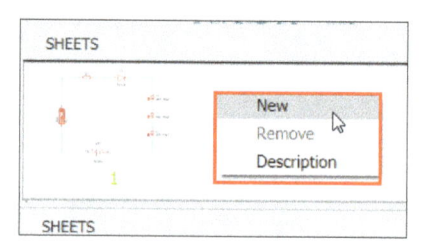

Figure-36. New option

Figure-35. SHEETS tile

WORKFLOW IN ELECTRONICS SCHEMATIC DESIGN

The workflow of electronics schematic design is not a unidirectional flow rather it is a bidirectional flow in which components can be placed before creating net and after creating net depending on the situations. So, the flow chart of electronics schematic design can be described as shown in Figure-37.

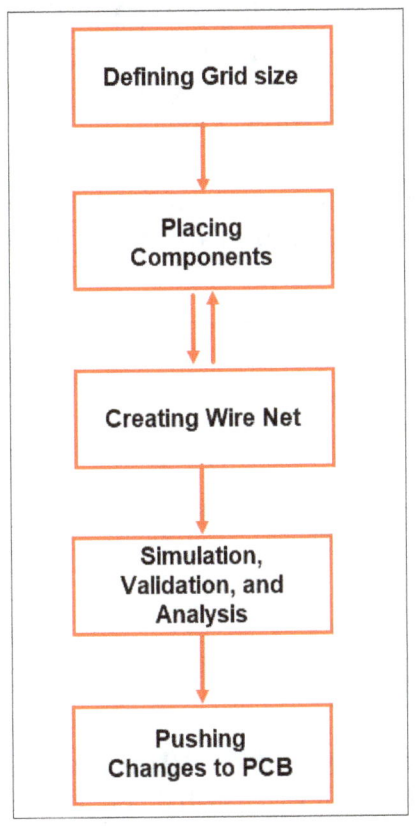

Figure-37. Electronics schematic design flowchart

The process of defining grid size has been discussed earlier. Now, you will learn about other processes involved in schematic design.

PLACING COMPONENTS AND BLOCKS

The tools in the **PLACE** drop-down of **DESIGN** tab in the **Ribbon** are used to insert electronic components, create pattern of components, and insert earlier created schematics, design blocks, and electronic design projects; refer to Figure-38. Various tools of this drop-down are discussed next.

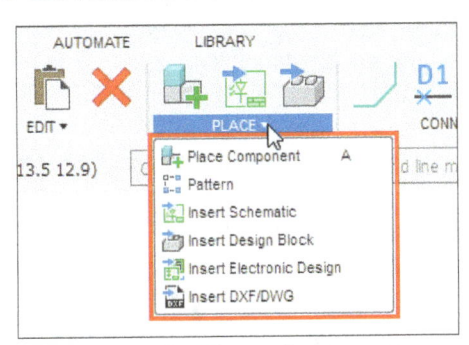

Figure-38. PLACE drop-down

Placing Components

The `Place Component` tool as name suggests is used to place various electronic components. On clicking this tool, the **PLACE COMPONENTS** Manager will be displayed. The procedure of using **PLACE COMPONENTS** Manager has been discussed earlier.

Creating Pattern of Objects

The **Pattern** tool in **PLACE** drop-down is used to create linear and circular patterns of selected objects (components, net segment etc.) in the schematic drawing. The procedure to use this tool is given next.

- Select the object(s) whose pattern is to be created and click on the **Pattern** tool from the **PLACE** drop-down of **DESIGN** tab in the **Ribbon**. The **PATTERN Dialog** will be displayed; refer to Figure-39.

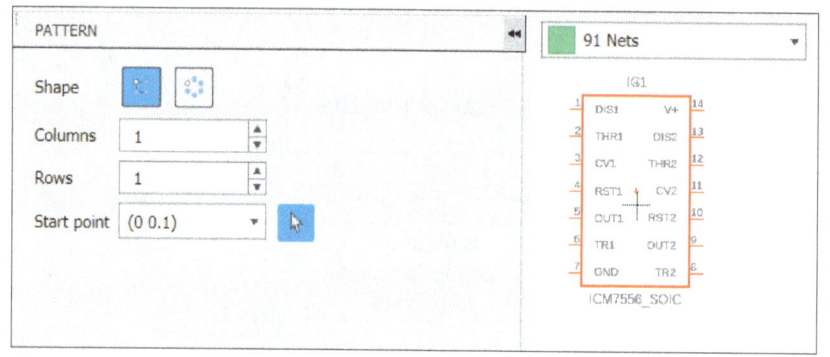

Figure-39. PATTERN Dialog

Creating Rectangular Pattern

- Select the **Rectangular Pattern** button from the **Shape** section of **Pattern Dialog** to create linear pattern. The options will be displayed as shown in Figure-39.
- Specify the number of instances to be created in the **Columns** and **Rows** edit boxes of the **Dialog**. Preview of the pattern will be displayed; refer to Figure-40.

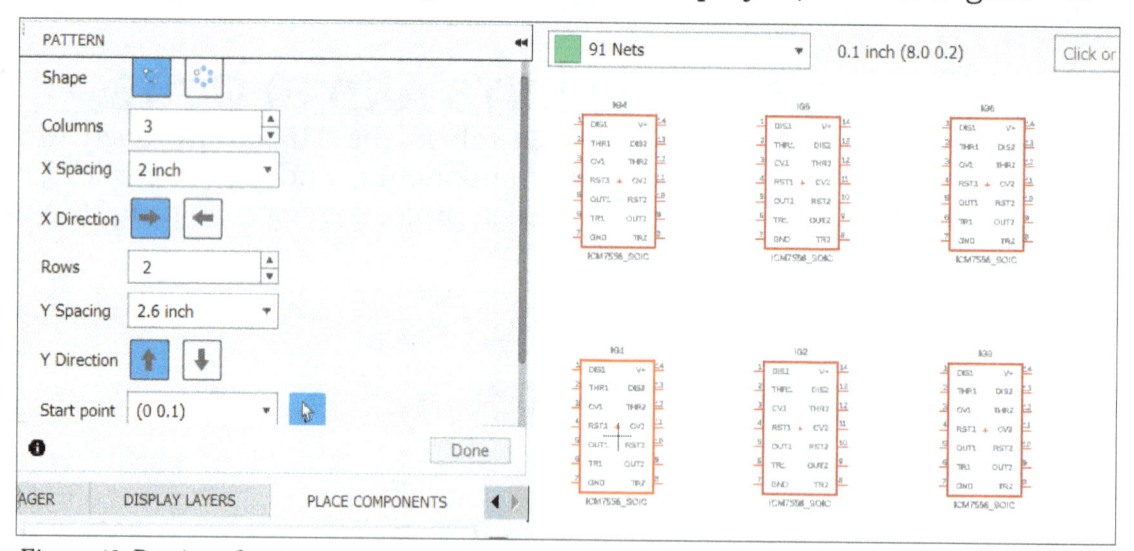

Figure-40. Preview of pattern

- Specify the distance between two instances of pattern in X and Y directions in the **X Spacing** and **Y Spacing** edit boxes, respectively.
- You can change the reference point for pattern by using the selection button for **Start point** option in the Dialog. After selecting this button, click at desired location to use it as start point. Alternatively, you can specify coordinates of start point in the **Start point** edit box.
- After setting desired parameters, click on the **Done** button. The pattern will be created.

Autodesk Fusion 360 PCB Black Book 2-23

Inserting Schematic in Drawing

The **Insert Schematic** tool is used to insert earlier created schematic design in current drawing. The procedure to use this tool is given next.

- Click on the **Insert Schematic** tool from the **PLACE** drop-down in the **DESIGN** tab of the **Ribbon**. The **Browse Team** dialog box will be displayed; refer to Figure-41.

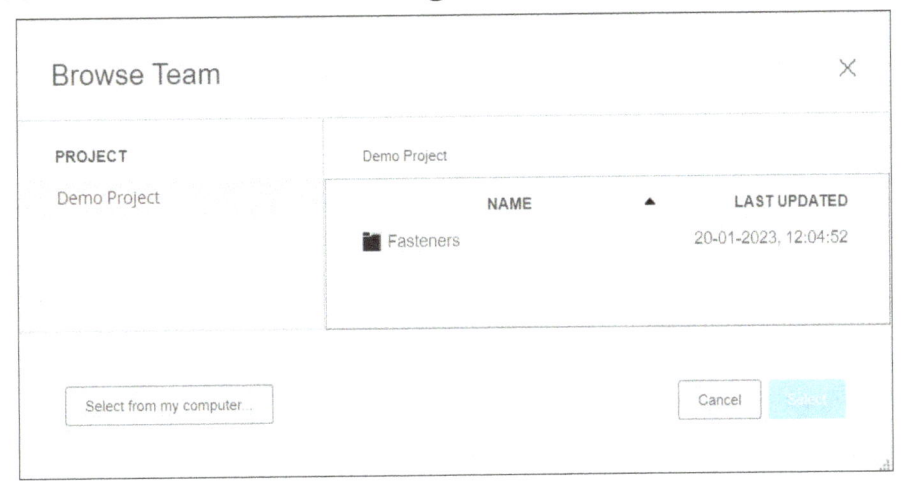

Figure-41. Browse Team dialog box

- Select the schematic design file from the dialog box if the file is on cloud server. Click on the **Select from my computer** button to use file saved in your local drive. The **Select File** dialog box will be displayed; refer to Figure-42.

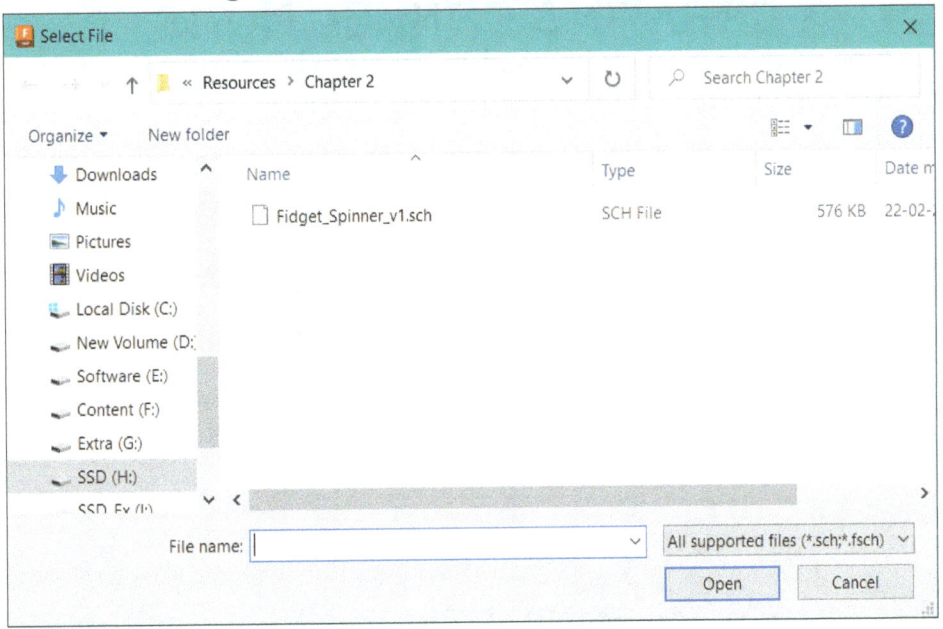

Figure-42. Select File dialog box

- Select desired file for schematic and click on the **Open** button. The schematic design will be attached to cursor; refer to Figure-43.

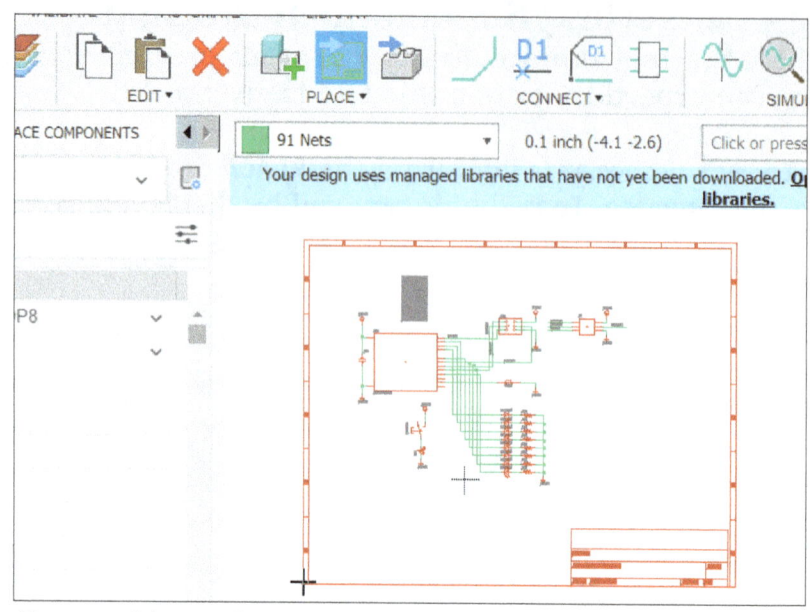

Figure-43. Schematic design attached to cursor

- Click at desired location to place the design.

Similarly, you can use **Insert Design Block**, **Insert Electronics Design**, and **Insert DXF/DWG** tools to insert design blocks, electronic design projects, and dxf/dwg files.

CREATING CONNECTIONS

The tools in **CONNECT** drop-down are used to create wire networks, junctions, labels, and other network segments; refer to Figure-44. Various tools in this drop-down are discussed next.

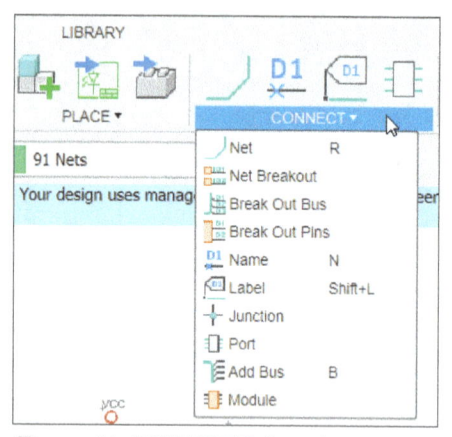

Figure-44. CONNECT drop-down

Creating Net

Net in electronics design is the web of electricity conducting thin copper strips used for connecting various components to perform operations. Net should always be created on a separate layer named Nets. This allows easy modification of all the net segments in the design. The procedure to create a net is given next.

- Click on the **Net** tool from the **CONNECT** drop-down in the **DESIGN** tab of **Ribbon**. The **NET Dialog** will be displayed; refer to Figure-45.

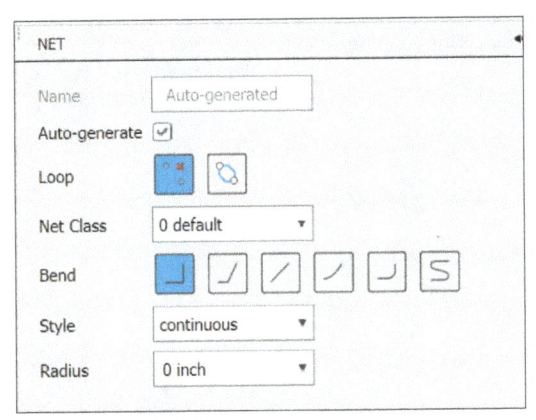

Figure-45. NET Dialog

- By default, the name of net segment is auto generated. To manually specify the name of net segment, clear the **Auto-generate** check box and specify desired name in the **Name** edit box.
- Select the **Loop Remove** button to remove extra loops from the net segment which do not have any component attached. Select the **Loop Preserve** button to keep extra loops of net segment.
- Select desired option from the **Net Class** drop-down to define the category of net you want to use for creating net segment. Note that net classes allow you to create nets of different properties. You will learn to create net classes later.
- Select desired button from the **Bend** section of Dialog to define the shape of bends to be created while drawing net.
- Select desired option from the **Style** drop-down to define the line style to be used for creating net.
- If you are using bend style which has fillet/radius at the intersection point then you can specify the radius value in the **Radius** edit box; refer to Figure-46.

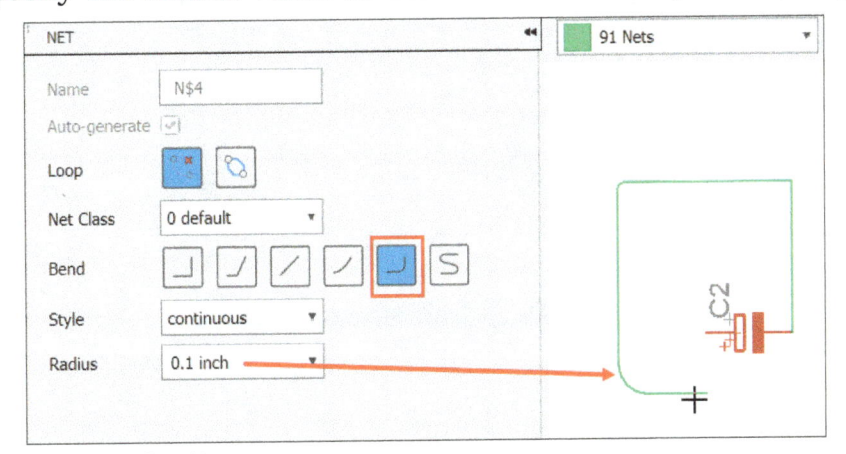

Figure-46. Bend in net

- After setting desired parameters, click at desired location/terminal point of component from the drawing area to specify start point of net segment. You will be asked to specify end point of the net segment; refer to Figure-47.

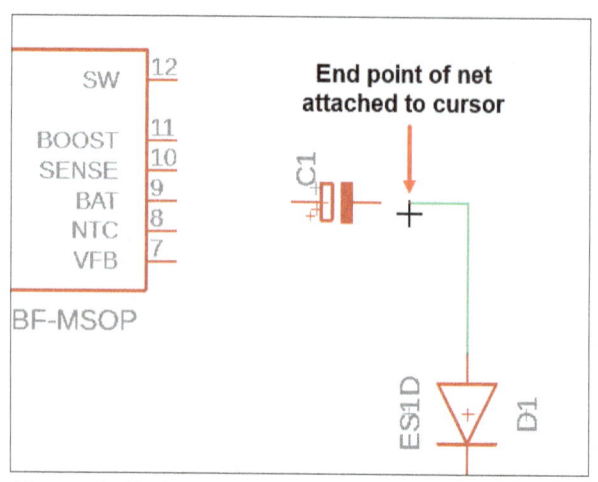

Figure-47. End point of net attached to cursor

- If you click at the terminal of another component then the current segment will be completed else you will be asked to specify next point of net segment.
- Click to specify the next point or press **ESC** to create current segment and start a new segment.
- Press **ESC** again to exit the tool.

Creating Net Class

The **DESIGN MANAGER** is used to create a new net class and set related parameters. The procedure to create net class is given next.

- Click on the **DESIGN MANAGER** from the left area in the application window and select **Nets** option from the **View** drop-down at the top in the **Manager**; refer to Figure-48.

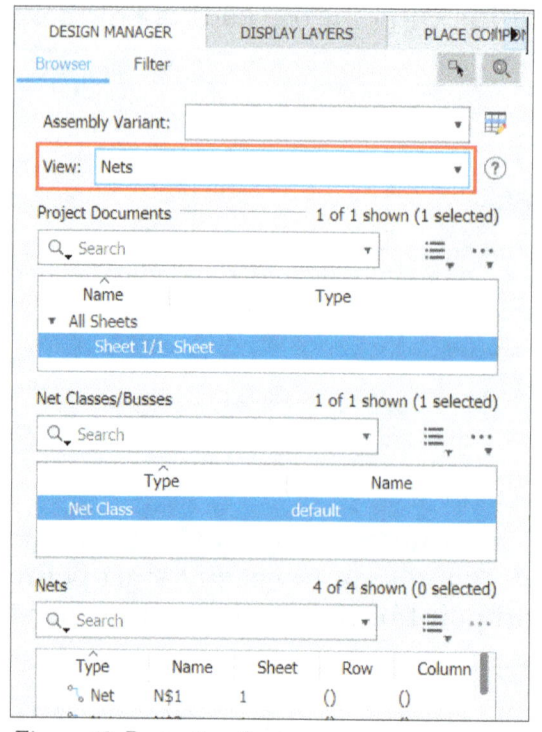

Figure-48. Properties of nets

- Click on the **New Netclass** option from the **Net Classes** drop-down; refer to Figure-49. The **Net classes** dialog box will be displayed; refer to Figure-50.

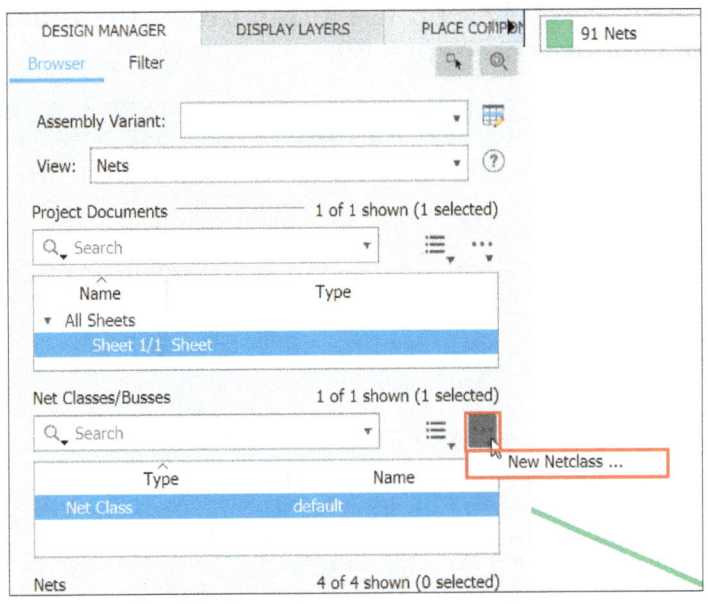

Figure-49. New Netclass option

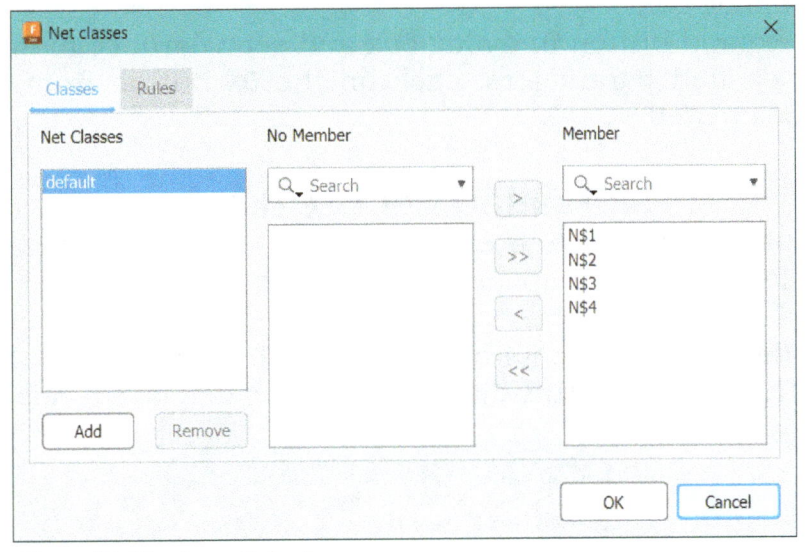

Figure-50. Net classes dialog box

- Click on the **Add** button from the dialog box. You will be asked to specify name of the class.
- Type desired name; refer to Figure-51 and press **ENTER**.

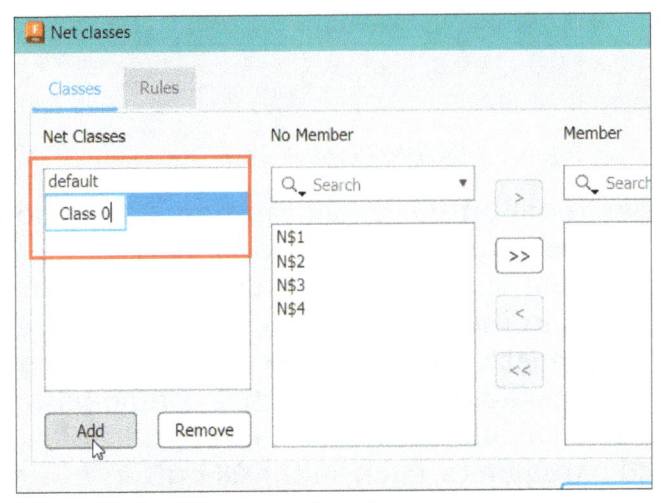

Figure-51. Specifying name of net class

- After creating the classes, click on the **Rules** tab of the dialog box. The options will be displayed as shown in Figure-52.

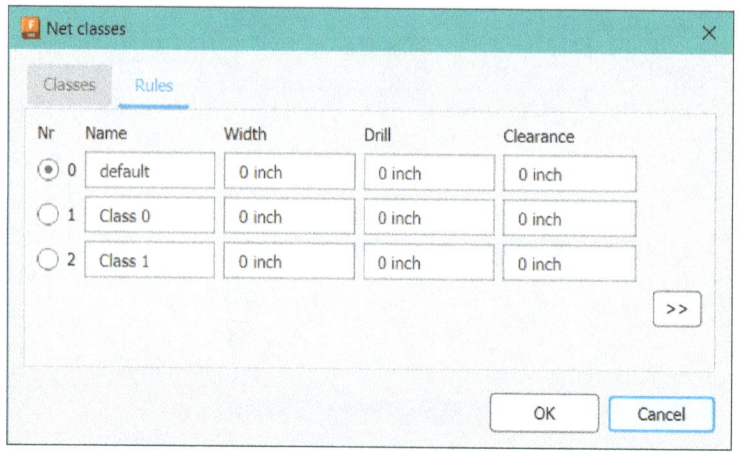

Figure-52. Rules tab of Net classes dialog box

- Specify desired name, width, drill, and clearance parameters in respective edit boxes of the dialog box.
- Click on the expand button in dialog box and set clearance matrix parameters.
- After setting desired parameters, click on the **OK** button from dialog box. The classes will be created.

Creating Net Breakout

The **Net Breakout** tool is used to break out pin and create small wire segment with annotation to selected pin of component. The procedure to use this tool is given next.

- Click on the **Net Breakout** tool from the **CONNECT** drop-down in the **DESIGN** tab of the **Ribbon**. The **Breakout Pins** dialog box will be displayed; refer to Figure-53.

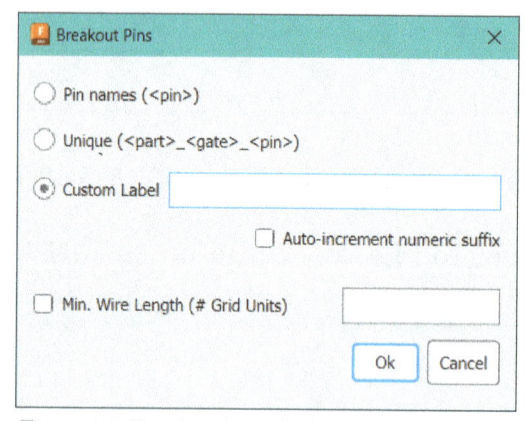

Figure-53. Breakout Pins dialog box

- Select the **Pin names** radio button to set pin name as annotation with extended wire. Select the **Unique** radio button to add part name-gate-pin number as annotation with extended wire. Select the **Custom Label** radio button and type desired annotation text in the edit box.
- Select the **Min. Wire Length (# Grid Units)** check box to specify the length of wire segment that will be attached to selected component pin. Note that length of wire segment will be calculated by number of grid units.
- After setting desired parameters, click on the **OK** button. You will be asked to select unoccupied pins of the component.

- Click on desired pin of the component to create net breakout; refer to Figure-54.

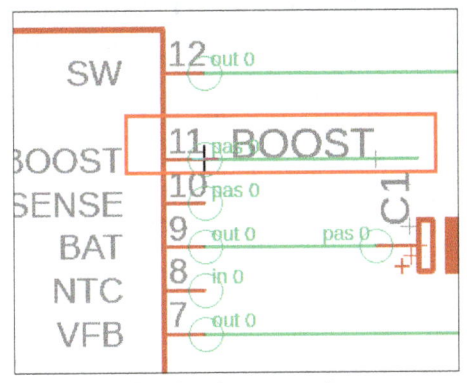

Figure-54. Pin breakout created

- After creating desired net breakouts, press **ESC** to exit the tool.

Drawing Bus

The **Add Bus** tool is used to create conductor lines in the schematic which are collections of net segments. If you assume wire as net then bus can be assumed as cable. The procedure to create bus is given next.

- Click on the **Add Bus** tool from the **CONNECT** drop-down in the **DESIGN** tab of the Ribbon. The **ADD BUS Dialog** will be displayed; refer to Figure-55.

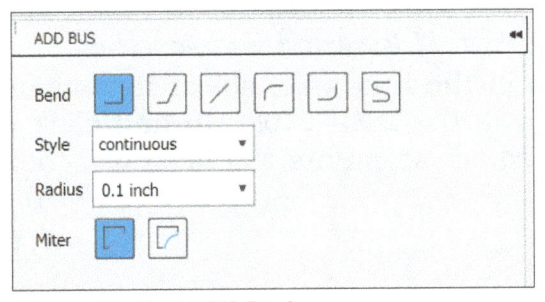

Figure-55. ADD BUS Dialog

- Select desired button from the **Bend** section to define shape of bend in the bus.
- Set the style and radius parameters as discussed earlier.
- Select the **Straight miter wire** button to create a straight line miter corner and select the **Round miter wire** button to create a round at the corner in bus wire; refer to Figure-56.

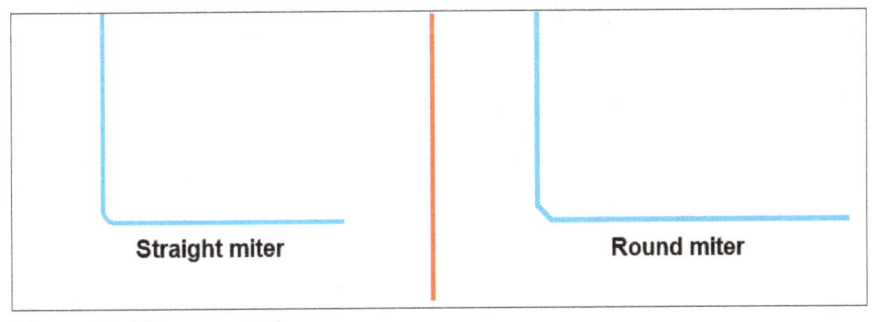

Figure-56. Miter types in bus

- After setting desired parameters, click at desired location to define start point and end point of the bus as discussed for net.

Adding Net Segments to Bus Wire

- After creating bus wire, select the bus wire from the drawing area. The parameters in **INSPECTOR** will be displayed as shown in Figure-57.

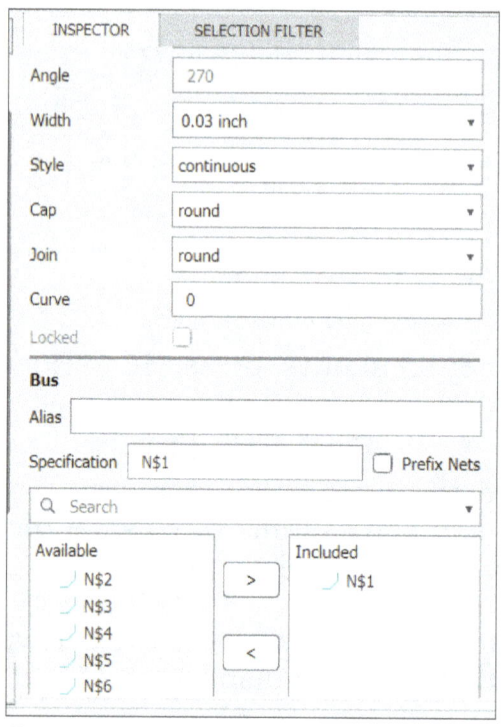

Figure-57. Inspector parameters for bus wire

- To add net segments in the bus wire, select the segments from the **Available** list box of **Bus** section in the **INSPECTOR** tab and click on the **>** button; refer to Figure-58. The selected net segments will be added to bus wire.

Figure-58. Adding net segment to Bus

Breaking Out Bus

The **Break Out Bus** tool is used to break out net segments from selected bus. The procedure to use this tool is given next.

- Click on the **Break Out Bus** tool from the **CONNECT** drop-down in the **DESIGN** tab of **Ribbon**. You will be asked to select a bus.
- Select desired bus wire from the drawing area. All the net segments will be displayed attached to cursor; refer to Figure-59.

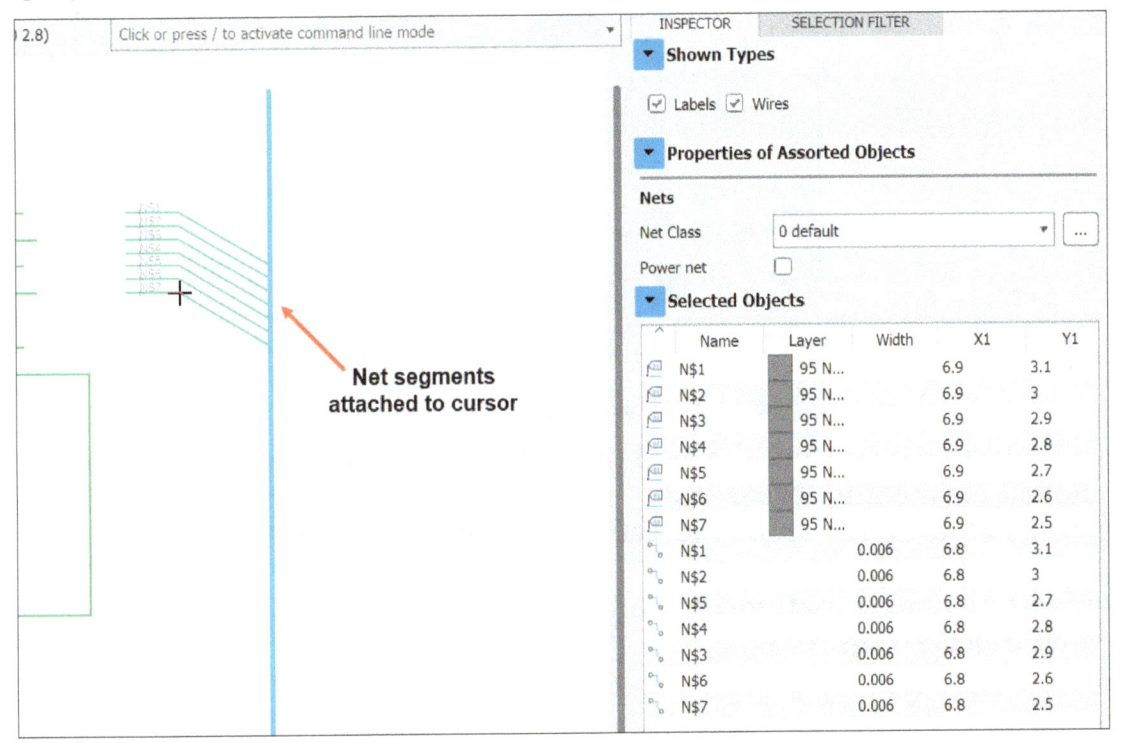

Figure-59. Net segments attached to cursor

- You can modify individual net segments of wire bus using the **INSPECTOR**.
- After setting desired parameters, click at desired location to create the break out.

Creating Part Pin Break Out

The **Break Out Pins** tool is used to create pin break outs of selected part. The procedure to use this tool is given next.

- Click on the **Break Out Pins** tool from the **CONNECT** drop-down in the **DESIGN** tab of the **Ribbon**. The **BREAK OUT PINS** dialog box will be displayed; refer to Figure-60. You will be asked to select the part/component for which you want to create breakouts.

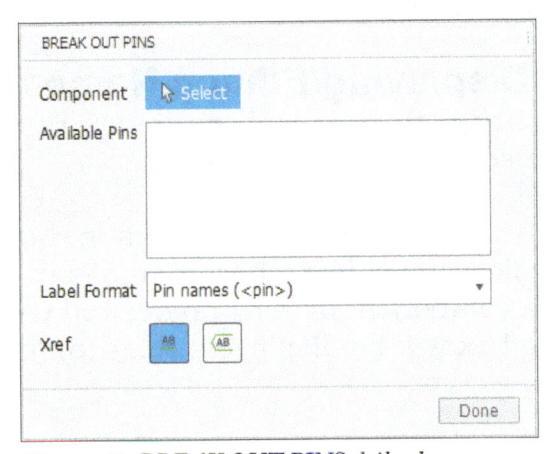

Figure-60. BREAK OUT PINS dailog box

- Select desired parts or components. The pins will be displayed in the dialog box; refer to Figure-61.

Figure-61. Pins available in the dialog box

- Select multiple pins from the **Available Pins** area of the dialog box to add them in the list for creating break out.
- Select desired option from the **Label format** drop-down to define the naming convention for pin break outs.
- After setting desired parameters, click on the **Done** button. The break outs will be created; refer to Figure-62.

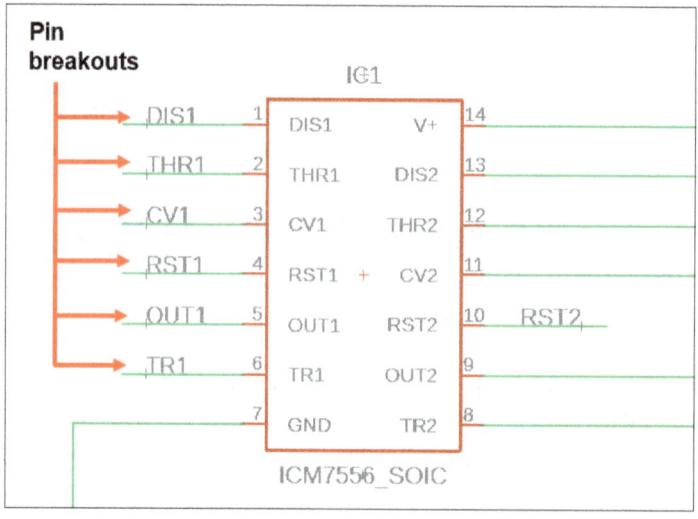

Figure-62. Created breakouts

Displaying/Editing Names

The **Name** tool is used to change the name of selected part, element, pins, gates, and so on. The procedure to use this tool is given next.

- Click on the **Name** tool from the **CONNECT** drop-down in the **DESIGN** tab of the **Ribbon**. You will be asked to select the object whose name is to be changed.
- Click on the **+** sign of a component or click on desired wire/bus/net to change its name. The **Name** dialog box will be displayed; refer to Figure-63.

Autodesk Fusion 360 PCB Black Book

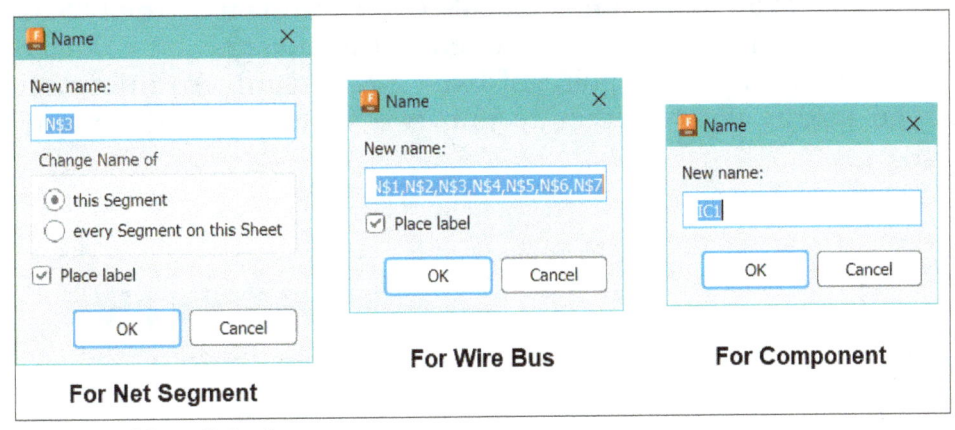

Figure-63. Name dialog box

- Note that in case of net segment, you can define whether you want to change the name of current segment or all the segments on the sheet.
- Change the name of component/net in the **New name** edit box and click on the **OK** button.

Assigning Label to Bus/Net

The **Label** tool is used to assign label to selected bus/net. The procedure to use this tool is given next.

- Click on the **Label** tool from the **CONNECT** drop-down in the **DESIGN** tab of **Ribbon**. The **LABEL Dialog** will be displayed at the left in the application window; refer to Figure-64 and you will be asked to select bus/net.

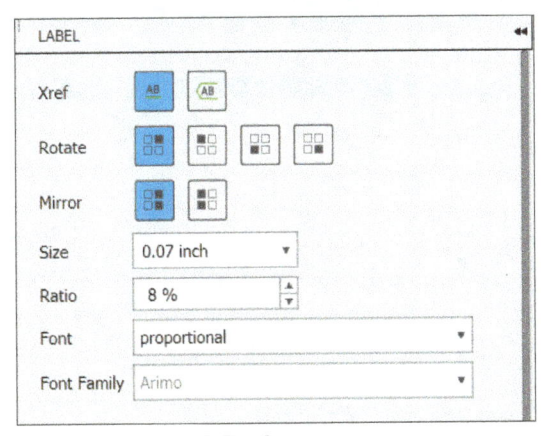

Figure-64. LABEL Dialog

- Click on desired net/bus from the drawing area. The label will get attached to cursor.
- Set desired parameters in the **Dialog** as discussed earlier and click at desired location to place the label.
- Press **ESC** to exit the tool.

Creating Junction

The **Junction** tool is used to place a junction dot at intersection of nets. The procedure to use this tool is given next.

- Click on the **Junction** tool from the **CONNECT** drop-down in the **DESIGN** tab of the **Ribbon**. The junction dot will get attached to the cursor.
- Click at desired intersection point of two net segment. An information box will be displayed asking you whether to join two segments and rename it; refer to Figure-65.

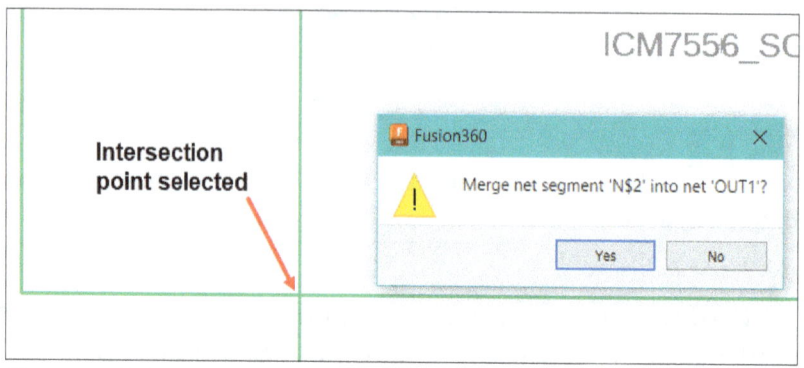

Figure-65. Information box

- Click on the **Yes** button to apply changes. Press **ESC** to exit the tool.

Creating Modules

The **Module** tool is used to custom modules that represent components which are not available in the library. The procedure to create module is given next.

- Click on the **Module** tool from the **CONNECT** drop-down in the **DESIGN** tab of the **Ribbon**. The **Select** dialog box will be displayed; refer to Figure-66.

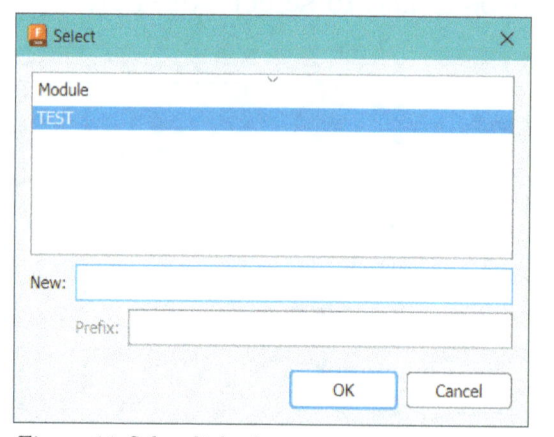

Figure-66. Select dialog box

- Type desired name of the module in **New** edit box, prefix name in the **Prefix** edit box, and click on the **OK** button. The module will get attached to cursor; refer to Figure-67.

Autodesk Fusion 360 PCB Black Book

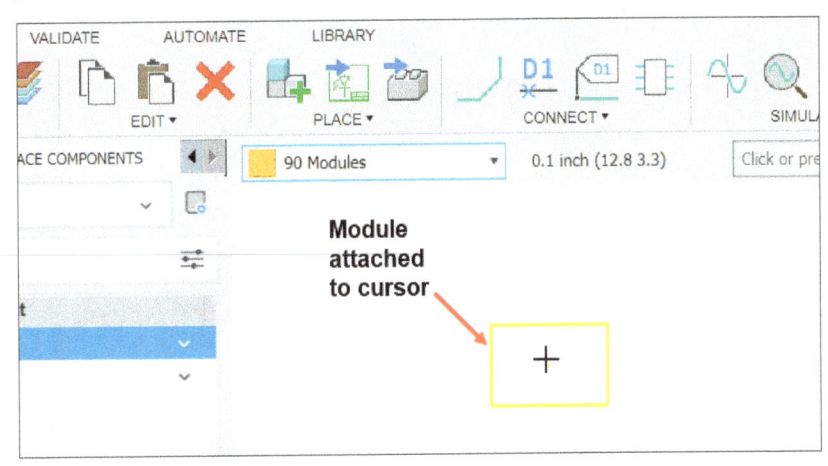

Figure-67. Module attached to cursor

- Click at desired location to place the module. Press **ESC** to exit the tool.

Creating Port in Module

The **Port** tool is used to add connection ports to selected module. The procedure to use this tool is given next.

- Click on the **Port** tool from the **CONNECT** drop-down in the **DESIGN** tab of **Ribbon**. The **Port Dialog** will be displayed; refer to Figure-68.

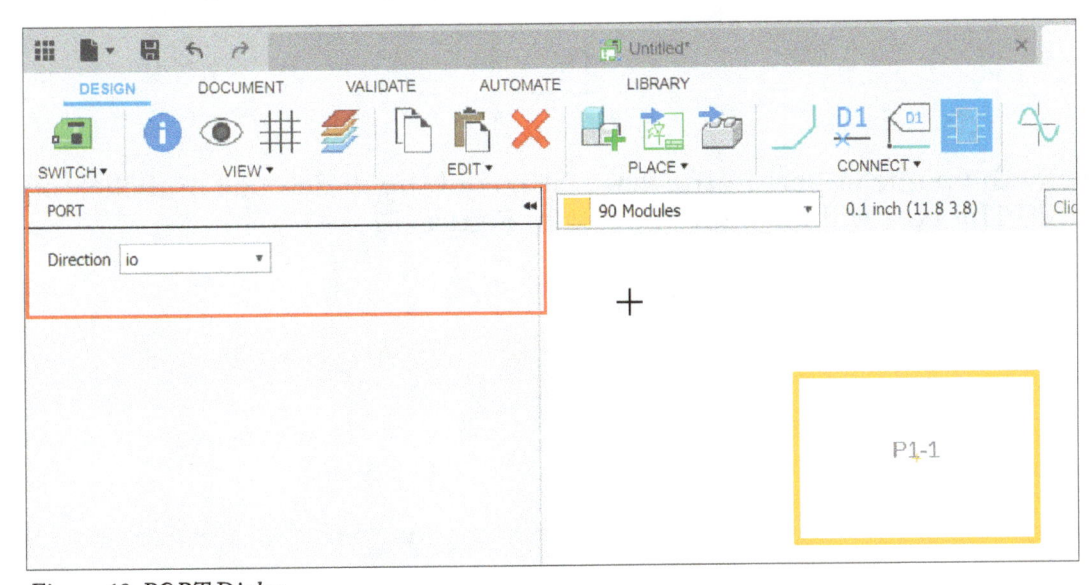

Figure-68. PORT Dialog

- Select desired port direction from the **Direction** drop-down. Select the **in**, **out**, **io**, **oc**, **pwr**, **nc**, **pass**, or **hiz** option from the drop-down to create respective port.
- After selecting desired option, click on the module at desired location to place the port. The port will get attached to cursor; refer to Figure-69.

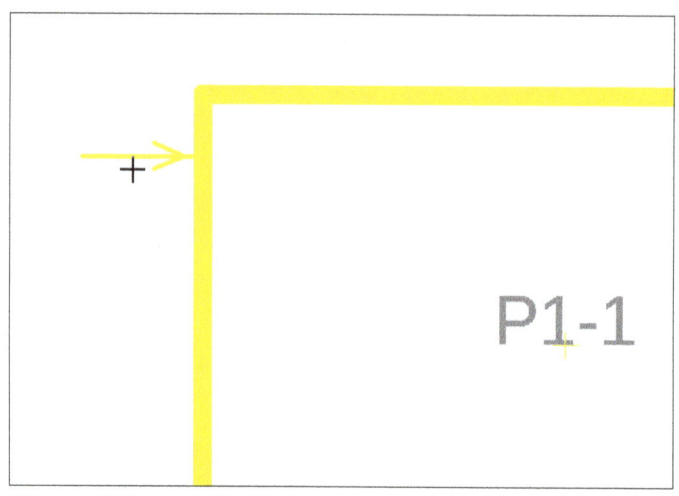

Figure-69. Port attached to cursor

- Click to place the port. The **Select** dialog box will be displayed; refer to Figure-70.

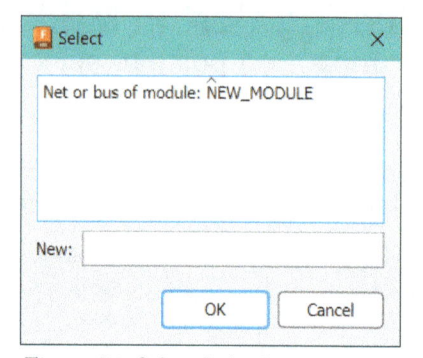

Figure-70. Select dialog box

- Specify the name of port in the **New** edit box of the dialog box and click on the **OK** button. The port will be created; refer to Figure-71.

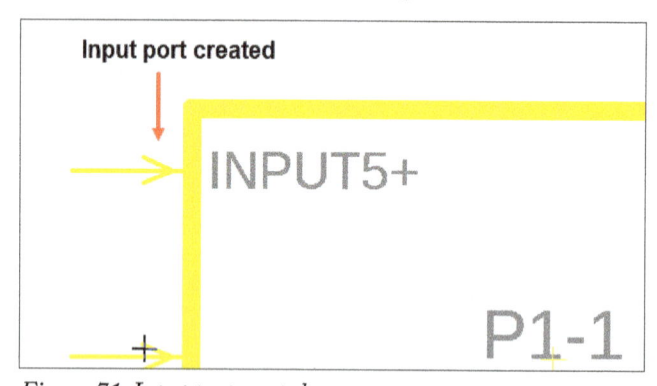

Figure-71. Input port created

- You can create more ports as discussed earlier or press **ESC** to exit the tool.

Self Assessment

Q1. Transistor is a passive electronic component. (T/F)

Q2. PMOS is a type of transistor. (T/F)

Q3. Magnetic devices use electrical field to store energy and release them as electrical discharge when required. (T/F)

Q4. What is the main purpose of a diode in circuit?

Q5. What is Integrated Circuit and what type of operations are performed by Integrated Circuits?

Q6. What is the workflow of Electronics Schematic Design in Autodesk Fusion PCB?

Q7. The .... is used to create re-configurable digital circuits.

Q8. The capacitors are used to store .... charge and release it when required in the circuit.

Q9. Which of the following is a fixed value resistor?

a) Rheostat   b) Power Resistor

c) Potentiometer   d) Trim pot

Q10. Which of the following tools is used to create a new net class?

a) Design Manager   b) Net Breakout

c) Junction   d) Module

Answers: **Ans 1.** F **Ans 2.** T **Ans 3.** F **Ans 7.** Programmable Logic Devices **Ans 8.** Electric **Ans 9.** b **Ans 10.** a

For Student Notes

Chapter 3

Modifying and Simulating Electronics Schematic

Topics Covered

The major topics covered in this chapter are:

- *Schematic Modification tools*
- *Rework tools*
- *Simulation tools*

INTRODUCTION

In the previous chapter, you learned about various tools related to creating schematic design. In this chapter, you will learn about various tools to modify the schematic design and perform simulations to test the design.

REWORKING ON WIRE/NET SEGMENTS

The tools in **REWORK** drop-down of **Ribbon** are used to modify various wire joints; refer to Figure-1. These tools are discussed next.

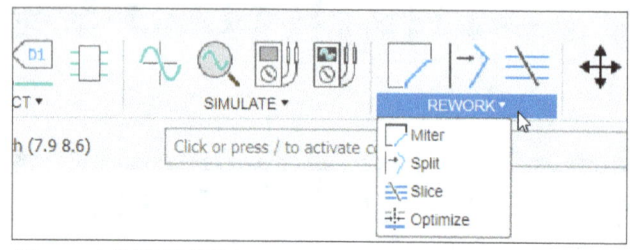

Figure-1. REWORK drop-down

Applying Miter Wire Joints

The `Miter` tool is used to change the corner point of two intersecting wires to miter cut. The procedure to use this tool is given next.

- Click on the `Miter` tool from the **REWORK** drop-down in the **DESIGN** tab of the **Ribbon**. The **MITER Dialog** will be displayed; refer to Figure-2.

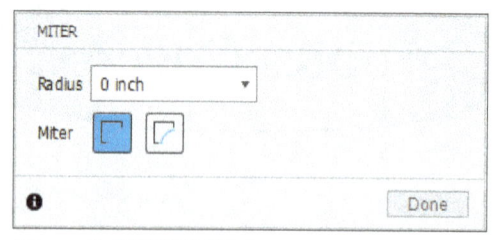

Figure-2. MITER Dialog

- Select desired button from the `Miter` section to define how miter cut will be created. Select the `Straight miter wire` button to create a chamfer at corner intersection of wires and select the `Round miter wire` button to create a fillet at the corner intersection of wires.
- Specify the size of chamfer/round in the **Radius** edit box.
- After setting desired parameters, click on the intersection point wires. The miter will be created; refer to Figure-3.

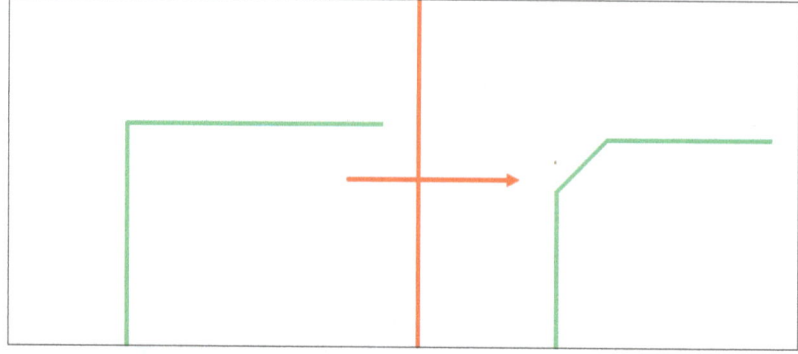

Figure-3. Miter joint created

- Press **ESC** to exit the tool.

Splitting Wire Segments/Polygon Edges

The **Split** tool is used to split selected wire segment/polygon edge into two segments at selected point. The procedure to use this tool is given next.

- Click on the **Split** tool from the **REWORK** drop-down in the **DESIGN** tab of the **Ribbon**. The **SPLIT Dialog** will be displayed; refer to Figure-4 and you will be asked to select wire to be split.

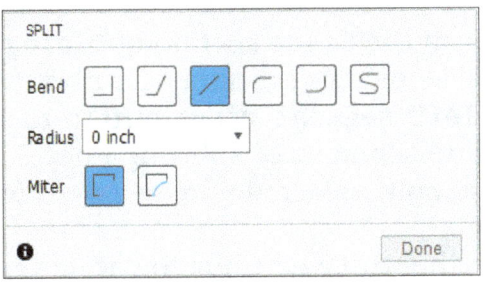

Figure-4. SPLIT Dialog

- Set the parameters in dialog box as discussed earlier and click at desired location on the wire to create segment. The split point of wire/net will get attached to cursor; refer to Figure-5.

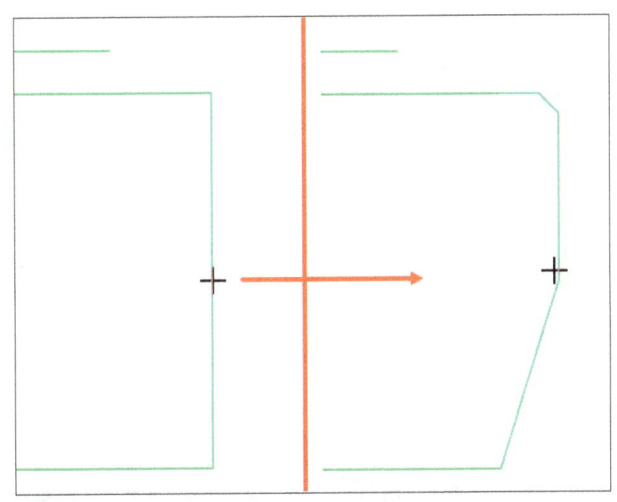

Figure-5. Splitting wire

- Click at desired location to apply the changes. You will be asked to specify split point for further splitting of line.
- Click at desired location to perform split or press **ESC** twice to exit the tool.

Slicing Wire Segments

The **Slice** tool is used to trim selected wire segment upto predefined width. The procedure to use this tool is given next.

- Click on the **Slice** tool from the **REWORK** drop-down in the **DESIGN** tab of the **Ribbon**. The **SLICE Dialog** will be displayed; refer to Figure-6 and you will be asked to create a slicing line.

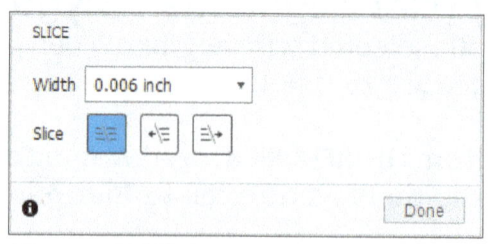

Figure-6. SLICE Dialog

- Select desired button from the **Slice** section of **Dialog**. Select the **Slice without ripup** button to split selected wire segment without deleting segment of wire. Select the **Slice with left segment ripup** button to split selected wire segment while deleting the left side segment of splitting line. Select the **Slice with right segment ripup** button to split selected wire segment while deleting the right side segment of splitting line.
- Specify desired value in the **Width** edit box to define the distance upto which wire segment will be deleted when using the **Slice** tool.
- After setting desired parameters, create a split line starting from bottom to top; refer to Figure-7. The slice will be created; refer to Figure-8.

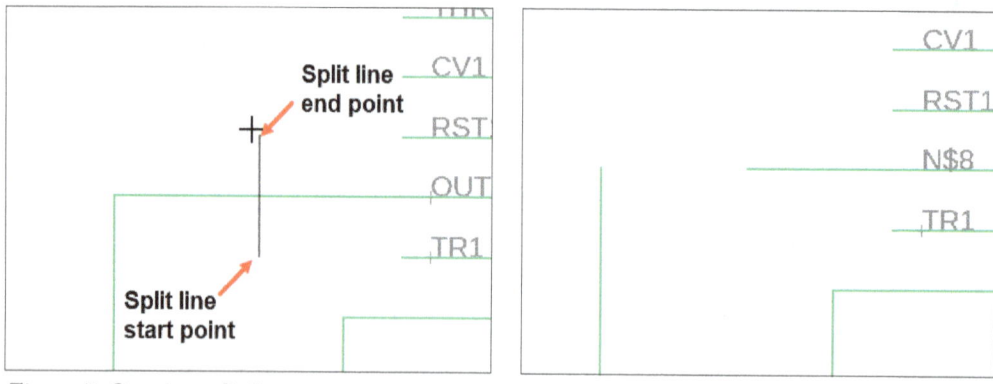

Figure-7. Creating split line *Figure-8. After applying slice*

- Press **ESC** to exit the tool.

Optimizing/Joining Wire Segments

The **Optimize** tool is used to join wire segments that lie in straight line. The procedure to use this tool is given next.

- Click on the **Optimize** tool from the **REWORK** drop-down in the **DESIGN** tab of **Ribbon**. You will be asked to select bus or net segment.
- Select the wire segments that lie on same layer, have same width, and lie in straight line. The wire segments will be joined.

MODIFYING SCHEMATIC DESIGN

The tools in the **MODIFY** drop-down are used to perform modifications like move, rotate, change, arrange, align, and so on; refer to Figure-9. Various tools of this drop-down are discussed next.

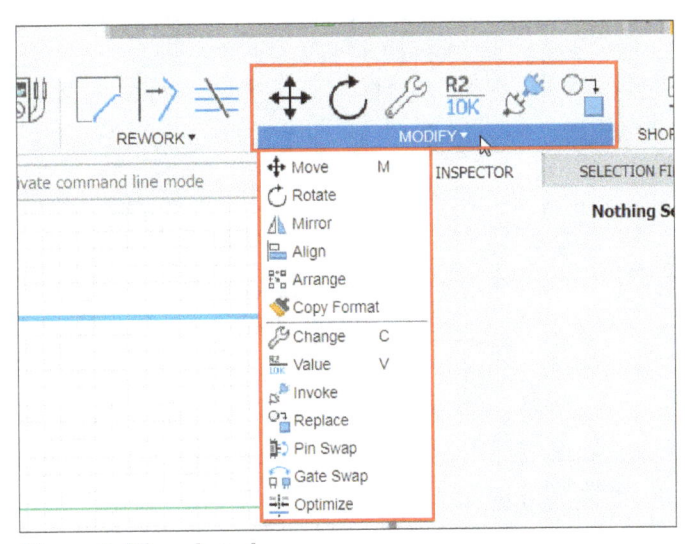

Figure-9. Move drop-down

Moving Selected Objects

The **Move** tool is used to move selected component, net, wire segment, pin, gate, and so on. The procedure to use this tool is given next.

- Click on the **Move** tool from the **MODIFY** drop-down in the **DESIGN** tab of the **Ribbon**. The **MOVE Dialog** will be displayed; refer to Figure-10.

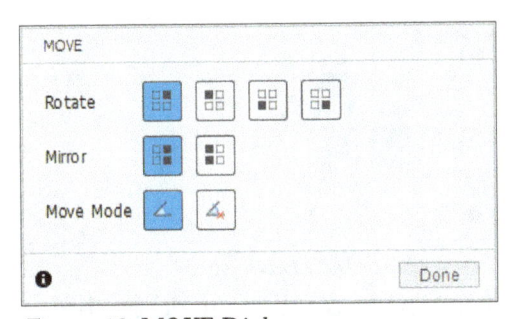

Figure-10. MOVE Dialog

- Select desired buttons from the **Dialog** to perform respective modifications while moving the component. Select the **Preserve Angles** button to keep the orientation of connected wire segments same. Select the **Free** button to freely move the component in drawing area.
- Click on the **+** sign of the component to be moved. The selected component will get attached to cursor.
- Click at desired location to place the component; refer to Figure-11.

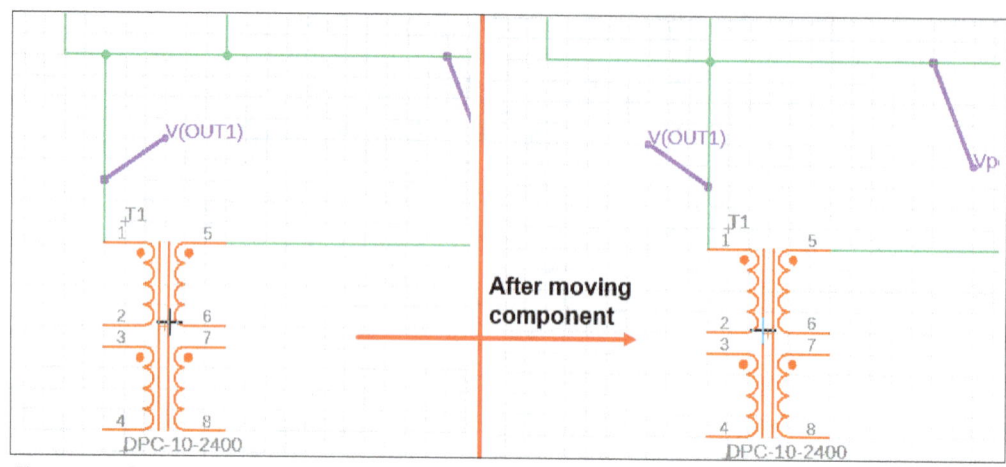

Figure-11. Moving component

- Press **ESC** to exit the tool.

Rotating Objects

The **Rotate** tool is used to rotate selected objects. The procedure to use this tool is given next.

- Click on the **Rotate** tool from the **MODIFY** drop-down in the **DESIGN** tab of the **Ribbon**. The **ROTATE Dialog** will be displayed; refer to Figure-12.

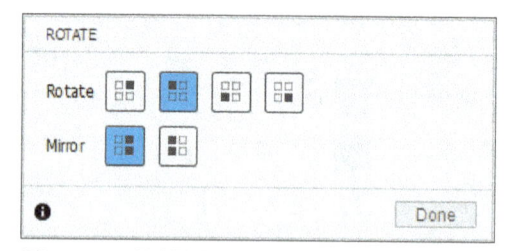

Figure-12. ROTATE Dialog

- Set desired buttons in the **Dialog** as discussed earlier and click on the component to rotate it.
- Press **ESC** to exit the tool.

Mirroring Objects

The **Mirror** tool is used to flip the component as mirror image. The procedure to use this tool is given next.

- Click on the **Mirror** tool from the **MODIFY** drop-down in the **DESIGN** tab of the **Ribbon**. You will be asked to select the component to be flipped.
- Select desired component from the drawing area. The component will be flipped; refer to Figure-13.
- Press **ESC** to exit the tool.

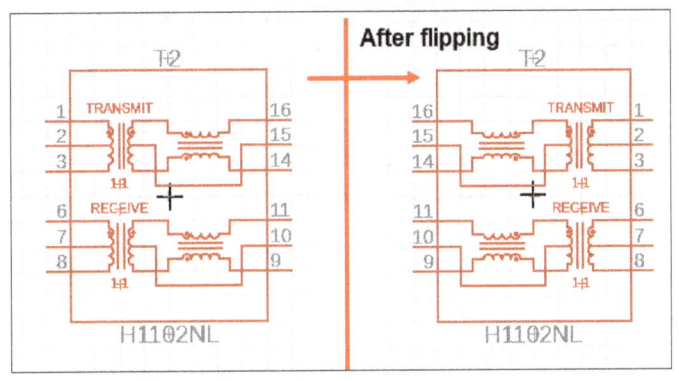

Figure-13. After applying mirror

Aligning Objects

The **Align** tool is used to align two or more selected objects. The procedure to use this tool is given next.

- Select all the objects to be aligned while holding the **CTRL** key and click on the **Align** tool from the **MODIFY** drop-down in the **DESIGN** tab of the **Ribbon**. The **ALIGN Dialog** will be displayed; refer to Figure-14.

Figure-14. ALIGN Dialog

- Select desired buttons from the **Dialog** to apply alignment operation. Press **ESC** to exit the tool.

Arranging Components

The **Arrange** tool is used to arrange selected components in linear or circular pattern. The procedure to use this tool is given next.

- Select the components to be arranged from the drawing area and click on the **Arrange** tool from the **MODIFY** drop-down in the **DESIGN** tab of the **Ribbon**. The **ARRANGE Dialog** will be displayed; refer to Figure-15.

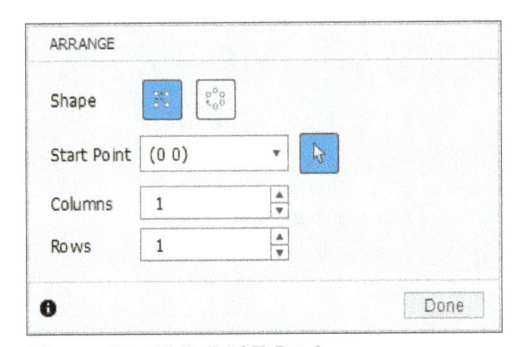

Figure-15. ARRANGE Dialog

Rectangular Arrange

- Select the **Rectangular Arrange** button from the **ARRANGE Dialog** to arrange components in rectangular pattern.
- Click at desired location to specify start point for arranging components.
- Set desired values in **Columns** and **Rows** edit box to define how many columns and rows can be created. Preview of arrangement will be displayed; refer to Figure-16.

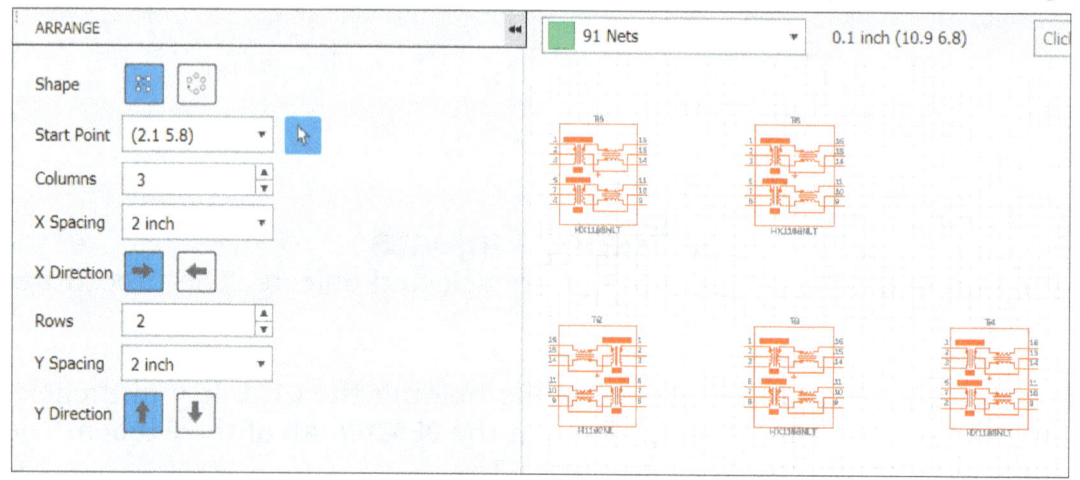

Figure-16. Rectangular arrangement of components

- Click on the **Done** button to create the arrangement.

Circular Arrange

- Select the **Circular Arrange** button from the **Shape** section of **ARRANGE Dialog**. The options to arrange components in circular pattern will be displayed.
- Set the parameters as discussed earlier. The circular arrangement will be created; refer to Figure-17.

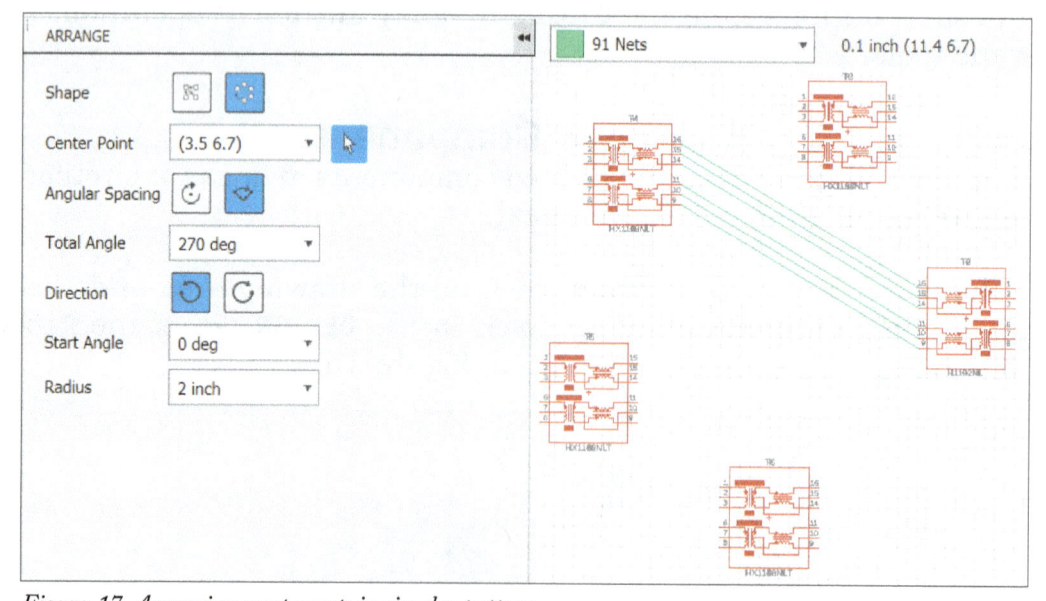

Figure-17. Arranging components in circular pattern

- Press **ESC** to exit the tool.

Copying Format

The **Copy Format** tool is used to copy properties of selected object and transfer those properties to similar type of other component. The procedure to use this tool is given next.

- Click on the **Copy Format** tool from the **MODIFY** drop-down in the **DESIGN** tab of the **Ribbon**. You will be asked to select the source object whose properties are to be copied.
- Select desired object from the drawing area. The **Select properties to copy** dialog box will be displayed; refer to Figure-18.

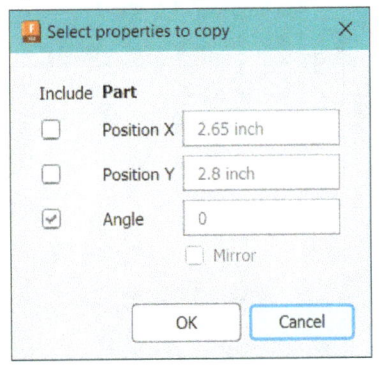

Figure-18. Select properties to copy dialog box

- Select check boxes for parameters to be included when copying and click on the **OK** button. You will be asked select to the component to which properties will be copied.
- Select desired component. The properties will be applied to destination component; refer to Figure-19.

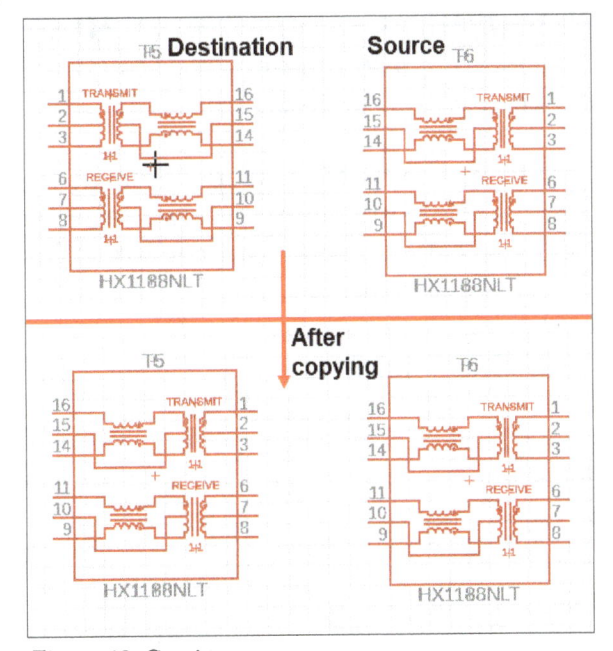

Figure-19. Copying

- Press **ESC** to exit the tool.

Changing Properties of Selected Object

The **Change** tool is used to change various properties of selected object. The procedure to use this tool is given next.

- Click on the **Change** tool from the **MODIFY** drop-down in the **DESIGN** tab of **Ribbon**. The **Change** menu will be displayed; refer to Figure-20.

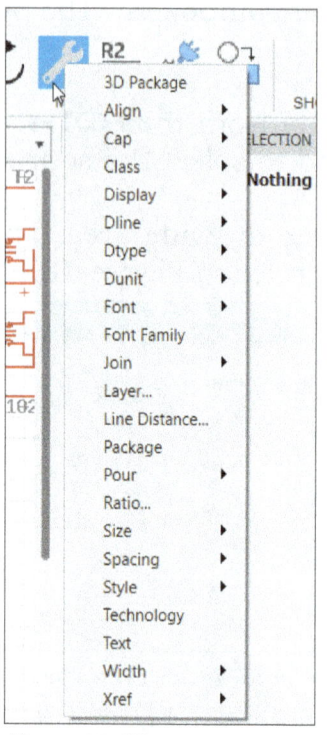

Figure-20. Change menu

- Select desired option from the menu and apply changes to selected object.
- Press **ESC** to exit the tool.

Modifying Component Values

The **Value** tool is used to modify value of components like resistor, capacitors, and so on. The procedure to use this tool is given next.

- Click on the **Value** tool from the **MODIFY** drop-down in the **DESIGN** tab of **Ribbon**. You will be asked to select the component whose value is to be changed.
- Select desired component (at its + icon) from the graphics area. The **Value** dialog box will be displayed; refer to Figure-21.

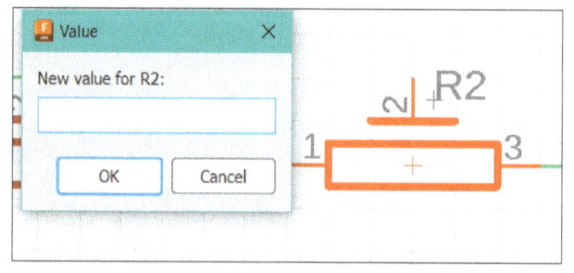

Figure-21. Value dialog box

- Specify desired value in the edit box of **Value** dialog box and click on the **OK** button to apply the value. Press **ESC** to exit the tool.

Invoking/Fetching Gates from Device

The **Invoke** tool is used to fetch the power connection details of selected logic IC. You can use this tool to place power symbols for the logic IC components. The procedure to use this tool is given next.

- Click on the **Invoke** tool from the **MODIFY** drop-down in the **DESIGN** tab of the **Ribbon**. The **INVOKE Dialog** will be displayed; refer to Figure-22 and you will be asked to select the part.

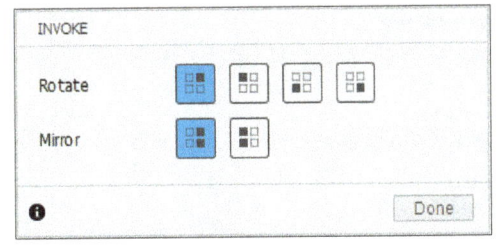

Figure-22. INVOKE Dialog

- Set desired parameters in the **Dialog** and select the IC component with logic gate. The **Invoke** dialog box will be displayed; refer to Figure-23.

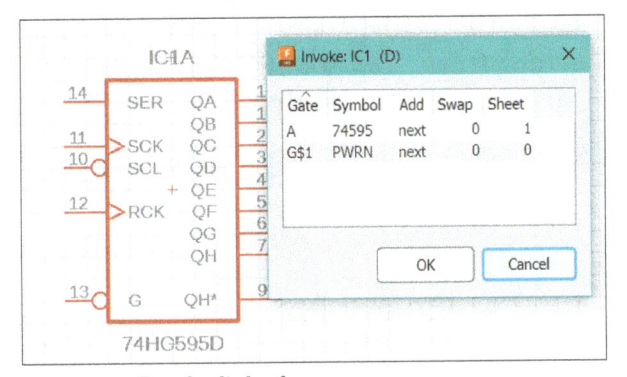

Figure-23. Invoke dialog box

- Select the power symbol PWRN from the list in dialog box and click on the **OK** button. The symbol will get attached to cursor; refer to Figure-24.

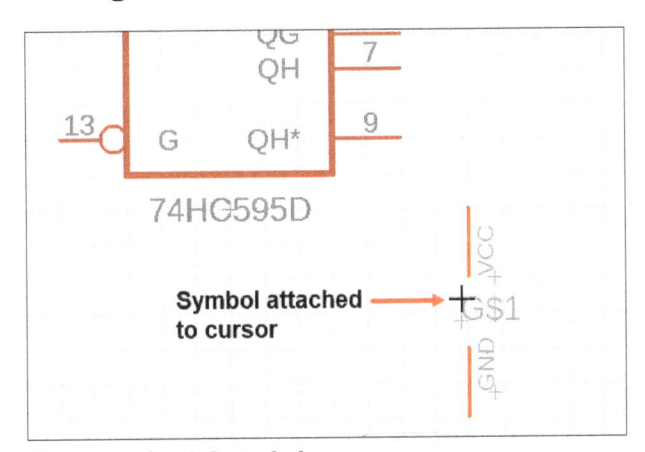

Figure-24. Symbol attached to cursor

- Set desired orientation of the symbol by using buttons in the dialog box and place it at appropriate location on the component; refer to Figure-25.

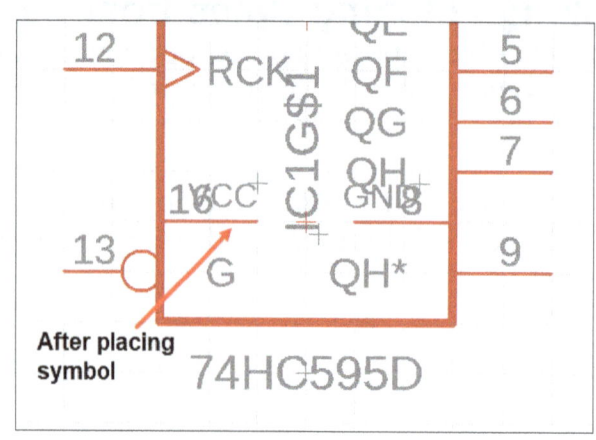

Figure-25. After placing symbol

- Press **ESC** to exit the tool.

Replacing Part

The **Replace** tool is used to replace selected part with another part. The procedure to use this tool is given next.

- Click on the **Replace** tool from the **MODIFY** drop-down in the **DESIGN** tab of the **Ribbon**. The **REPLACE** dialog box will be displayed; refer to Figure-26.

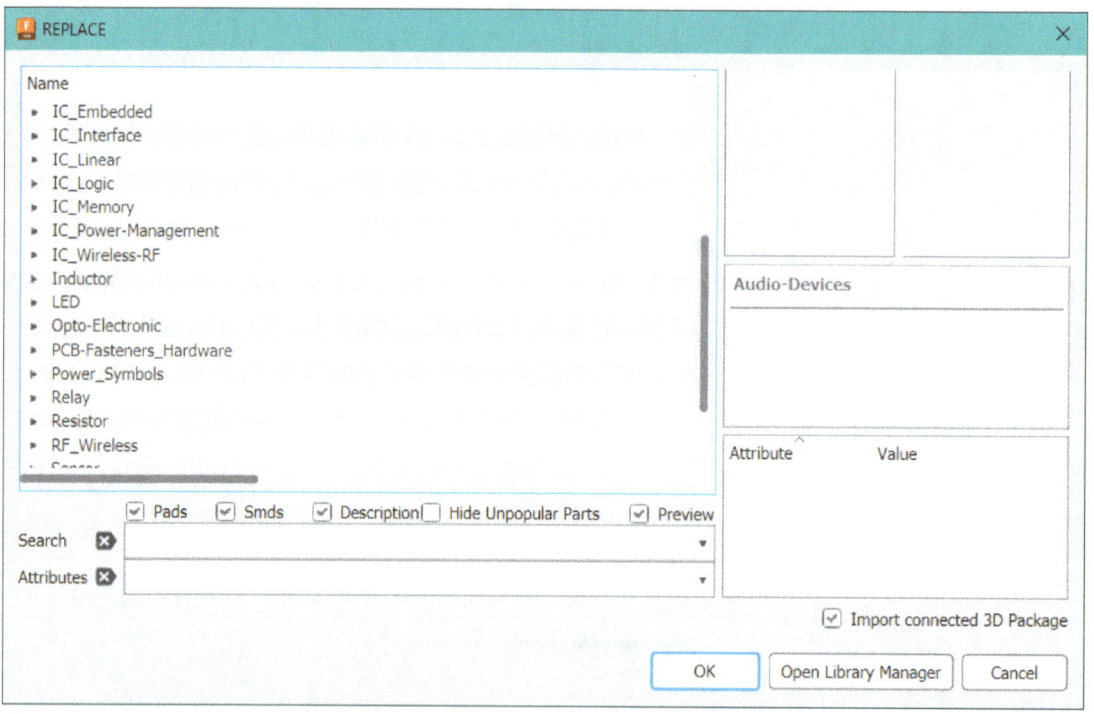

Figure-26. REPLACE dialog box

- Expand category and sub-category of component that you want to use as replacement component and double-click on desired component; refer to Figure-27. You will be asked to select component to be replaced from the drawing area.

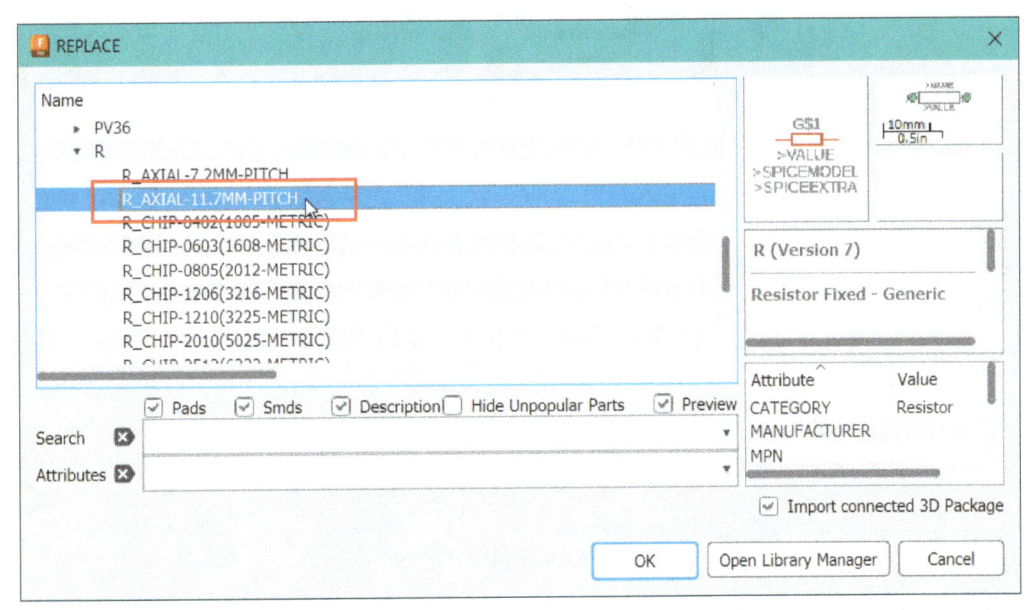

Figure-27. Selecting replacement component

- Click on the component to be replaced; refer to Figure-28.

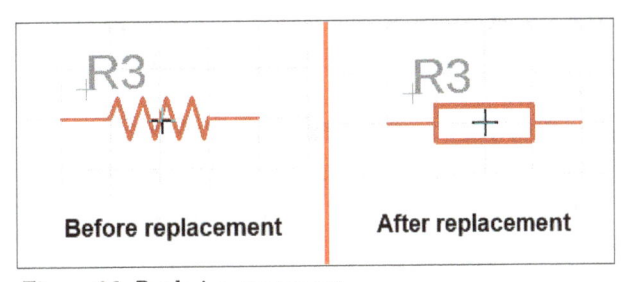

Figure-28. Replacing component

- Press **ESC** to exit the tool.

Performing Pin Swap

The **Pin Swap** tool is used to interchange selected pins of a component. Note that there are some components like resistor, non-polarized capacitors, and so on. Only the pins which are set to swap level other than 0 can be swapped by the tool. The procedure to use this tool is given next.

- Click on the **Pin Swap** tool from the **MODIFY** drop-down in the **DESIGN** tab of the **Ribbon**. You will be asked to select the first pin to be swapped.
- Select the first pin of component. You will be asked to select the second pin.
- Select the other pin of component. The pins will be swapped; refer to Figure-29.

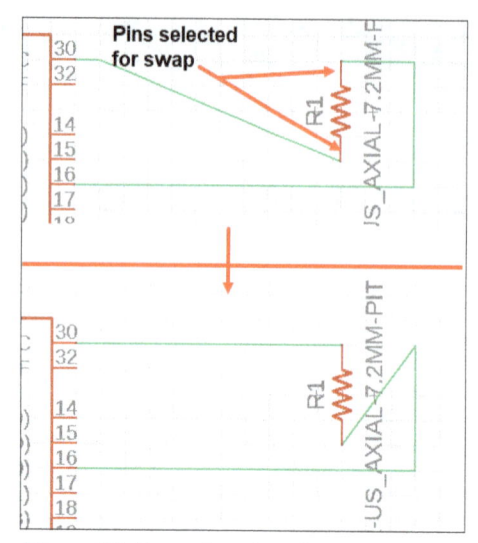

Figure-29. Swapping pins of component

- Press **ESC** to exit the tool.

Applying Gate Swap

The **Gate Swap** tool, as name suggests, is used to swap two selected gate components in schematic. The procedure to use this tool is given next.

- Click on the **Gate Swap** tool from the **MODIFY** drop-down in the **DESIGN** tab of the **Ribbon**. You will be asked to select the gates to be swapped.
- One by one select the two gates to be swapped to perform operation; refer to Figure-30.

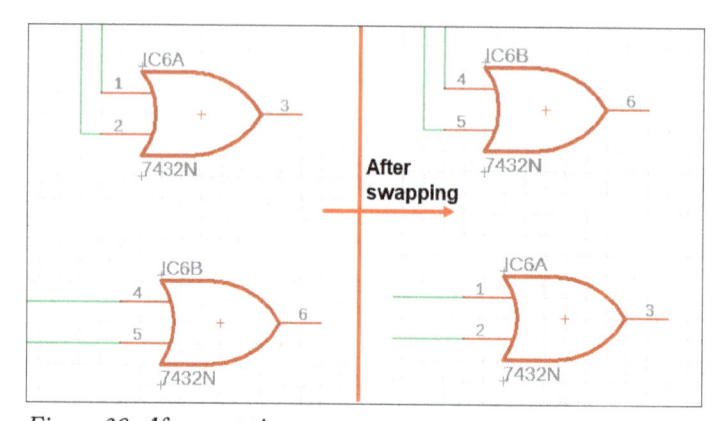

Figure-30. After swapping gate

- Press **ESC** to exit the tool.

Optimizing Wire Nets

The **Optimize** tool is used to connect various wire net segments which lie in straight line and have same properties. The procedure to use this tool is given next.

- Click on the **Optimize** tool from the **MODIFY** drop-down in the **DESIGN** tab of the **Ribbon**. You will be asked to select the wire net to be optimized.
- Select the straight wire net which has multiple mid points to remove extra midpoints.
- Press **ESC** to exit the tool.

SIMULATION

The tools in **SIMULATE** drop-down are used to perform various signal and electronic simulations; refer to Figure-31. Autodesk Fusion uses SPICE to perform electronic simulations. Various tools of this drop-down are discussed next.

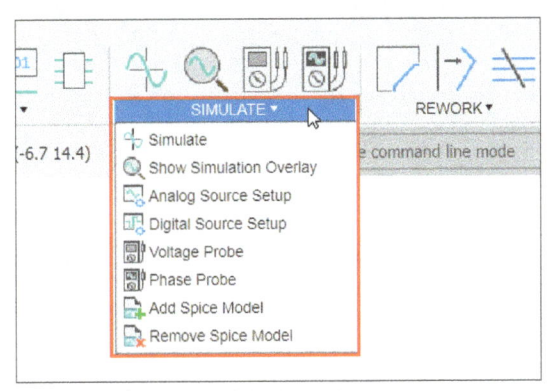

Figure-31. SIMULATE drop-down

Adding Spice Model

The schematic generally created in Fusion 360 PCB is not a SPICE model. To perform simulation study, we need spice model that represents the schematic. SPICE stands for Simulation Program with Integrated Circuit Emphasis. Check the Wikipedia link on SPICE (https://en.wikipedia.org/wiki/SPICE) for more information. The procedure to convert schematic component to SPICE model is given next.

- After creating schematic circuit, click on the **Add Spice Model** tool from the **SIMULATE** drop-down in the **DESIGN** tab of the **Ribbon**. You will be asked to select the component to be made SPICE simulation compatible.
- Select desired component from graphic area. The **Add Spice Model** dialog box will be displayed; refer to Figure-32.

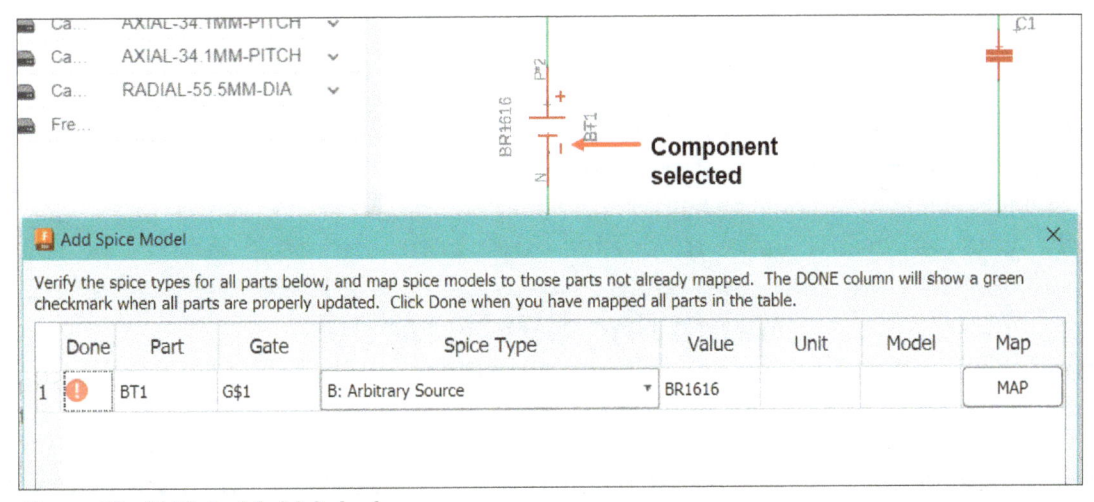

Figure-32. Add Spice Model dialog box

- Select desired option from drop-down in the **Spice Type** column to define the spice type of component. For example, in this case we want to define spice type for a battery so we will select **V: Independent Voltage Source** option from the drop-down.

- After selecting the option from **Spice Type** drop-down, click on the **MAP** button to map pins of schematic symbol with SPICE model. The **Map To Model** dialog box will be displayed; refer to Figure-33.

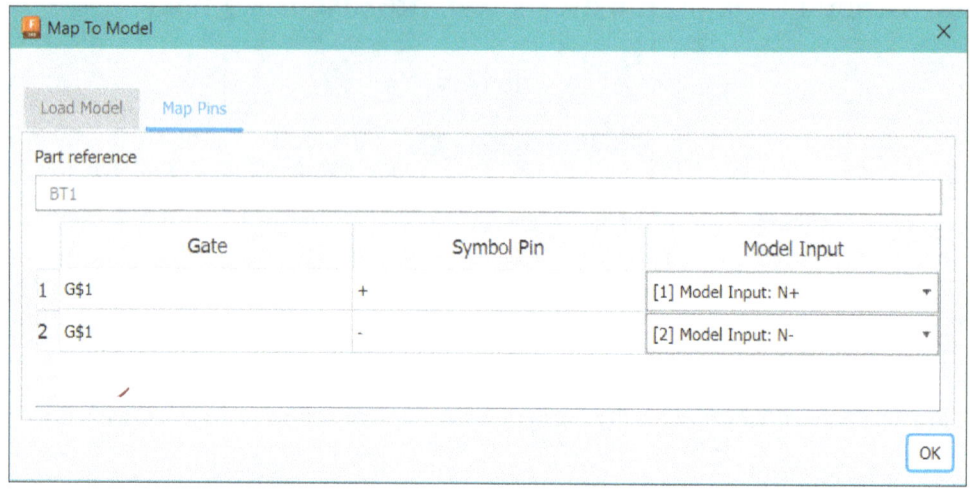

Figure-33. Map To Model dialog box

- Map the + terminal of component with + input and - terminal of component with - input.
- Click on the **OK** button to apply changes. The **Add Spice Model** dialog box will be displayed again with a green tick mark against the component which signifies that the component is properly mapped; refer to Figure-34.

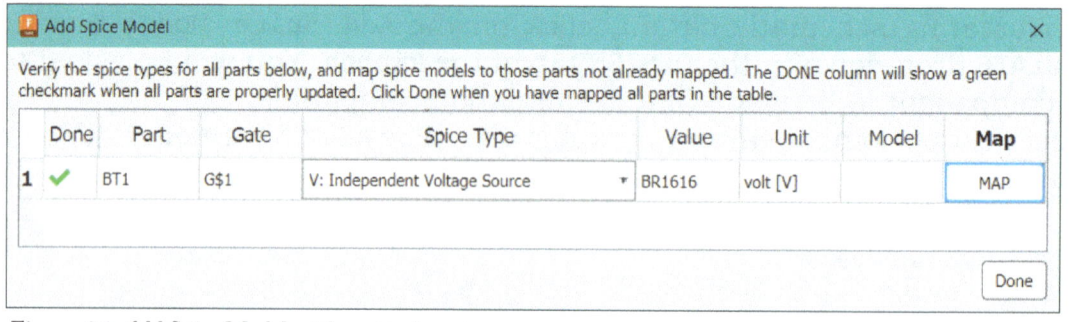

Figure-34. Add Spice Model with mapped component

- Click on the **Done** button to apply changes.

Similarly, you can create spice model for other components.

Removing SPICE Model

The **Remove Spice Model** tool is used to remove SPICE attributes of selected component. The procedure to use this tool is given next.

- Click on the **Remove Spice Model** tool from the **SIMULATE** drop-down in the **DESIGN** tab of the **Ribbon**. You will be asked to select the component whose SPICE attributes are to be removed.
- Select desired component from the graphics area. A message box will be displayed telling you that mapping has been removed; refer to Figure-35.

Figure-35. SPICE mapping removed message box

Analog Source Setup

The `Analog Source Setup` tool is used to define voltage/current data for selected SPICE compatible voltage/current source. The procedure to use this tool is given next.

- Click on the `Analog Source Setup` tool from the **SIMULATE** drop-down in the **DESIGN** tab of the **Ribbon**. You will be asked to select the voltage/current source.
- Select desired voltage/current source component from graphics area. The `Source Setup` dialog box will be displayed; refer to Figure-36.

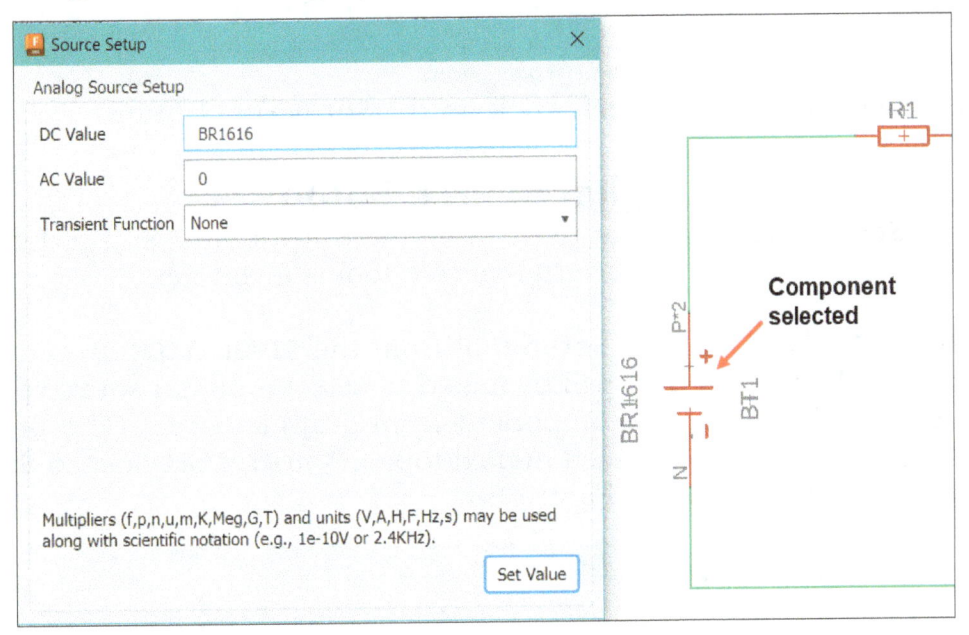

Figure-36. Source Setup dialog box

- Specify desired value in the `DC Value` edit box if the source of current/voltage is DC supply and specify desired value in `AC Value` edit box if the source is AC supply.
- Select desired option from the `Transient Function` drop-down to define whether the supply is time dependent or fixed. Select `Sinusoidal` option from the drop-down if you want to use sinusoidal waveform for current (generally in case of alternating current); refer to Figure-37. Select `Exponential` option if you want to create current/voltage source changing as per exponential curve parameters specified in the dialog box. Select `Pulse` option from the drop-down if the power is released in pulses from the source.

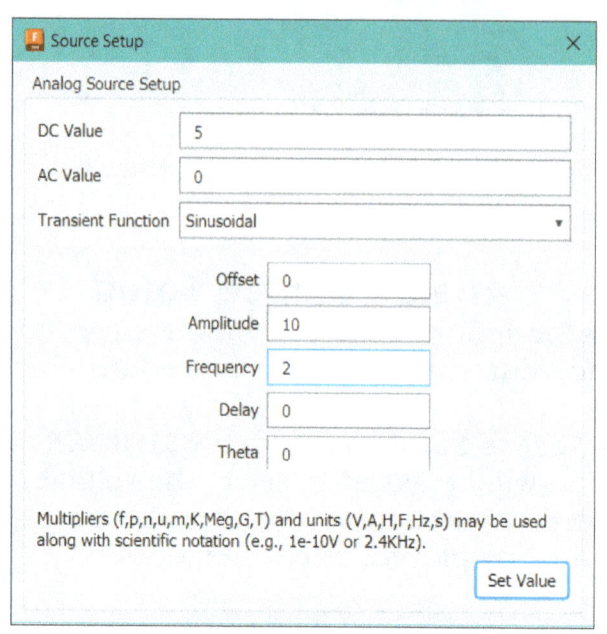

Figure-37. Sinusoidal voltage source

- After setting desired parameters, click on the **Set Value** button.

Digital Source Setup

The **Digital Source Setup** tool is used to define output values of a digital source with respect to time. The procedure to use this tool is given next.

- Click on the **Digital Source Setup** tool from the **SIMULATION** drop-down in the **DESIGN** tab of the **Ribbon**. You will be asked to select a digital source component.
- Select desired digital source component from graphics area. (The procedure to insert a digital source is discussed in next topic). The **Digital Source Setup** dialog box will be displayed; refer to Figure-38.

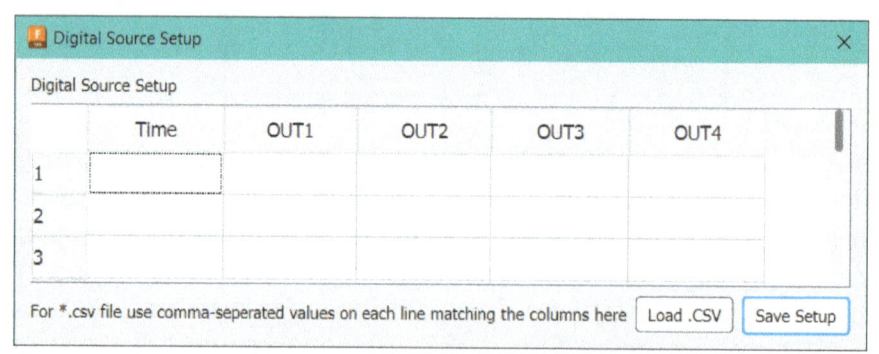

Figure-38. Digital Source Setup dialog box

- Specify the 0 or 1 values (digital output) for various outputs with respect to time in respective columns of the table; refer to Figure-39.

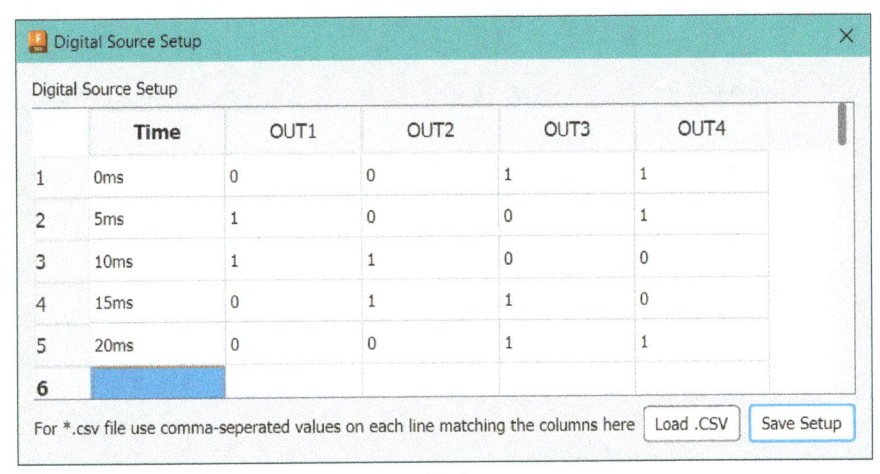

Figure-39. Digital source parameters specified

- Click on the **Load .CSV** button from dialog box if you want to use a CSV spreadsheet file for entries in the dialog box.
- After setting desired parameters, click on the **Save Setup** button to apply parameters and exit the dialog box.

Inserting Digital Source Component in Schematic

The procedure to insert a digital source component is similar to inserting any other component in schematic. For inserting SPICE compatible digital source, you need to activate the ngspice-digital library. The procedure to insert digital source component is given next.

- Click on the **Open Library Manager** button from the **PLACE COMPONENTS MANAGER**; refer to Figure-40. You can also access this tool from **LIBRARIES** panel in the **LIBRARY** tab of the **Ribbon**. The **Library Manager** will be displayed; refer to Figure-41.

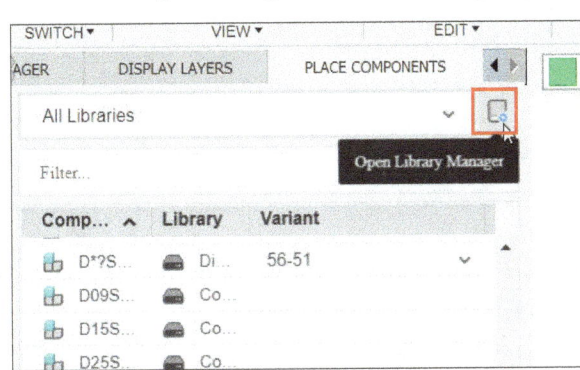

Figure-40. Open Library Manager button

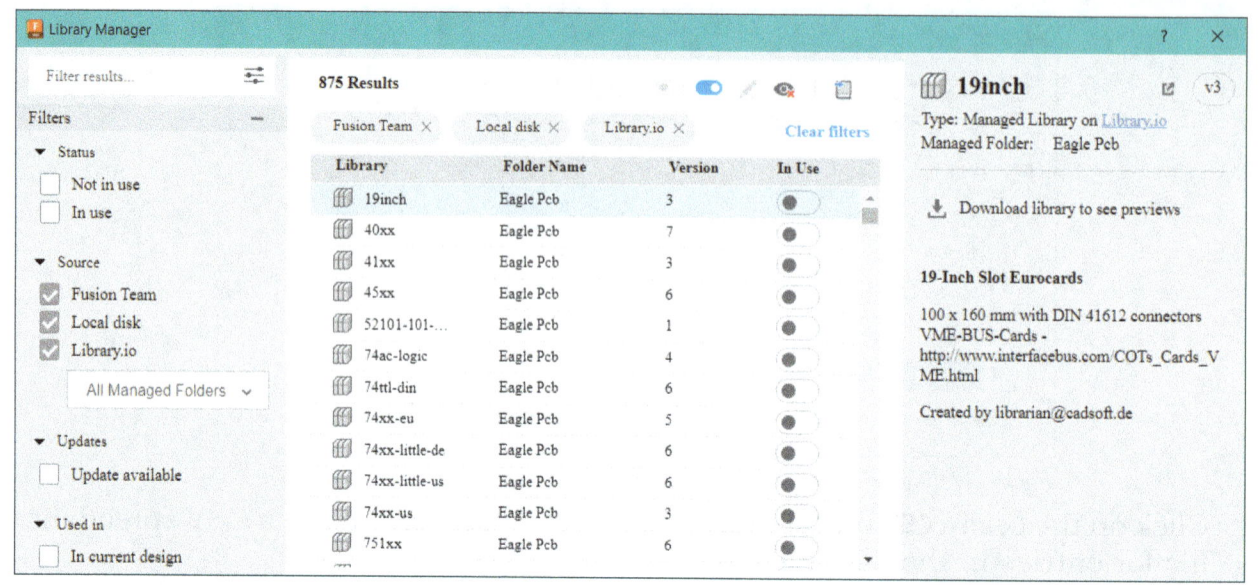

Figure-41. Library Manager

- Make sure the **Library.io** check box is selected in the **Source** section at the left in the dialog box and then type **spice** in **Filter results** edit box. The options will be displayed as shown in Figure-42.

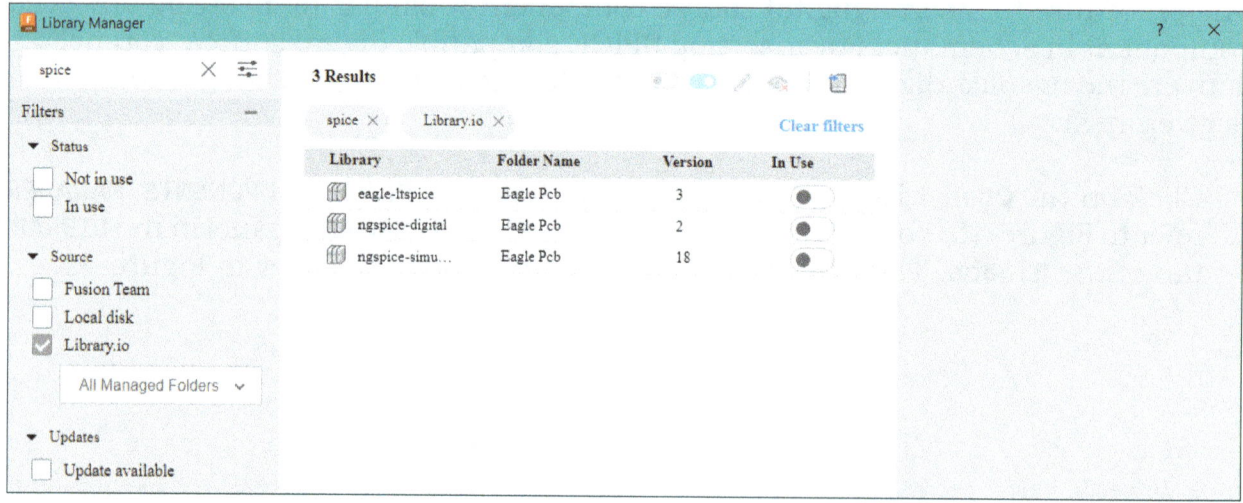

Figure-42. Spice libraries after filtering

- Select the **In Use** toggle button for ngspice-digital library from list to activate the respective library for component insertion.
- Similarly, select the **In Use** toggle button for ngspice-simulation library to use SPICE compatible analog electronic components.
- Close the **Library Manager** by using **Close** button at top-right corner.
- Click on the **Libraries** drop-down at the top in **PLACE COMPONENTS MANAGER** and select the **ngspice-digital** option from drop-down; refer to Figure-43. The components will be displayed as shown in Figure-44.

Autodesk Fusion 360 PCB Black Book 3-21

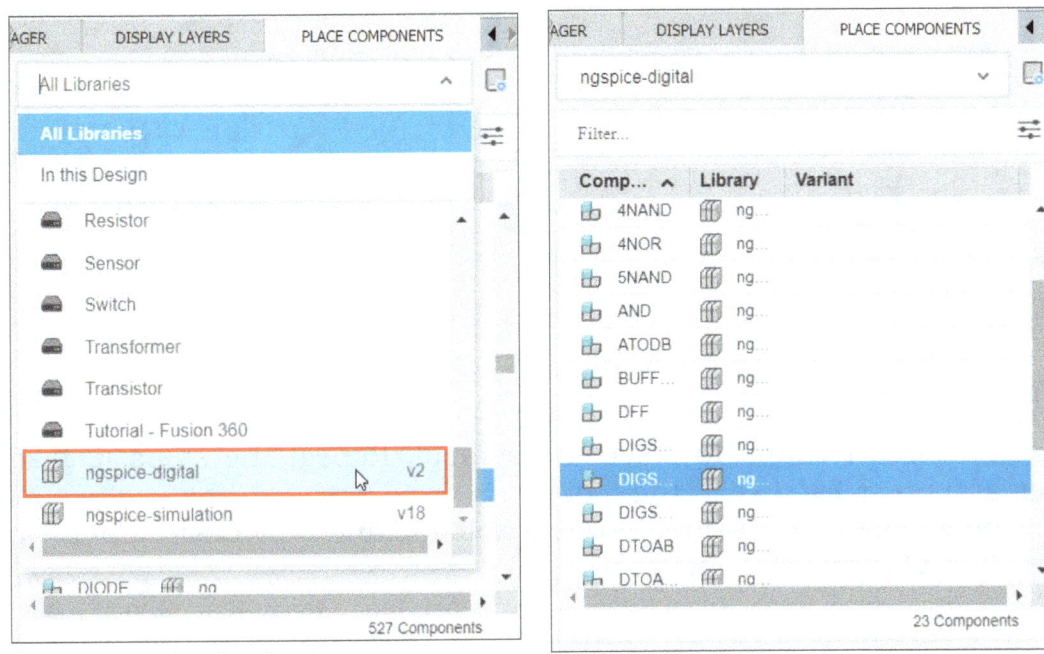

Figure-43. ngspice-digital option

Figure-44. ngspice-digital components

- Double-click on desired **DIGSRC** component from the component list in **MANAGER**. The component will be attached to cursor.
- Click at desired location in schematic to place the component; refer to Figure-45.

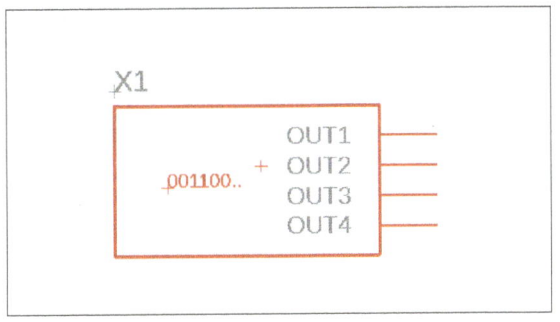

Figure-45. Digital source component

Similarly, you can place SPICE compatible analog components using the **ngspice-simulation** library.

Applying Voltage Probe

The **Voltage Probe** tool is used to place voltage measuring sensors at desired locations on circuit to output voltage in simulation output. The procedure to use this tool is given next.

- Click on the **Voltage Probe** tool from the **SIMULATE** drop-down in the **DESIGN** tab of the **Ribbon**. The **VOLTAGE PROBE Dialog** will be displayed; refer to Figure-46 and you will be asked to specify location of probe.
- Set the parameters as discussed earlier in the **VOLTAGE PROBE Dialog** and click on the circuit at the location where you want to check the voltage. The end point of voltage probe will get attached to cursor; refer to Figure-47.

Figure-46. VOLTAGE PROBE Dialog

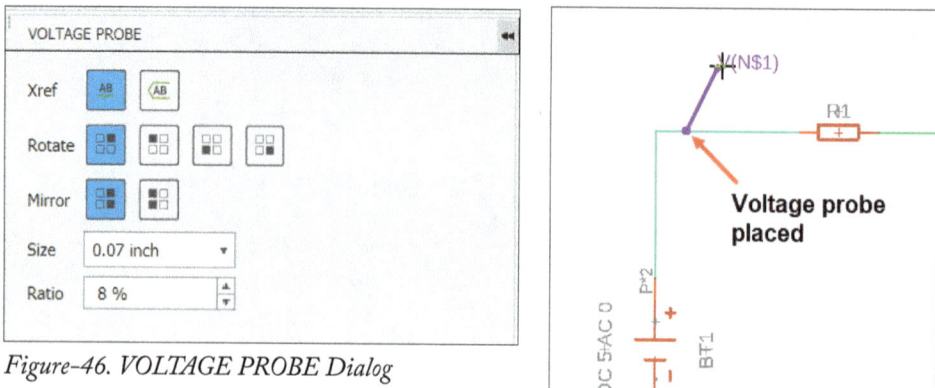

Figure-47. Placing voltage probe

- Click at desired location to place the symbol. Repeat the steps to place more probes and press **ESC** to exit the tool.
- Select the voltage probe symbol from schematic and type desired name of probe in the **INSPECTOR**; refer to Figure-48.

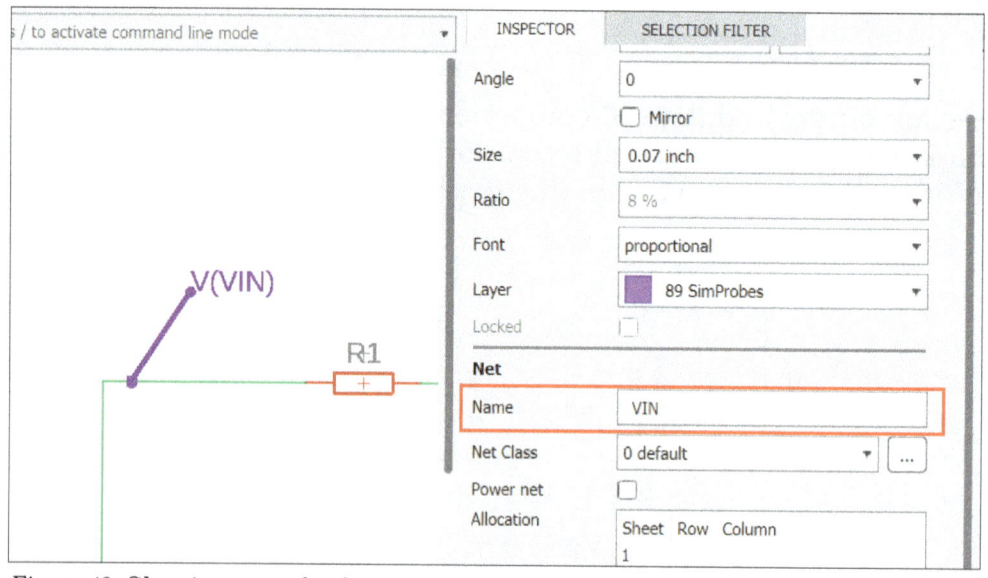

Figure-48. Changing name of probe

Applying Phase Probe

The **Phase Probe** tool is used to measure voltage phase relative to AC source in the circuit. This tool is useful for AC Sweep analysis. The procedure to use this tool is similar to the **Voltage Probe** tool.

PERFORMING SIMULATION

The **Simulate** tool is used to perform SPICE simulation in Fusion 360 PCB schematic. In Fusion 360, you can perform four types of SPICE simulation:

- Operation Point
- DC Sweep
- AC Sweep
- Transient Analysis

These simulations are discussed next.

Operating Point Simulation

The Operating Point simulation is performed to check voltage and current readings on various probes at the startup of circuit operation. Note that to perform any SPICE simulation, all the components in circuit must be mapped with SPICE compatible output types. The procedure to perform this simulation is discussed next.

- Click on the **Simulate** tool from the **SIMULATE** drop-down in the **DESIGN** tab of the **Ribbon**. The **SIMULATION** dialog box will be displayed; refer to Figure-49.

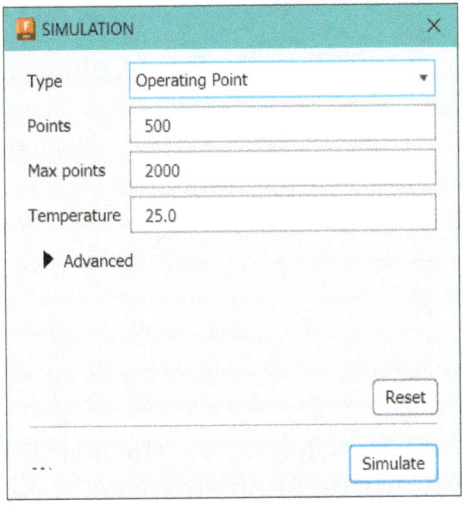

Figure-49. SIMULATION dialog box

- Select the **Operating Point** option from the **Type** drop-down to perform respective simulation.
- Specify the number of node points in circuit for which you want to calculate the voltage and current parameters.
- Specify desired value in **Max points** edit box of the dialog box to define maximum number of nodes upto which the parameters will be calculated during simulation if needed by analysis.
- Specify desired value in the **Temperature** edit box to define the operating temperature for circuit in real-world. Note that some components operate differently at various temperatures.
- Expand the **Advanced** node to modify advanced parameters of the analysis; refer to Figure-50.
- Specify desired value in the **ABSTOL** edit box to specify absolute current error tolerance allowed in simulation results. The default value is 1 picoamp.
- Specify desired value in **GMIN** edit box to specify the minimum conductance allowed by program. The default value for this parameter is 1e-12.
- Specify desired value in the **ITL1** edit box to specify maximum number of iterations allowed in DC simulation.
- Specify desired value in the **PIVREL** edit box to specify the relative ratio between the largest column entry and an acceptable pivot value. The default value is 1.0e-3. In the numerical pivoting algorithm, the allowed minimum pivot value is determined by EPSREL=AMAX1(PIVREL*MAXVAL, PIVTOL) where MAXVAL is the maximum element in the column where a pivot is sought (partial pivoting).
- Specify desired value in the **PIVTOL** edit box to specify the absolute minimum value for a matrix entry to be accepted as a pivot. The default value is 1.0e-13.
- Specify desired value in the **RELTOL** edit box to specify the relative error tolerance of the program. The default value is 0.001 (0.1%).

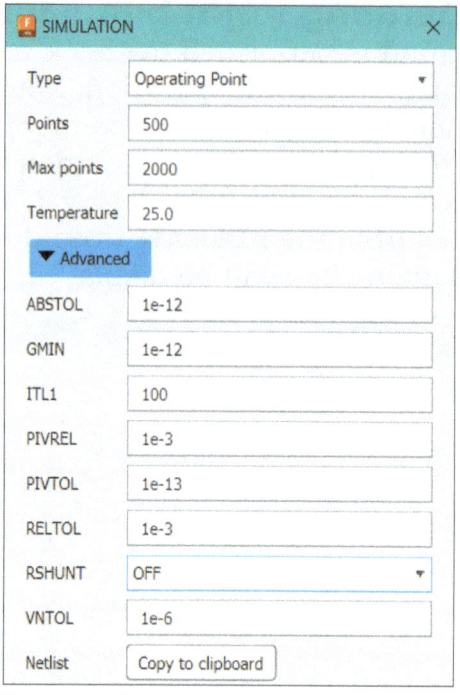

Figure-50. Advanced parameters of Simulation

- By default, the **OFF** option is selected in the **RSHUNT** drop-down, so no shunt resistor is added in circuit during analysis to avoid singularity in analysis. Select the **ON** option from drop-down if you want to use a shunt resistor to circuit during analysis.
- Specify desired value in the **VNTOL** edit box to define absolute voltage tolerance for the analysis.
- After setting desired parameters, click on the **Simulate** button from the dialog box. The results of analysis will be displayed; refer to Figure-51.

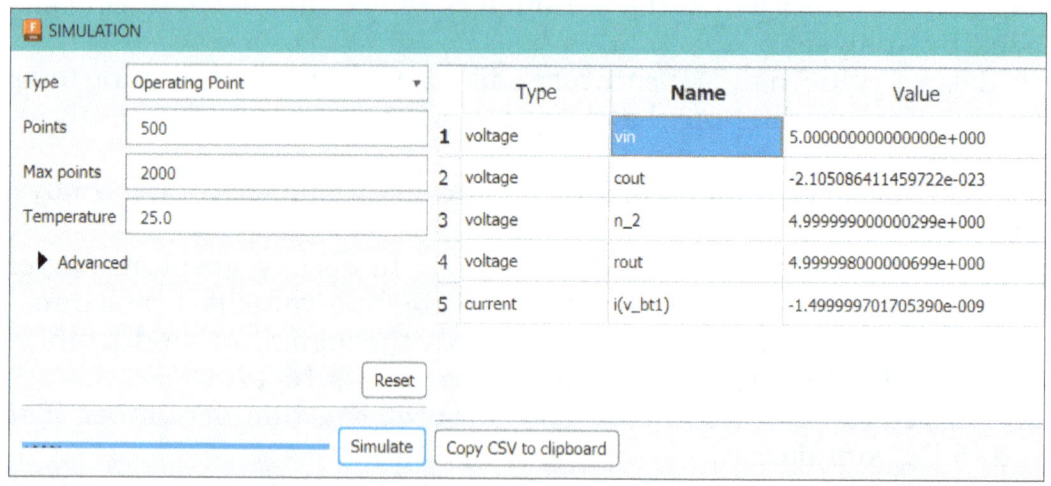

Figure-51. Simulation result

DC Sweep Simulation

The DC Sweep simulation is performed to check response of circuit in specified range of DC voltage. The procedure to perform this simulation is given next.

- Click on the **Simulate** tool from the **SIMULATE** drop-down in the **DESIGN** tab of the **Ribbon**. The **SIMULATION** dialog box will be displayed.

- Select the **DC Sweep** option from the **Type** drop-down to create a DC Sweep analysis. The options will be displayed as shown in Figure-52.

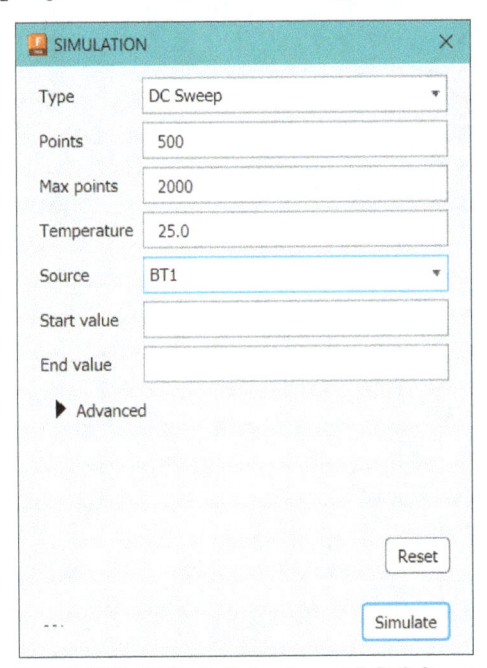

Figure-52. SIMULATION dialog box with DC Sweep options

- Specify the **Points**, **Max points**, and **Temperature** parameters as discussed earlier. Select desired option from the **Source** drop-down to define the voltage source whose value will vary during analysis.
- Specify desired value in the **Start value** edit box to define starting voltage of the source.
- Specify desired value in the **End value** edit box to define end voltage value of source for analysis.
- You can set the **Advanced** parameters as discussed earlier. After setting parameters, click on the **Simulate** button. Once the calculations are complete, the result of analysis will be displayed in a graph; refer to Figure-53.

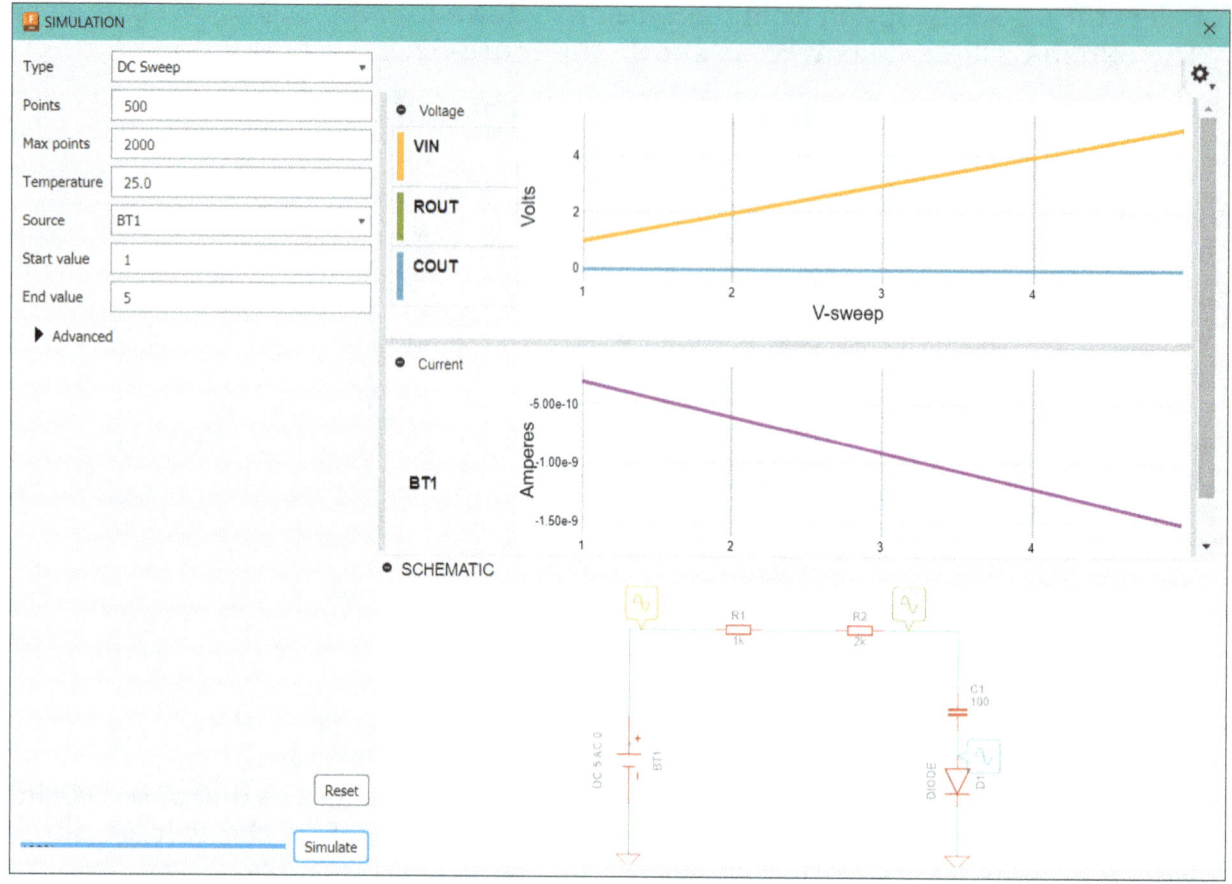

Figure-53. DC Sweep simulation result

- Click on the **Settings** button at top right corner of dialog box to switch between Dark mode or Light mode. You can also toggle between Logarithmic scale or Linear scale; refer to Figure-54.

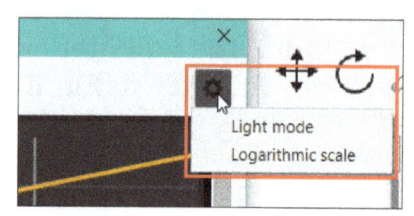

Figure-54. Settings menu

AC Sweep Simulation

The AC Sweep simulation is performed to check the response of circuit to specified range of frequencies. Note that you will need an AC supply source in circuit to perform this analysis. The procedure to perform this simulation is given next.

- Select the **AC Sweep** option from the **Type** drop-down in the **SIMULATION** dialog box. The options will be displayed as shown in Figure-55.

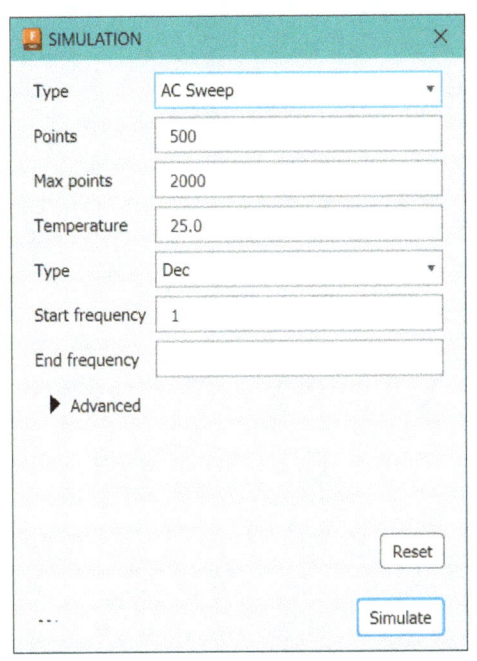

Figure-55. AC Sweep Simulation parameters

- Specify the **Points**, **Max points**, and **Temperature** parameters as discussed earlier.
- Select the **Dec** option from the **Type** drop-down to use logarithmic/decade scale for frequency. Select the **Lin** option from the **Type** drop-down to use linear scale for frequency when generating results.
- Specify desired values in the **Start frequency** and **End frequency** edit boxes to define starting frequency and end frequency for AC sweep analysis.
- After setting desired parameters, click on the **Simulate** button. The result of analysis will be displayed.

Transient Simulation

The Transient simulation is performed to check the response of circuit within specified time frame. You can check circuits like timer circuits to understand whether it will work as expected outside of operation time. The procedure to perform simulation is given next.

- Select the **Transient** option from the **Type** drop-down in the **SIMULATION** dialog box. The options will be displayed as shown in Figure-56.
- Specify desired values in the **Start time** and **End time** edit boxes to define the starting time and end time of circuit operation for simulation.
- Specify desired value in the **TMAX** edit box to define the time steps by which analysis will proceed.
- Set other parameters as discussed earlier and click on the **Simulate** button. The result will be displayed; refer to Figure-57.

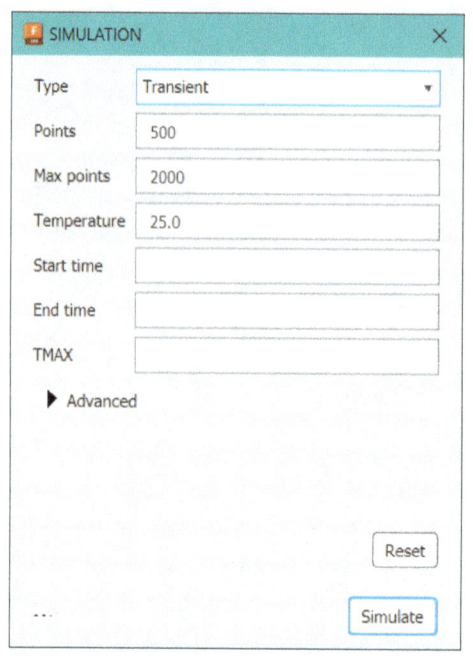

Figure-56. Transient type simulation options

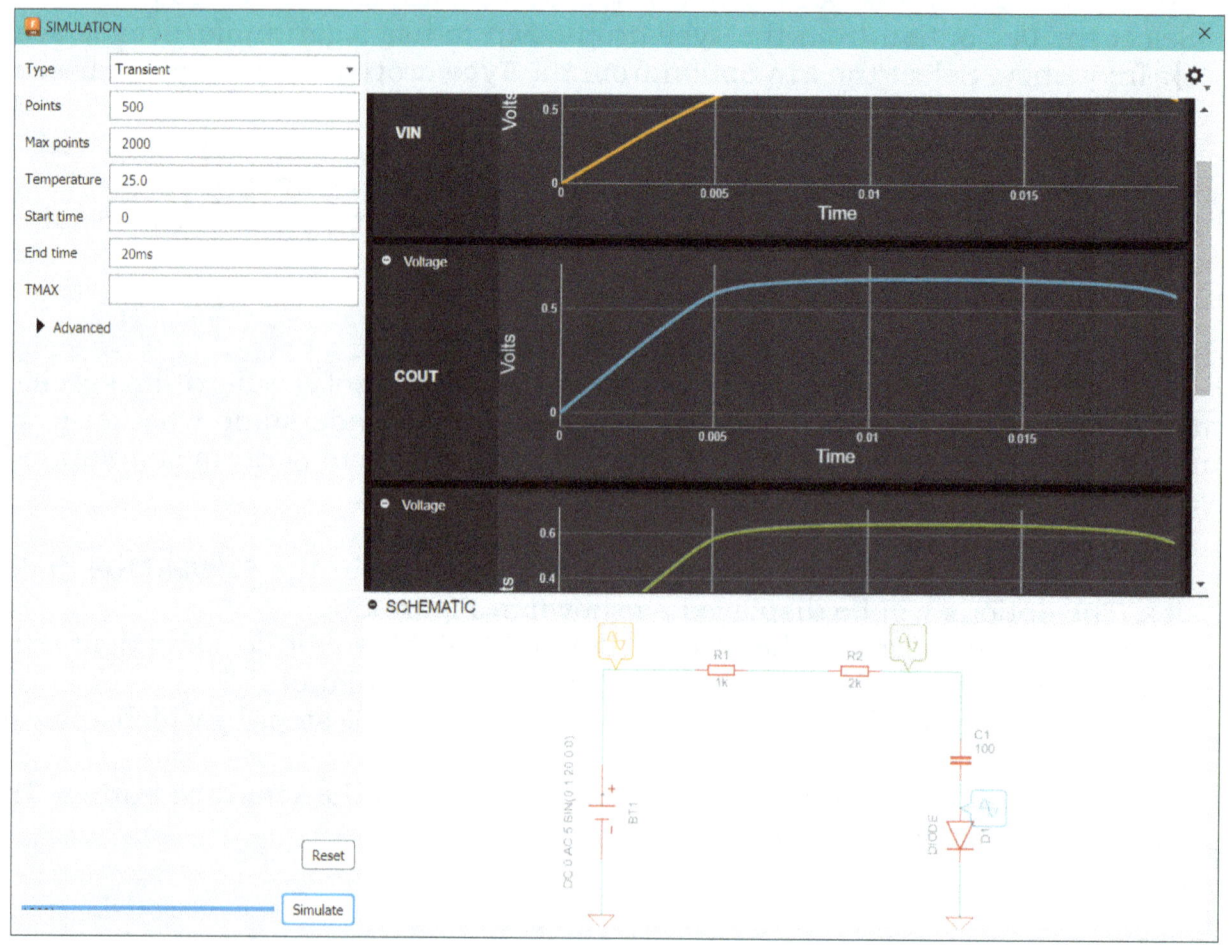

Figure-57. Result of transient simulation

- After checking results, close the dialog box.

SELF ASSESSMENT

Q1. What is the use of Miter tool?

Q2. How can you split wire segments?

Q3. The Slice tool is used to .... selected wire segment upto predefined width.

Q4. .... tool is used to place power symbols for the logic IC components.

Q5. For inserting SPICE compatible digital source, you need to activate the ngspice-digital library. (T/F)

Q6. The Digital Source Setup tool is used to define voltage/current data for selected SPICE compatible voltage/current source. (T/F)

Q7. Which of the following simulations is performed to check the response of circuit within specified time frame?

a) Operation Point Simulation
c) DC Sweep Simulation

b) AC Sweep Simulation
d) Transient Simulation

Answers: **Q3**. Trim **Q4**. Invoke **Q5**. T **Q6**. F **Q7**. d

FOR STUDENT NOTES

Chapter 4

Documentation, Validation, and Automation

Topics Covered

The major topics covered in this chapter are:

- ***Documentation***
- ***Validation***
- ***Automation***
- ***Library***

INTRODUCTION

In previous chapter, you have learned about electronic simulation. In this chapter, you will learn about tools related to documentation of schematic, validation of schematic design, and automation of schematic design. These tools are discussed next.

DOCUMENTATION TOOLS

The tools to generate documentation of schematic are available in the **DOCUMENT** tab of **Ribbon**; refer to Figure-1.

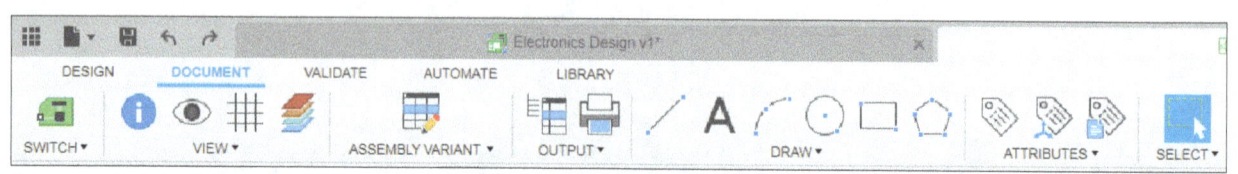

Figure-1. DOCUMENT tab

Assembly Variant

The **Assembly Variant** tool of the **ASSEMBLY VARIANT** drop-down in the **DOCUMENT** tab of the **Ribbon** is used to create and manage variations of the assembly. The process of creating and managing assembly variants has been discussed earlier in **DESIGN MANAGER** topic of Chapter 2.

Generating Bill of Materials

The **Bill of Materials** tool is used to generate table of all the components used in the schematics. The procedure to use this tool is given next.

- Click on the **Bill of Materials** tool from the **OUTPUT** drop-down in the **DOCUMENT** tab of the **Ribbon**. The **Fusion 360 Electronics: Bill Of Material** dialog box will be displayed; refer to Figure-2.

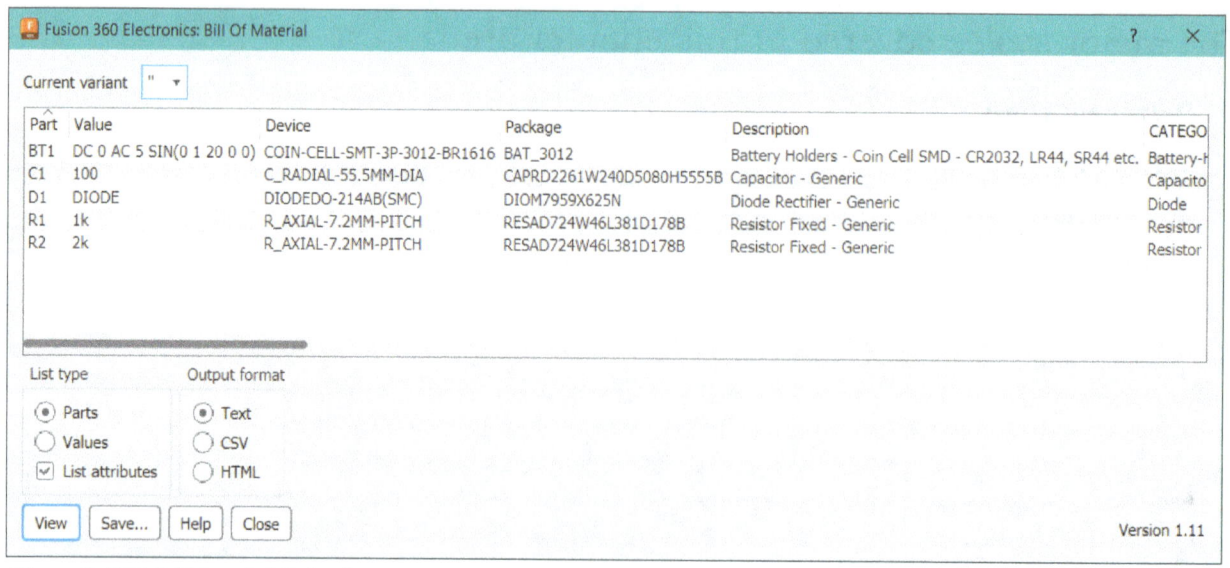

Figure-2. Bill Of Material dialog box

- If you have multiple assembly variants of the schematic then select desired variant from the **Current Variant** drop-down.

- Select desired radio button from the **List type** drop-down to define sorting order of table. Select **Parts** radio button to use alphabetical order of names of components or select the **Values** radio button to sort components by increasing order of their values.
- Select the **List attributes** check box to include attributes of components in the bill of materials.
- Select desired radio button from the **Output format** area of dialog box to define the format in which bill of materials table will be saved.
- Click on the **View** button to preview bill of materials before saving.
- Click on the **Save** button from the dialog box to save bill of materials in selected format. The **Save Bill Of Material** dialog box will be displayed; refer to Figure-3.

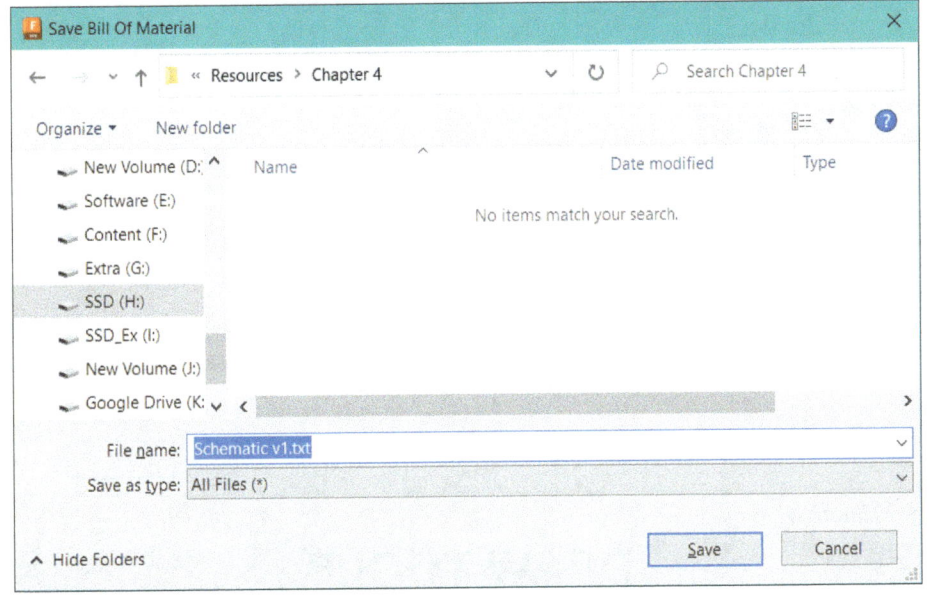

Figure-3. Save Bill Of Material dialog box

- Specify desired name in the **File name** edit box and click on the **Save** button. After saving file, press **ESC** to exit the tool.

Printing Schematic

The **Print** tool is used to print schematic drawing on paper. The procedure to use this tool is given next.

- Click on the **Print** tool from the **OUTPUT** drop-down in **DOCUMENT** tab of the **Ribbon**. The **Print** dialog box will be displayed; refer to Figure-4.
- Select desired option from the **Printer** drop-down to define the printer to be used for printing schematic on paper.
- Set the other parameters to printing in the dialog box and click on the **OK** button. The print will be generated.

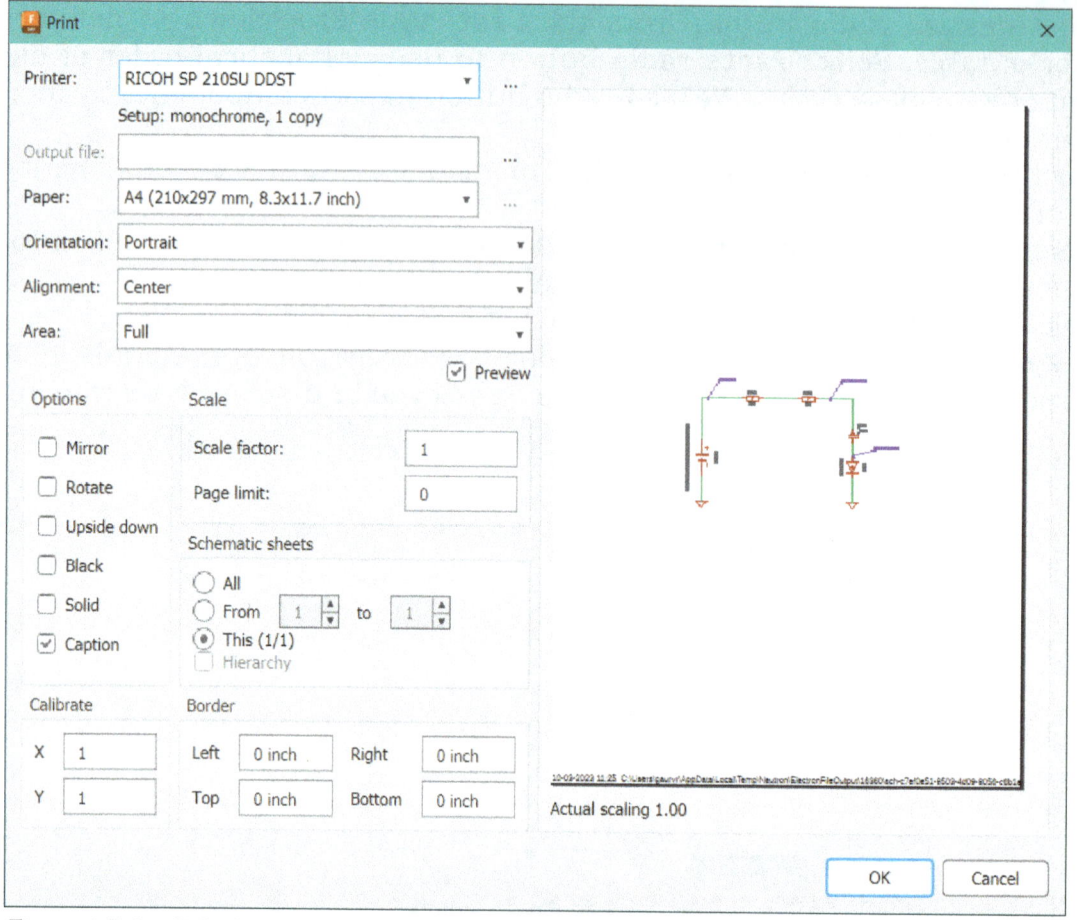

Figure-4. Print dialog box

DRAW TOOLS

The tools in **DRAW** drop-down are used to draw various geometric entities like line, arc, circle, polygon, and so on; refer to Figure-5. The tools in this drop-down are discussed next.

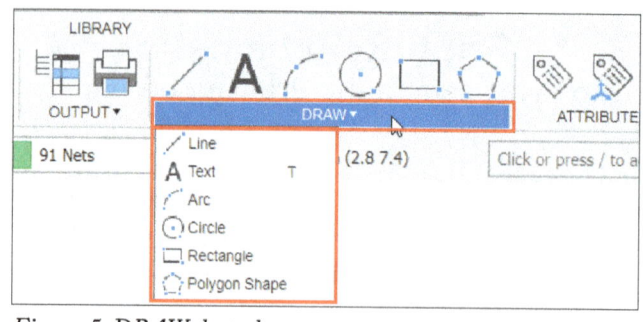

Figure-5. DRAW drop-down

Creating Line

The **Line** tool is used to create lines in the drawing. The procedure to use this tool is given next.

- Click on the **Line** tool from the **DRAW** drop-down in the **DOCUMENT** tab of the **Ribbon**. The **LINE Dialog** will be displayed; refer to Figure-6.

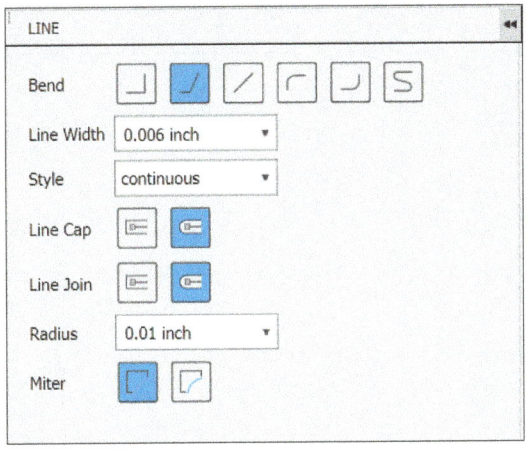

Figure-6. LINE Dialog

- Select desired option from the **Style** drop-down and set line type to be used for creating lines.
- Set the other parameters as discussed earlier.
- Click at desired location in the drawing area to specify starting point of line. You will be asked to specify end point of line.
- Click at desired location to create the line; refer to Figure-7.

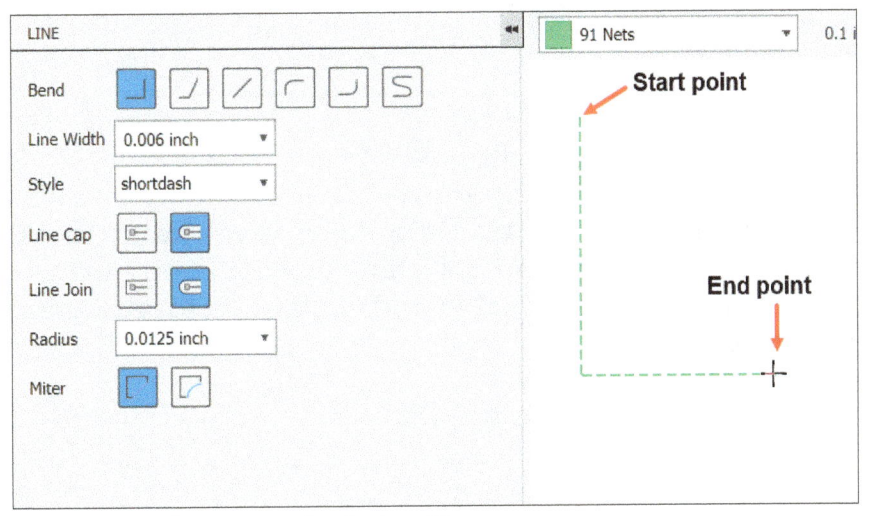

Figure-7. Specifying start and end point of line

- Press **ESC** to exit the current operation and press **ESC** again to exit the tool.

Writing Text

The **Text** tool is used to place user defined text in the drawing area. The procedure to use this tool is given next.

- Click on the **Text** tool from the **DRAW** drop-down in the **DOCUMENT** tab of the **Ribbon**. The **TEXT Dialog** will be displayed; refer to Figure-8.

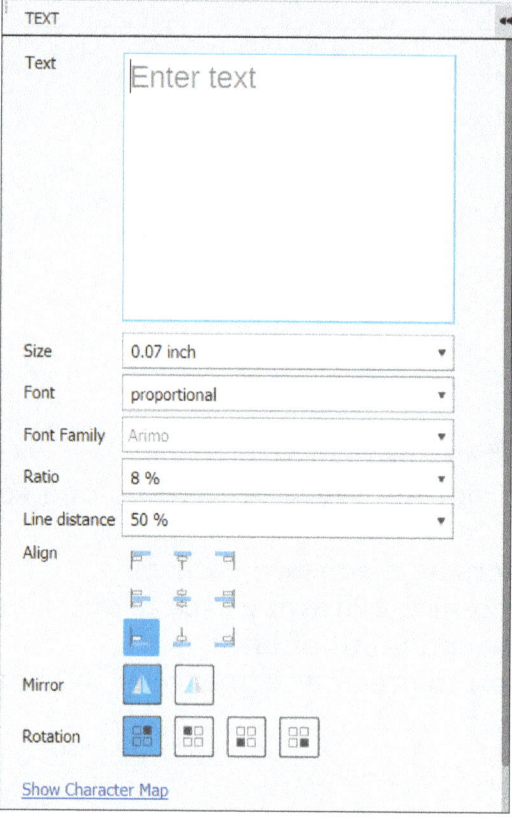

Figure-8. TEXT Dialog

- Specify desired text in the **Text** edit box.
- Set the size of text in the **Size** drop-down.
- Select custom option from the **Font** drop-down if you want to use custom font parameters or select standard options from the **Font** drop-down.
- Set the other parameters in the Dialog and click in the drawing area to place the text; refer to Figure-9.

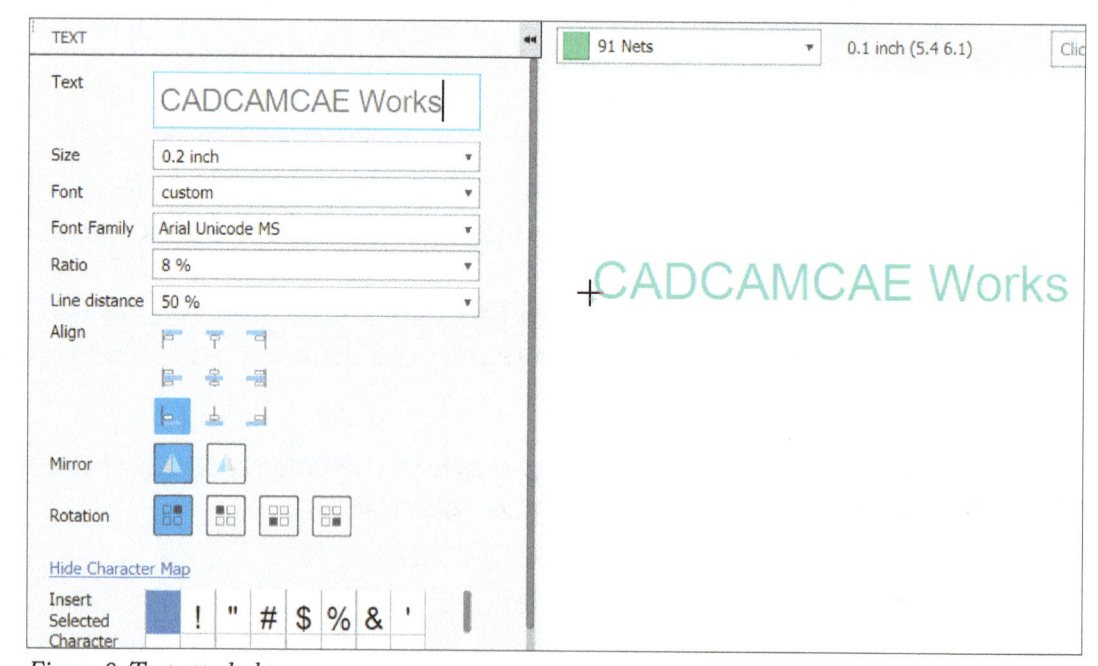

Figure-9. Text attached to cursor

- Press **ESC** to exit the tool.

Creating Arc

The **Arc** tool is used to create geometric arcs in the drawing area. The procedure to use this tool is given next.

- Click on the **Arc** tool from the **DRAW** drop-down in the **DOCUMENT** tab of the **Ribbon**. The **ARC Dialog** will be displayed; refer to Figure-10.

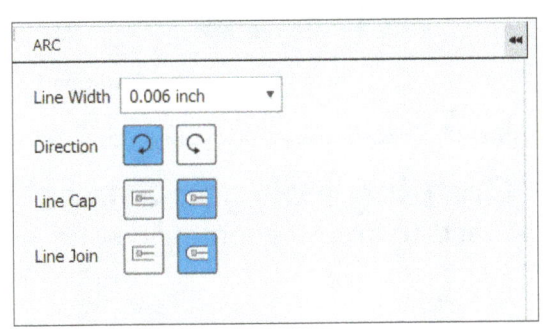

Figure-10. ARC Dialog

- Set the line width and other parameters in the **Dialog** as discussed earlier.
- Click in the drawing area to specify start point of arc, diametrical point of arc, and end point of arc.
- Press **ESC** to exit the tool.

Creating Circle

The **Circle** tool is used to create a circle in drawing area. The procedure to use this tool is given next.

- Click on the **Circle** tool from the **DRAW** drop-down in the **DOCUMENT** tab of the **Ribbon**. The **CIRCLE Dialog** will be displayed.
- Specify the line width parameter in the **Dialog** and click in the drawing area to specify center point of circle. You will be asked to specify diametrical point of circle.
- Click at desired location to specify diametrical point of circle to create the geometry.
- Press **ESC** to exit the tool.

Creating Rectangle

The **Rectangle** tool is used to create geometric rectangle in the drawing area. The procedure to use this tool is given next.

- Click on the **Rectangle** tool from the **DRAW** drop-down in the **DOCUMENT** tab of the **Ribbon**. You will be asked to specify start point of rectangle.
- Click at desired location to specify corner start point of rectangle. You will be asked to specify opposite corner end point of rectangle.
- Click at desired location. The rectangle will be created; refer to Figure-11.

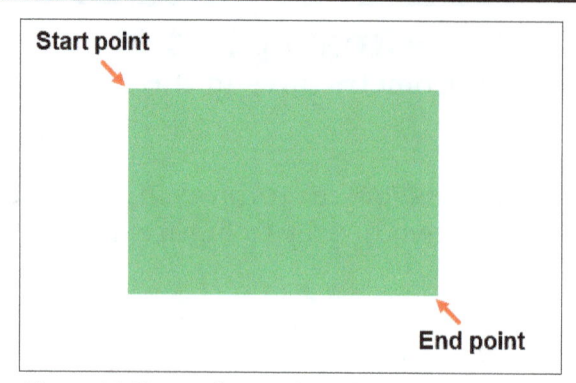

Figure-11. Rectangle created in drawing area

Creating Polygon Shape

The **Polygon Shape** tool is used to create geometric solid polygon. The procedure to use this tool is given next.

- Click on the **Polygon Shape** tool from the **DRAW** drop-down in the **DOCUMENT** tab of the **Ribbon**. The **POLYGON SHAPE Dialog** will be displayed; refer to Figure-12.

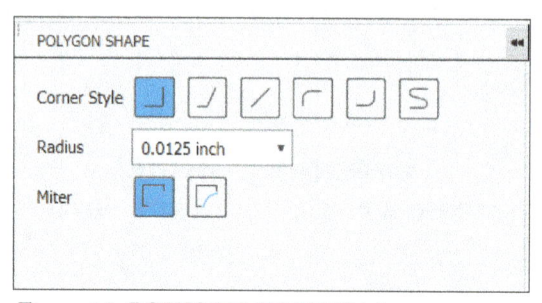

Figure-12. POLYGON SHAPE Dialog

- Set the parameters in **Dialog** as discussed earlier and click at desired location to specify start point of polygon.
- Click at desired locations to specify start point and consecutive points; refer to Figure-13.
- Double-click at desired location to specify end point of polygon and complete the geometry creation; refer to Figure-14.

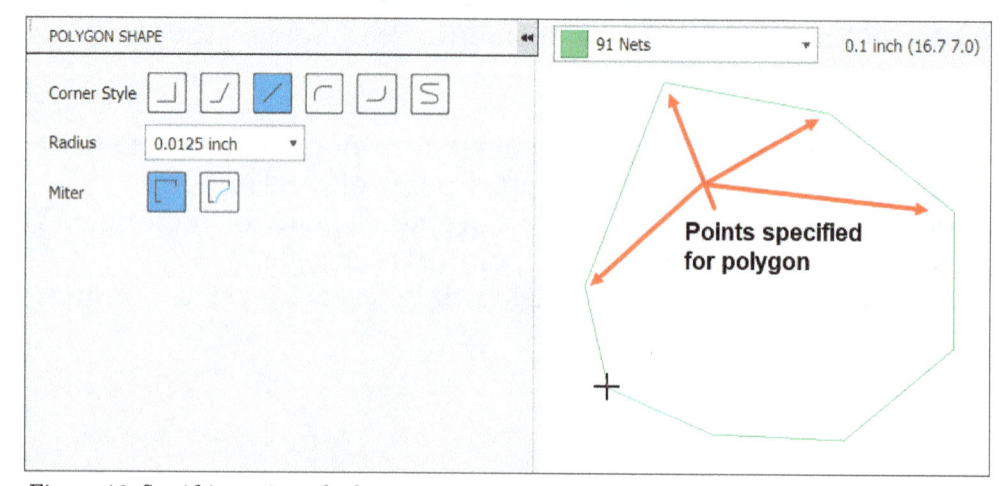

Figure-13. Specifying points of polygon

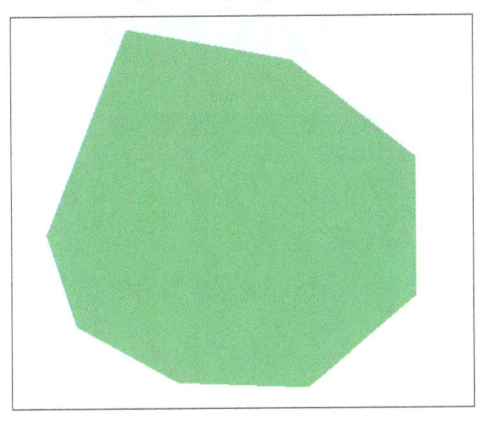

Figure-14. Polygon created

- Press **ESC** to exit the tool.

Defining Part Attributes

The **Attribute** tool is used to define various attributes/information for selected parts. The procedure to use this tool is given next.

- Click on the **Attribute** tool from the **ATTRIBUTES** drop-down in the **DOCUMENT** tab of the **Ribbon**. The **ATTRIBUTE Dialog** will be displayed; refer to Figure-15.

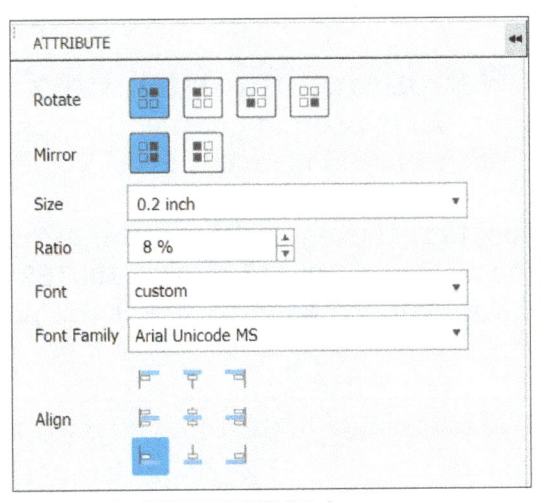

Figure-15. ATTRIBUTE Dialog

- Set the parameters in **Dialog** as discussed earlier and then click on the component whose attributes are to be defined. The **Attributes of part** dialog box will be displayed; refer to Figure-16.
- Double-click at desired parameter to change it. Respective **Change attribute** dialog box will be displayed; refer to Figure-17.

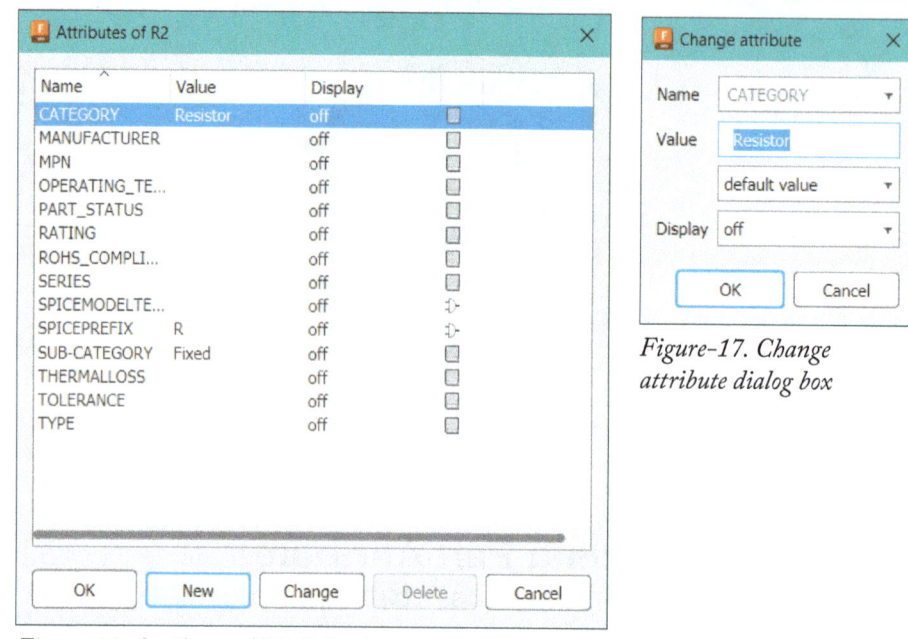

Figure-16. Attributes of R2 dialog box

Figure-17. Change attribute dialog box

- Set desired parameters in the **Change attribute** dialog box and click on the **OK** button.
- Press **ESC** to exit the tool.

Repositioning Attributes

The **Reposition Attributes** tool is used to change the position of selected attribute of part. The procedure to use this tool is given next.

- Click on the **Reposition Attributes** tool from the **ATTRIBUTES** drop-down in the **DOCUMENT** tab of the **Ribbon**. The **REPOSITION ATTRIBUTES Dialog** will be displayed; refer to Figure-18 and you will be asked to select the part.

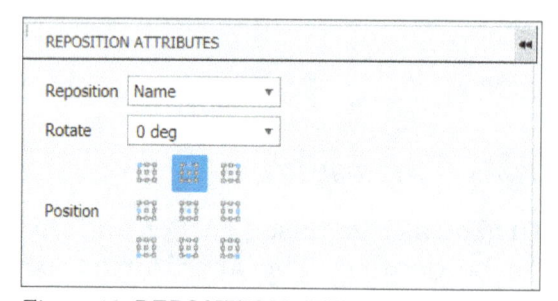

Figure-18. REPOSITION ATTRIBUTES Dialog

- Select desired option from the **Reposition** drop-down of **Dialog** to define the attribute of part to be repositioned.
- Select desired option from the **Rotate** drop-down to orient the attribute at specified angle.
- Select desired button from the **Position** section of **Dialog** to define position of attribute.
- After setting desired parameters, click on the part. The attribute will be repositioned; refer to Figure-19.

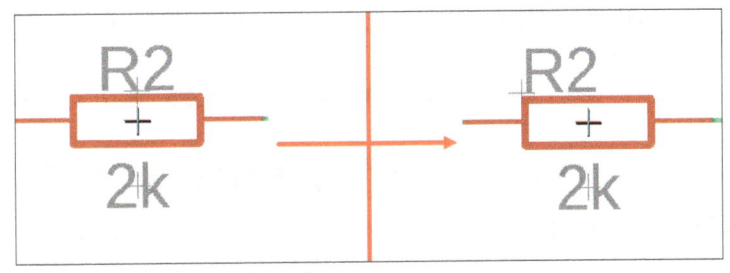

Figure-19. Repositioned attribute

- Press **ESC** to exit the tool.

Defining Document Attributes

The **Document Attributes** tool is used to modify attributes of current schematic. The procedure to use this tool is given next.

- Click on the **Document Attributes** tool from the **ATTRIBUTES** drop-down in the **DOCUMENT** tab of the **Ribbon**. The **Document Attributes** dialog box will be displayed; refer to Figure-20.

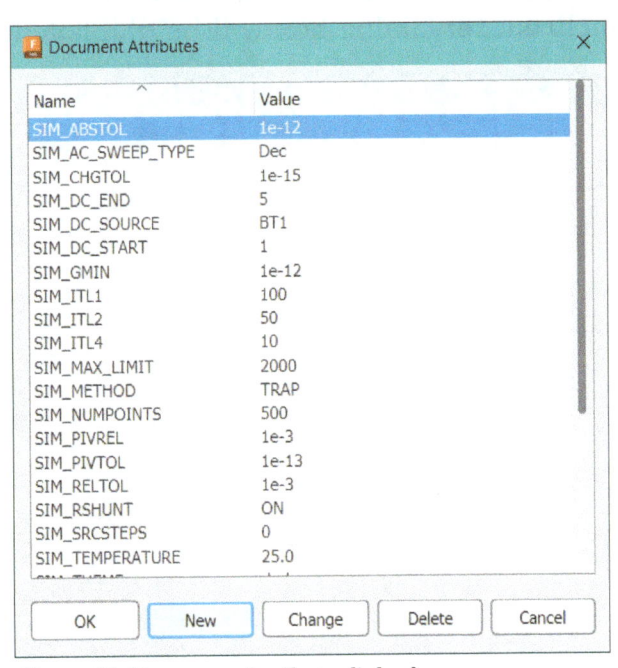

Figure-20. Document Attributes dialog box

- Double-click on desired parameter to modify it.
- Click on the **New** button to create a new attribute for document. The **New attribute** dialog box will be displayed; refer to Figure-21.

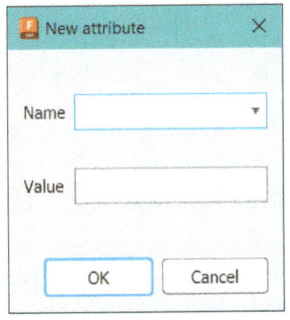

Figure-21. New attribute dialog box

- Specify the name and value of new attribute and click on the **OK** button to create the attribute.
- Click on the **OK** button from the **Document Attributes** dialog box to apply changes.

VALIDATION TOOLS

The tools in the **VALIDATE** tab are used to perform various validation tests on the circuit to check proper function of circuit; refer to Figure-22. These tools are discussed next.

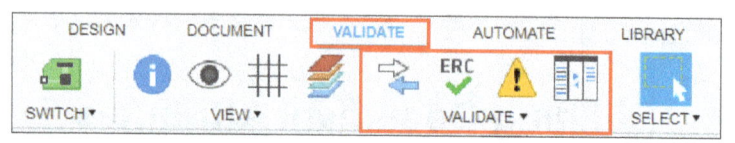

Figure-22. VALIDATE tab

Synchronizing Documents

The **Synchronize** tool is used to detect and compare two documents for differences. The procedure to use this tool is given next.

- Click on the **Synchronize** tool from the **VALIDATE** drop-down of **VALIDATE** tab in the **Ribbon**. The **Synchronizer** dialog box will be displayed; refer to Figure-23.

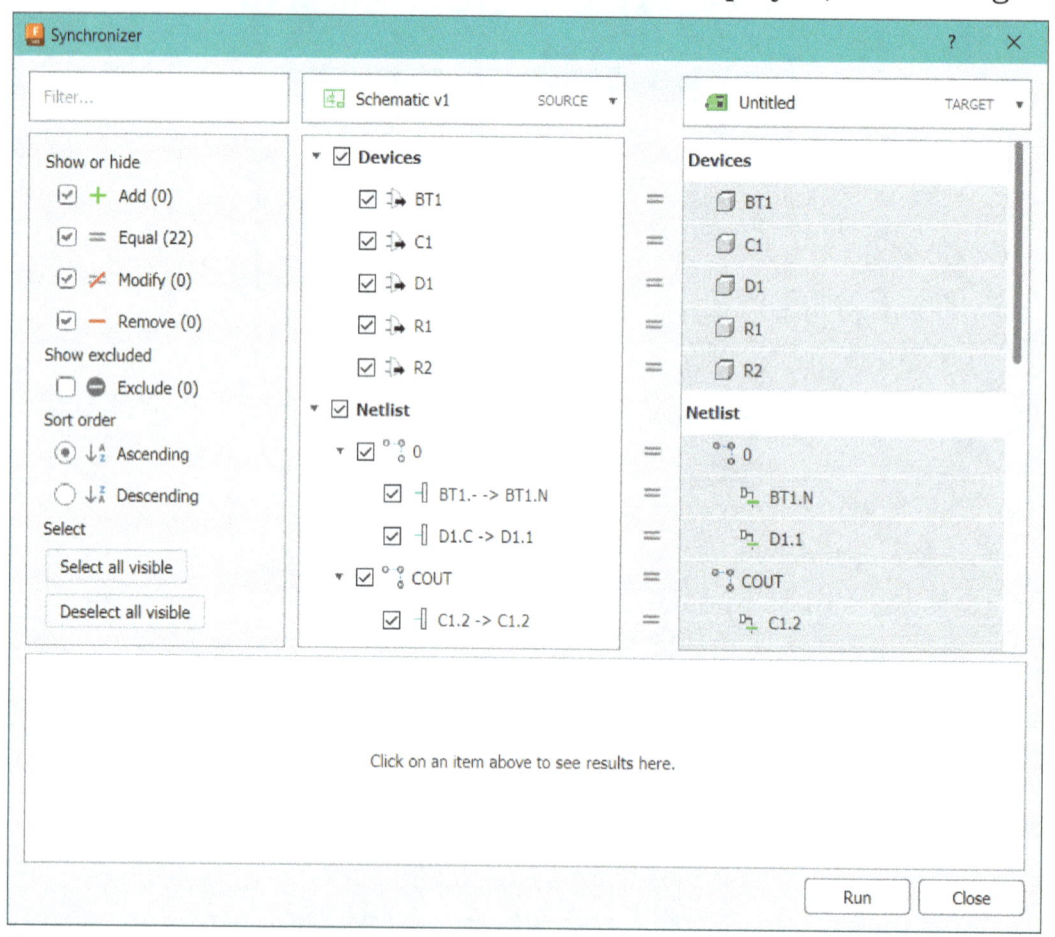

Figure-23. Synchronizer dialog box

- Select desired options from the left area of dialog box to perform different comparisons.
- Click on the **Run** button from the dialog box to perform synchronization.

Performing Electrical Rules Check

The **ERC** tool is used to perform various electrical rules check on the schematic. The procedure to use this tool is given next.

- Click on the **ERC** tool from the **VALIDATE** drop-down in the **VALIDATE** tab of the **Ribbon**. The ERC message will be displayed in the **Status Bar** as shown in Figure-24 when there is no error in the schematic.

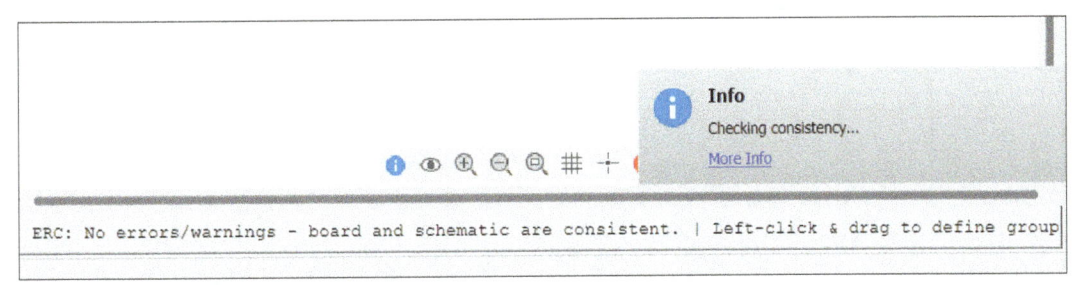

Figure-24. ERC no error message

- If there are errors in the schematic then **ERC Errors** dialog box will be displayed; refer to Figure-25.

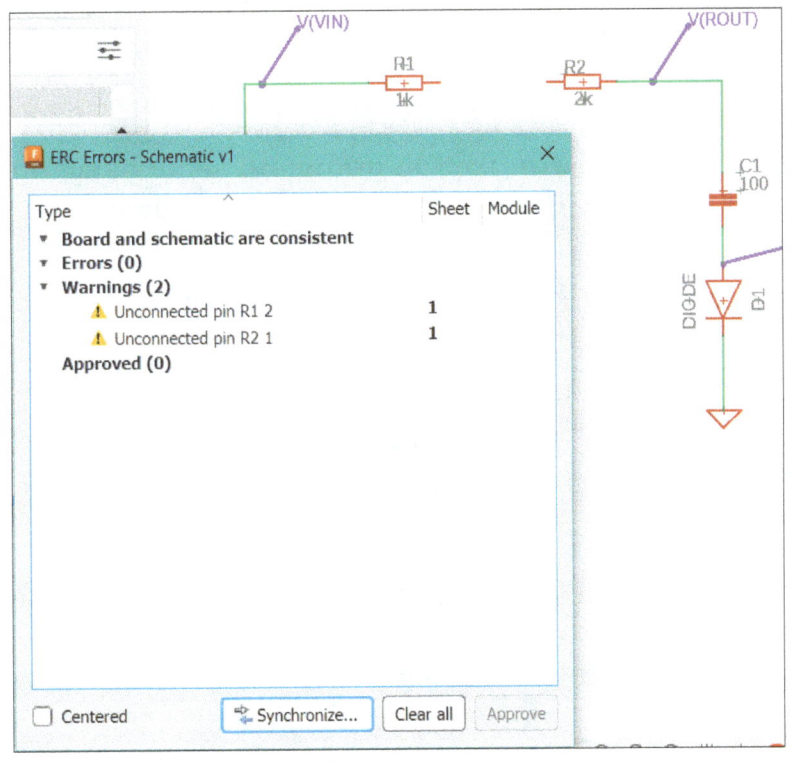

Figure-25. ERC Errors dialog box

- Select desired warning from the list to check the problem in drawing area. If the warning is acceptable then click on the **Approve** button. Otherwise, close the error dialog box and modify the design. Note that you can check the error anytime using the **Errors** tool.

Checking Net Classes

The **Class** tool is used to check data related to net class. The procedure to use this tool is given next.

- Click on the **Class** tool from the **VALIDATE** drop-down in the **VALIDATE** tab of the **Ribbon**. The **Net classes** dialog box will be displayed; refer to Figure-26.

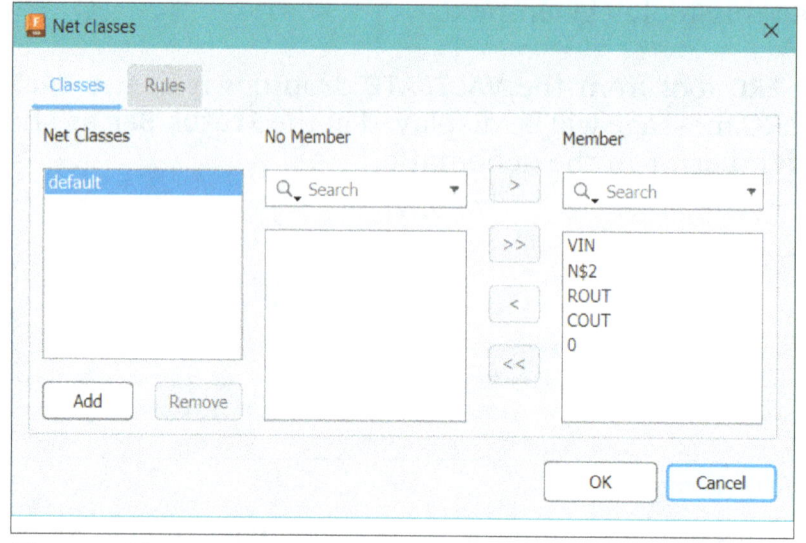

Figure-26. Net classes dialog box

- Set the parameters in **Classes** and **Rules** tab as discussed earlier in previous chapter and then click on the **OK** button.

AUTOMATION

The tools in the **AUTOMATE** tab are used to automate some basic operations created by scripts; refer to Figure-27. These tools are discussed next.

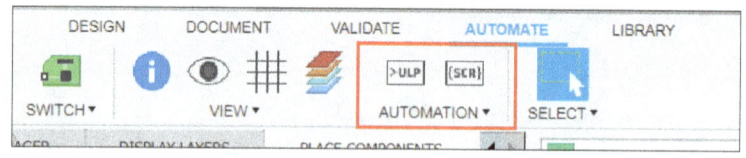

Figure-27. AUTOMATION drop-down

Running ULPs

The **Run ULP** tool is used to run a User Language Program. The procedure to run the ULP is given next.

- Click on the **Run ULP** tool from the **AUTOMATION** drop-down in the **AUTOMATE** tab of **Ribbon**. The **ULP** dialog box will be displayed; refer to Figure-28.

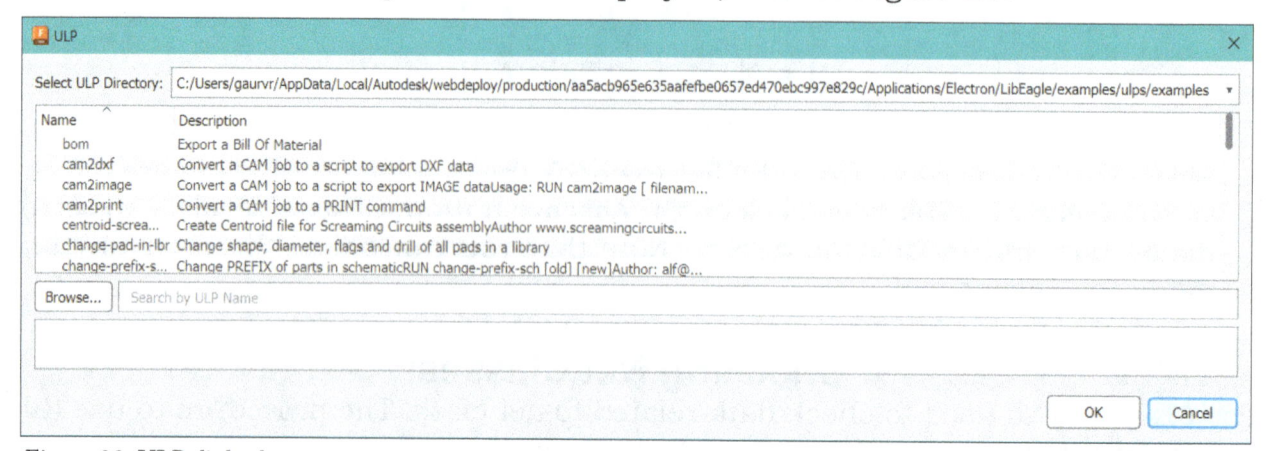

Figure-28. ULP dialog box

- Select desired ULP from the list and click on the **OK** button. If applicable, the related dialog box will be displayed; refer to Figure-29.

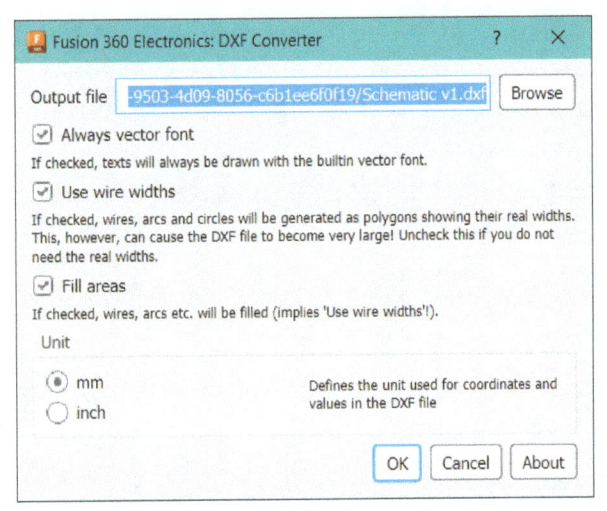

Figure-29. DXF Converter ULP dialog box

Running Scripts

The **Run Script** tool is used to run predefined scripts. The procedure to use this tool is given next.

- Click on the **Run Script** tool from the **AUTOMATION** drop-down in the **AUTOMATE** tab of **Ribbon**. The **Script** dialog box will be displayed; refer to Figure-30.

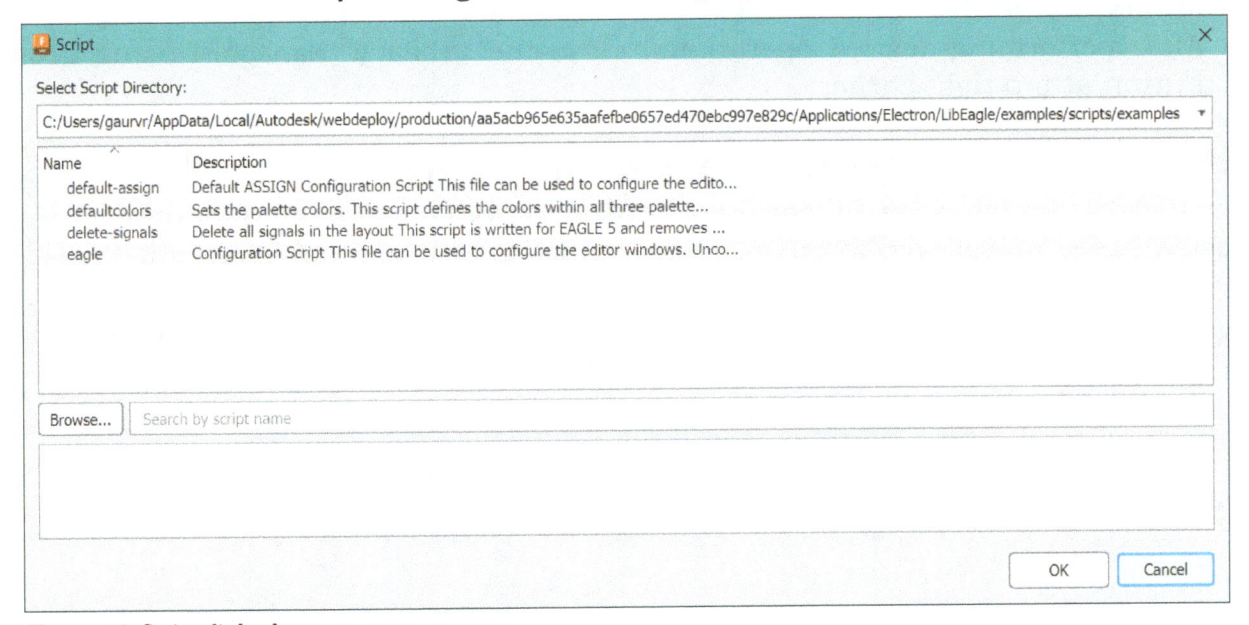

Figure-30. Script dialog box

- Select desired script from the list.
- Click on the **Browse** button to load script from local drive.
- After selecting script, click on the **OK** button to run the script.

LIBRARY MANAGER

The **Open Library Manager** tool is used to display **Library Manager** which is used to manage libraries of components. The procedure to use **Library Manager** is discussed next.

- Click on the **Open Library Manager** tool from the **LIBRARIES** drop-down in the **LIBRARY** tab of **Ribbon**. The **Library Manager** will be displayed; refer to Figure-31.

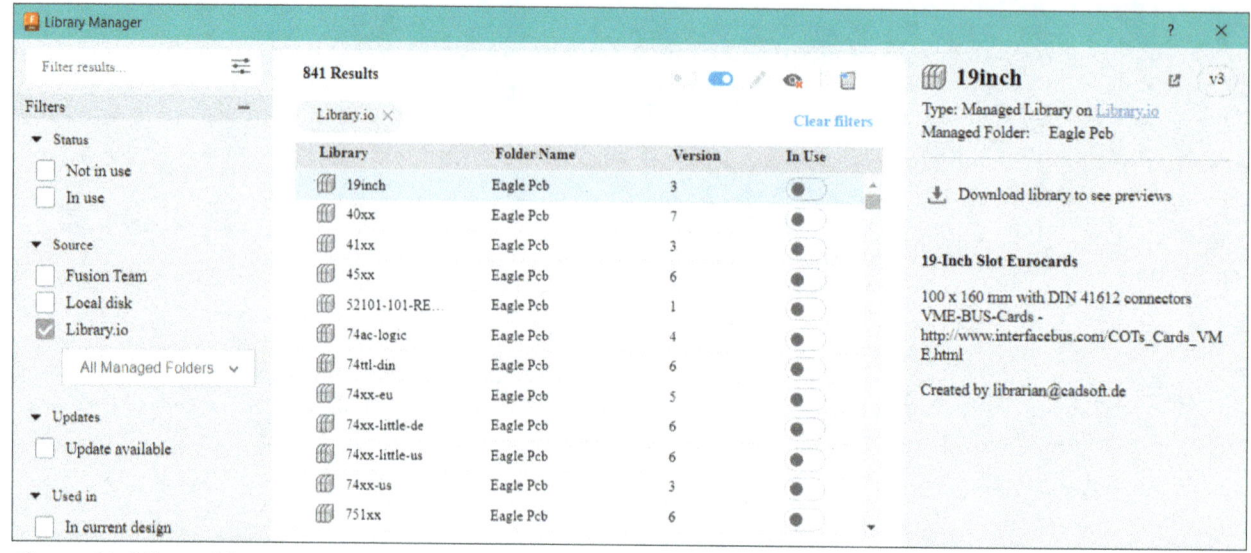

Figure-31. Library Manager

- Set the **In Use** toggle button to **ON** for libraries to be included in the current project.
- You can filter the list of available libraries by specifying keywords in the **Filter results** edit box.
- Click on the **Download library to see previews** button from the right area of the **Library Manager** to download preview images of the components in selected library.
- After performing desired operations, close the **Library Manager** using **Close** button at top right corner.

Updating Design from Library

The **Update design from library** tool is used to update components in the current design based on selected library. The procedure to use this tool is given next.

- Click on the **Update design from library** tool from the **LIBRARIES** drop-down in the **LIBRARY** tab of the **Ribbon**. The **Update** dialog box will be displayed; refer to Figure-32.

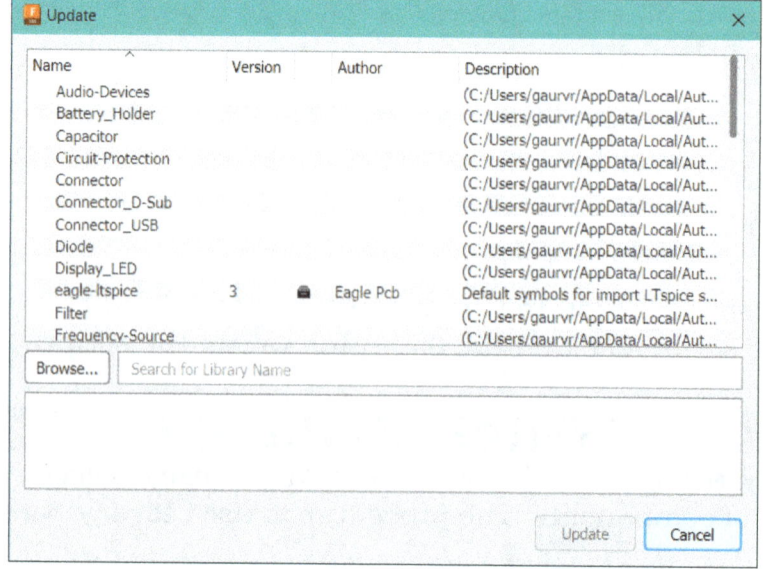

Figure-32. Update dialog box

- Select desired library component from the list to be updated. Note that you need to select the component of desired library to be used for updation. For example in Figure-33, you can see two libraries which have resistors. So, you need to select resistor which is from desired library.

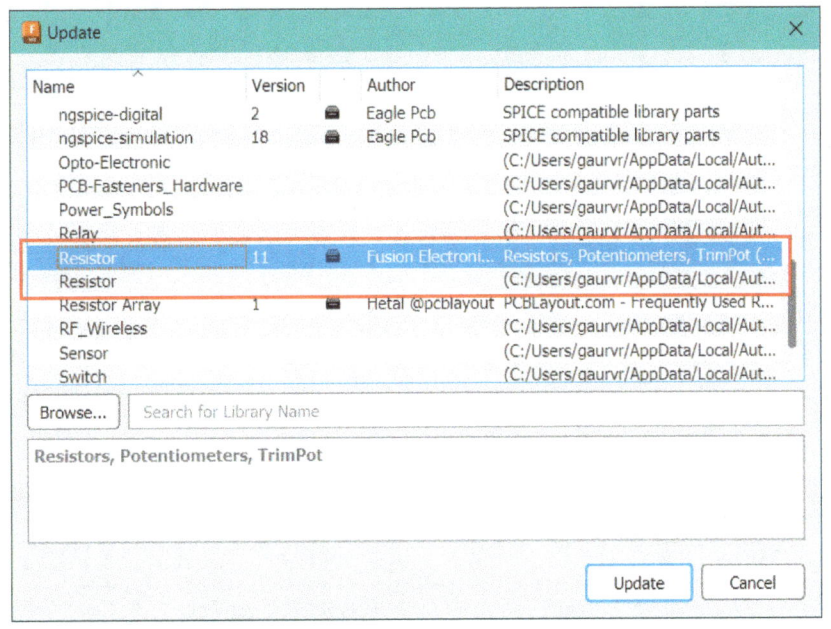

Figure-33. Resistors of Update dialog box

- After selecting the component to be updated, click on the **Update** button from the dialog box. A message will be displayed at the bottom right corner of drawing area telling you that design has changed and you need to perform DRC (Design Rule Check) again. Make sure to run the checks.

Updating Design From All Libraries Used in Design

The `Update design from all libraries` tool used in design tool is used to update all the components of designs based on current libraries.

Exporting Libraries

The `Export Libraries` tool is used to export all the libraries of components which are used in the current design. The procedure to use this tool is given next.

- Click on the `Export Libraries` tool from the **LIBRARIES** drop-down in the **LIBRARY** tab of the **Ribbon**. The `Export Electronics Libraries from Schematic` dialog box will be displayed; refer to Figure-34.
- Select check boxes for the source libraries to be exported.
- Select the `Export multiple libraries` radio button to export each library individually.
- Select the `Merge into one library` radio button to export all the selected libraries into one library. After selecting this radio button, select the `Prefix parts with library name` check box to add prefix of related library to component names when merging them in single library.
- Select the `Only export footprints which are used in the design` check box to export only those components which are used in current design.
- Select the `Export only footprints (No Devices or Symbols)` check box to export only footprints of the components in library.

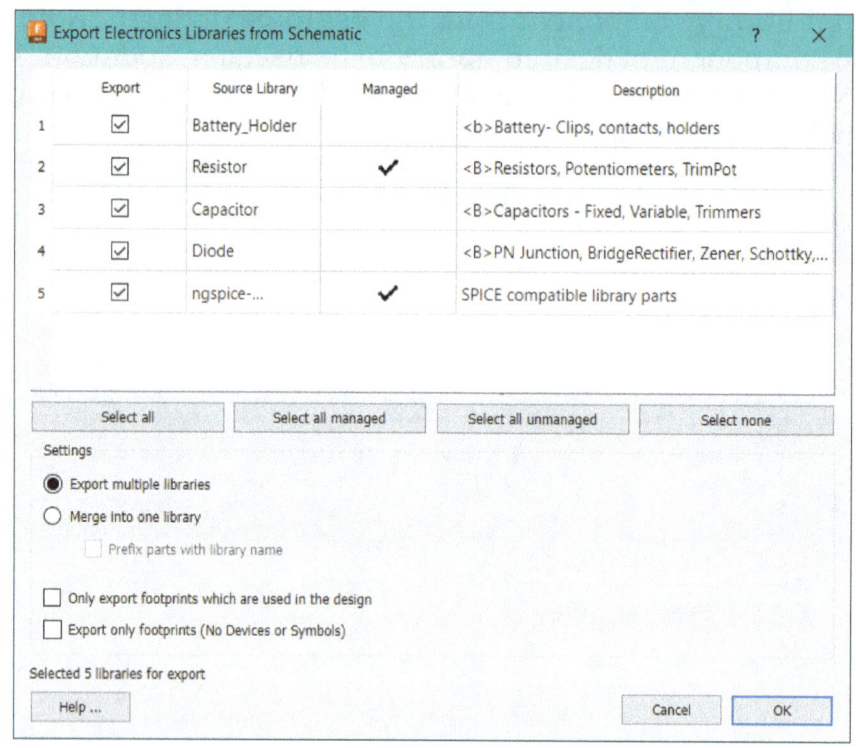

Figure-34. Export Electronics Libraries from Schematic dialog box

- After setting desired parameters, click on the **OK** button from the dialog box. The **Upload** dialog box will be displayed; refer to Figure-35.

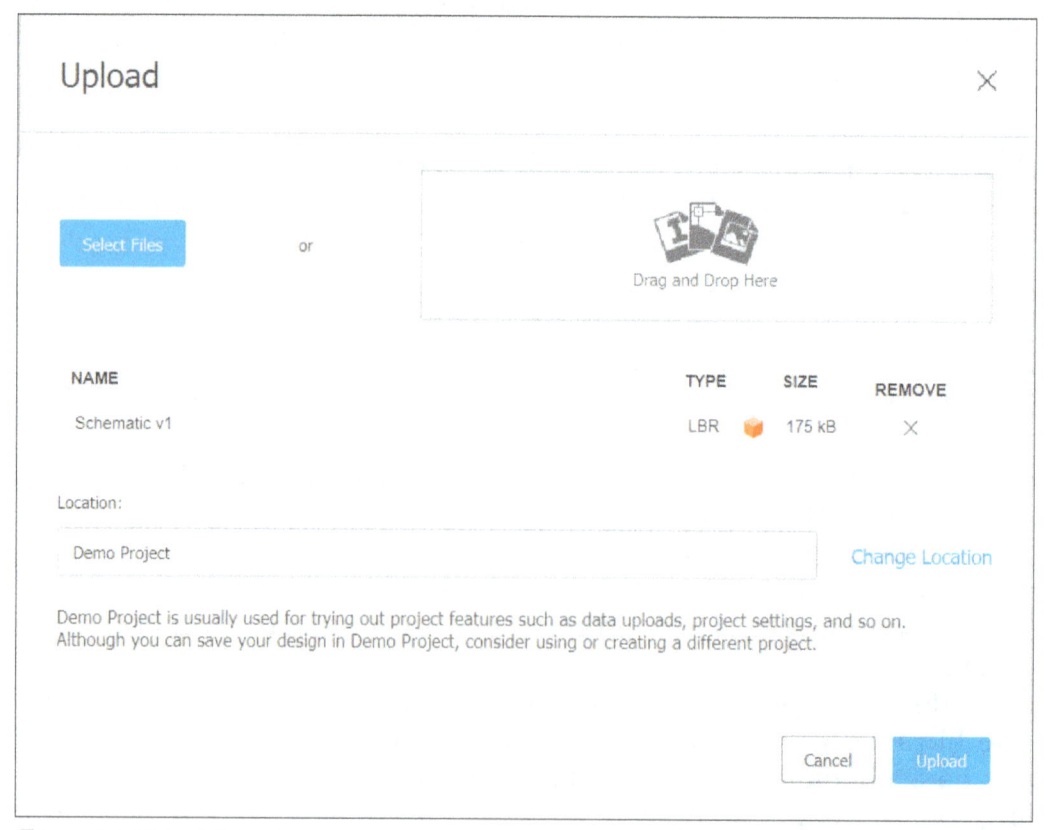

Figure-35. Upload dialog box

- If you want to change the location of exported library file then click on the **Change Location** button and set desired location.

- Click on the **Upload** button to upload library file in cloud. The status of uploading will be displayed; refer to Figure-36.

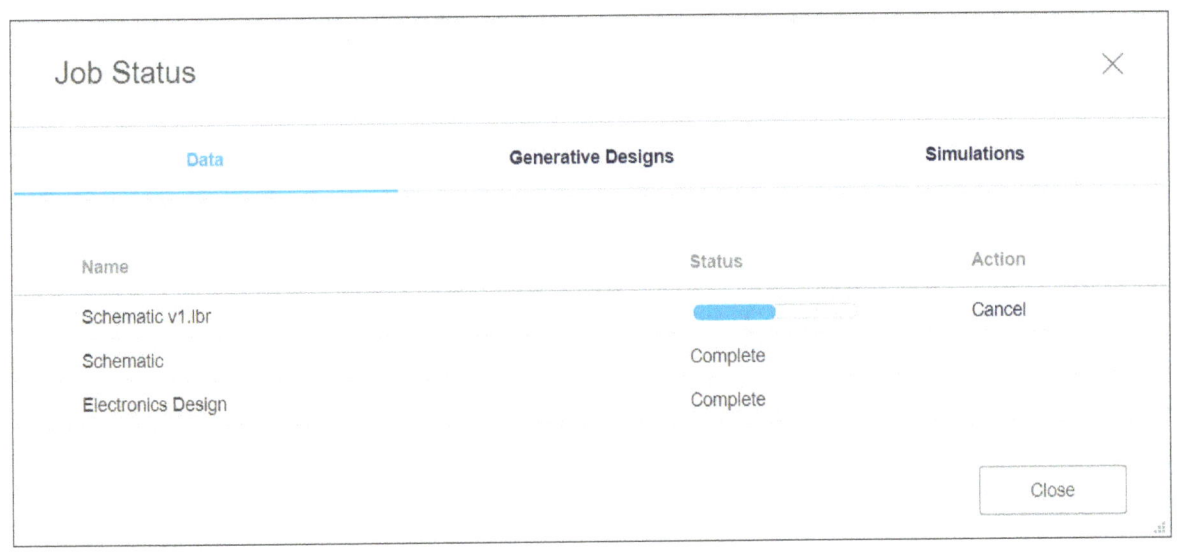

Figure-36. Job Status dialog box

- Once the status is complete, click on the **Close** button to exit.

SELF ASSESSMENT

Q1. What is the use of Bill of Materials?

Q2. What is the use of ERC tool?

Q3. The Attribute tool is used to define various attributes/information for selected parts. (T/F)

Q4. The Synchronize tool is used to detect and compare two documents for differences. (T/F)

Q5. The .... tool is used to perform various electrical rules check on the schematic.

Q6. The .... tool is used to run predefined scripts.

Answers: **Ans 3.** T **Ans 4.** T **Ans 5.** ERC **Ans 6.** Run Script

For Student Notes

Chapter 5

Introduction to PCB

Topics Covered

The major topics covered in this chapter are:

- ***Overview of PCB Design***
- ***Setting Layers***
- ***Creating/Editing Board Shape***
- ***Placing/Managing Components***

INTRODUCTION

The Electronics Design in Autodesk Fusion 360 PCB consists of 3 files: Schematic design, 2D PCB design, and 3D PCB model. In previous chapters, you have learned to create schematic designs. Now, we will discuss the process and procedures of creating PCB designs.

STARTING PCB DESIGN

The **New PCB** tool is used to start a new PCB design file. The two files of Electronics Design: Schematic Design and PCB Design have bidirectional associativity which means any object updated in either design will also be reflected in another design. The procedure to start a new PCB design file is given next.

- After starting new electronics design file or opening an electronics design file, click on the **New PCB** tool from the **CREATE** drop-down in the **COMMON** tab of the **Ribbon**; refer to Figure-1. The interface of PCB will be displayed with air wires connected to components; refer to Figure-2. Most of the interface elements have been discussed earlier in previous chapters.

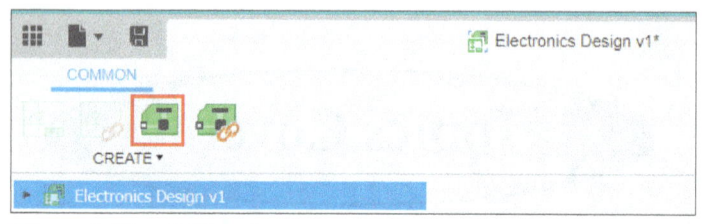

Figure-1. New PCB tool

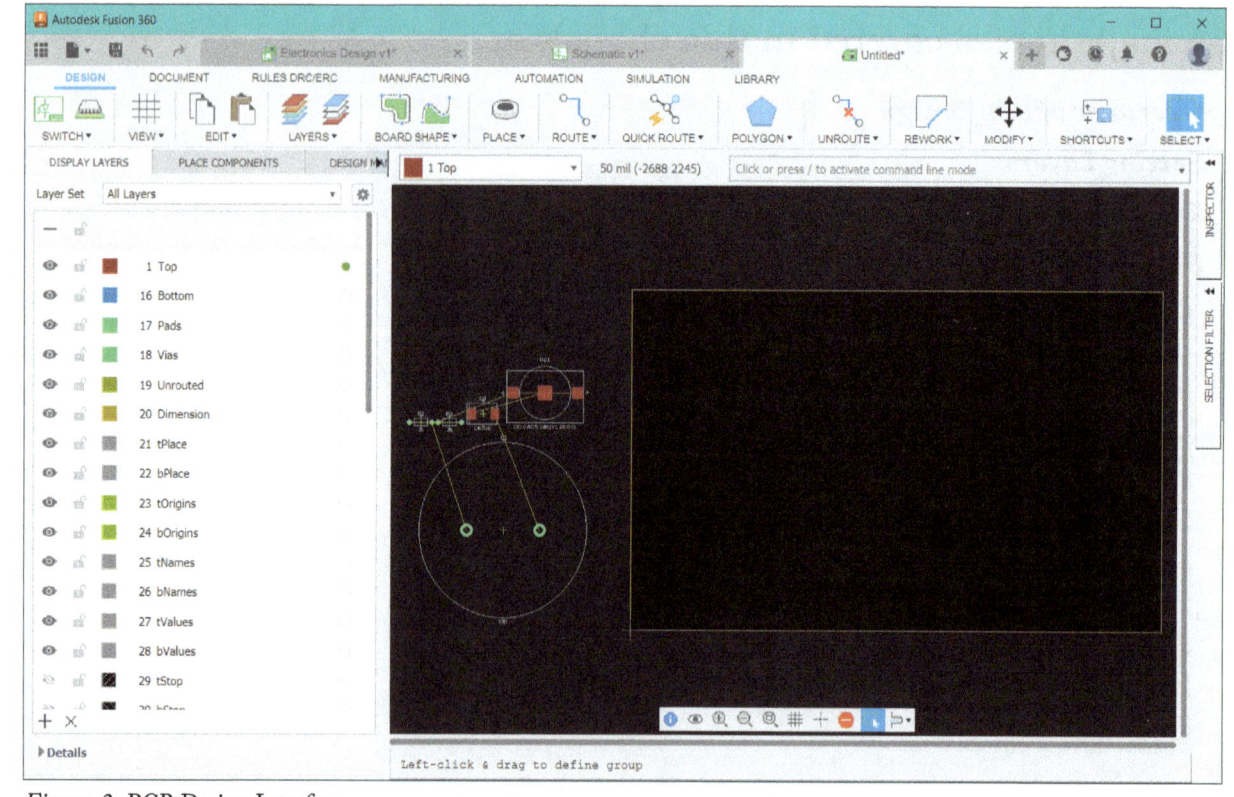

Figure-2. PCB Design Interface

LAYERS

The tools to create and manage layers are available in the **LAYERS** drop-down. Each layer can be distinguished by its color. The tools of this drop-down are discussed next.

Displaying Layers

The `Display Layers` tool is used to display/hide layers. On selecting this tool, the **DISPLAY LAYERS MANAGER** will be displayed which has been discussed earlier.

Flipping Board

The `Flip Board` tool is used to flip the PCB so that you can connect the components to the other side of board. Note that when you place the components on top layer then they are shown with green padding but when you place the components on bottom layer of PCB then they are displayed with red padding.

Single Layer/Multi Layer View

The `Toggle Single Layer View` tool is used to set whether all the layers of design will be displayed or only current layer will be displayed.

Layer Stack Manager

The `Layer Stack Manager` tool is used to set the parameters related to layers of PCB. The procedure to use this tool is given next.

- Click on the `Layer Stack Manager` tool from the **LAYERS** drop-down in the **DESIGN** tab of the **Ribbon**. The `Layer Stack Manager` will be displayed; refer to Figure-3.

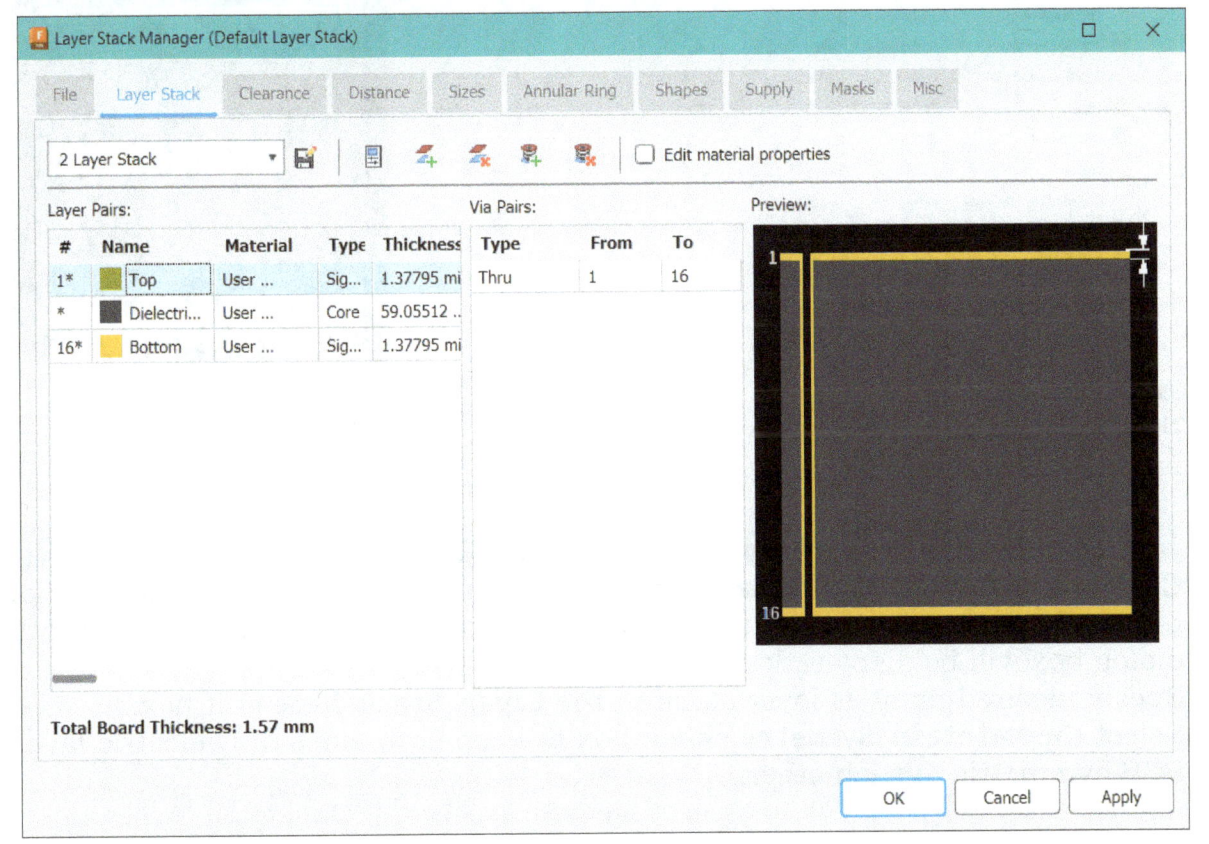

Figure-3. Layer Stack Manager

- Select desired option from the **Layer Stack** drop-down at the left corner in **Layer Stack** tab of the dialog box. A list of predefined layer setups will be displayed; refer to Figure-4.

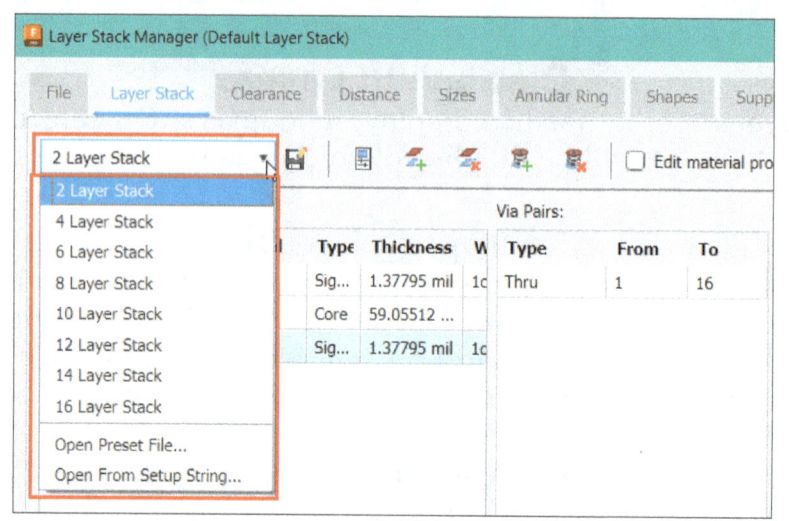

Figure-4. Layer Stack drop-down

- Select desired option from the drop-down to define the type of layer stacks to be used for current PCB design.
- Set desired parameters in the table to modify the PCB layers; refer to Figure-5.

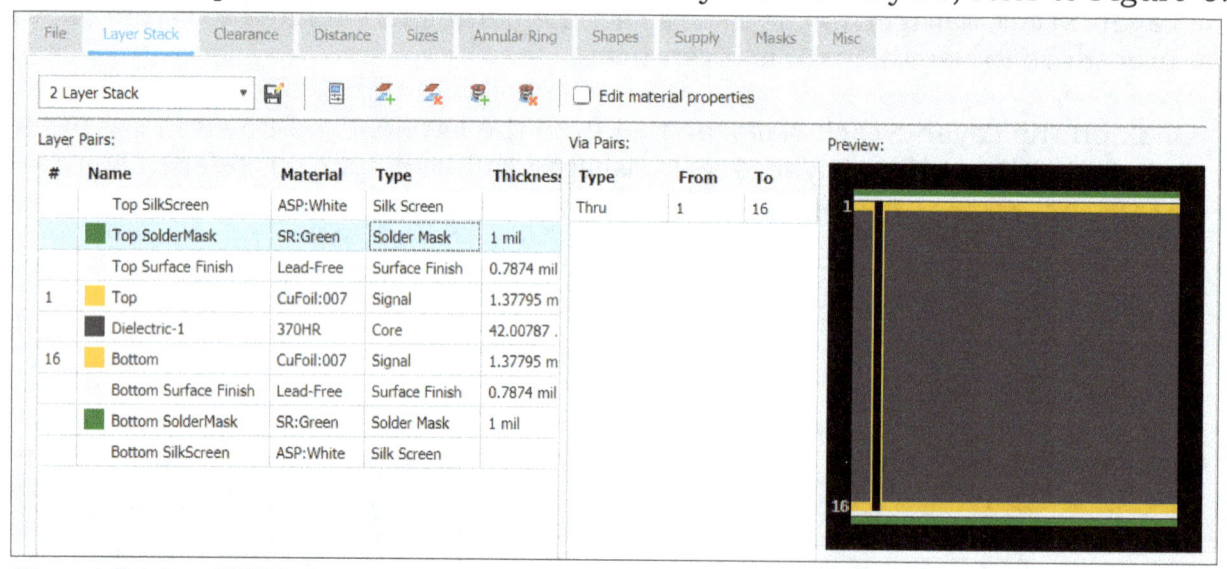

Figure-5. Preview of PCB layers

- Note that the thickness of layers is specified in `mil` which is 1 thousandth of an inch (1/1000 inch) that is 0.0254 mm.

Layer Stack Properties

- Click on the **Change layer stack properties** button from the toolbar in dialog box to change properties of current selected layer. The **Layer Stack Properties** dialog box will be displayed; refer to Figure-6.
- Specify desired name of layer stack in the **Layer Stack Name** edit box.
- Select the **Maintain Symmetry** check box to keep both sides of Dielectric layer of PCB symmetric to each other.

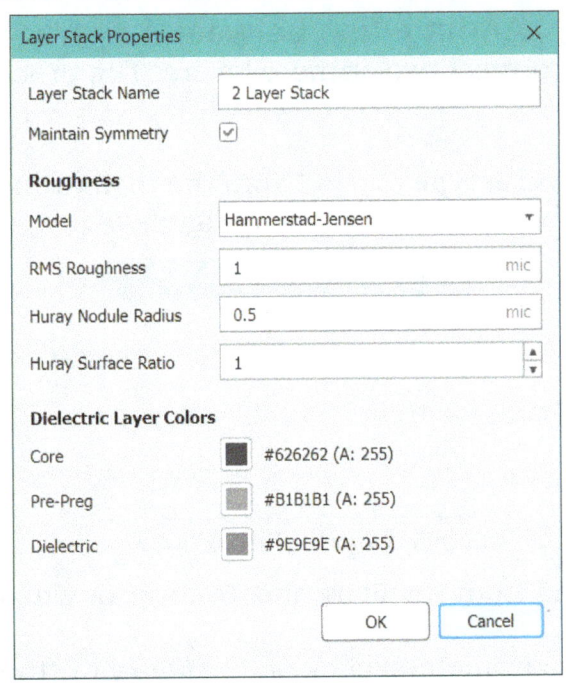

Figure-6. Layer Stack Properties dialog box

- Select desired option from the **Model** drop-down to define the method to be used for defining roughness of conductors and set the other parameters.
- After setting desired parameters, click on the **OK** button.

Adding Next Layer

- To add a new layer pair, click on the `Add next available layer pair` button from the toolbar in the dialog box. A new layer pair will be added in the list; refer to Figure-7.

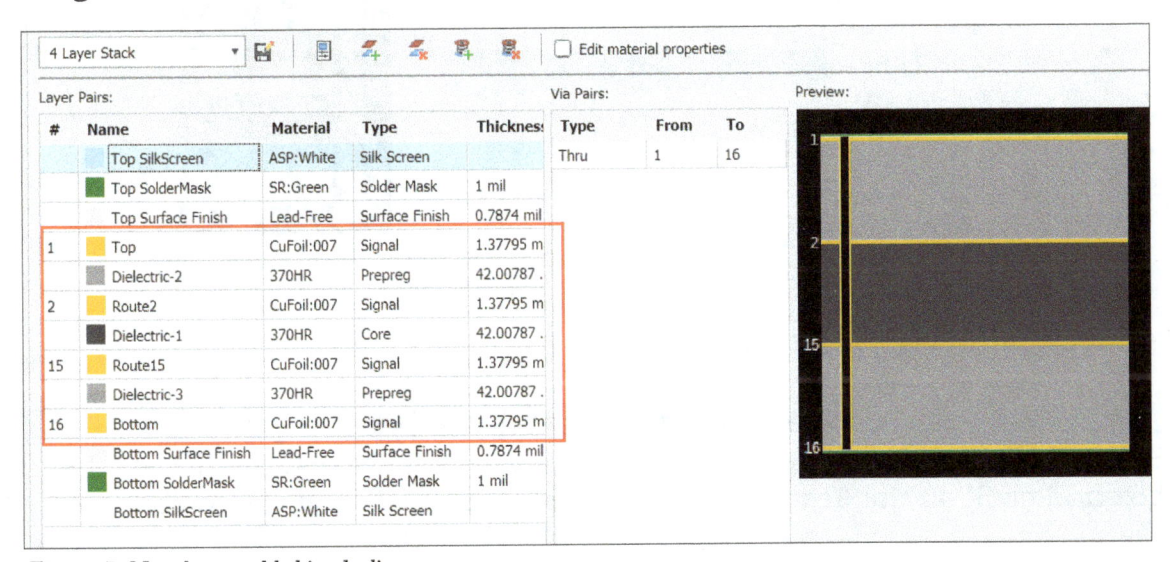

Figure-7. New layers added in the list

Removing Last Layer Pair

- To remove last added pair of layer, click on the `Remove last layer pair` tool from the toolbar in the dialog box. The last added layer pair will be removed.

Adding Vias between Layers

Via is a hole created to connect two or more layers. The procedure to add via is given next.

- Click on the **Add new via type** tool from the toolbar in dialog box. The **Add Via Pair** dialog box will be displayed; refer to Figure-8.

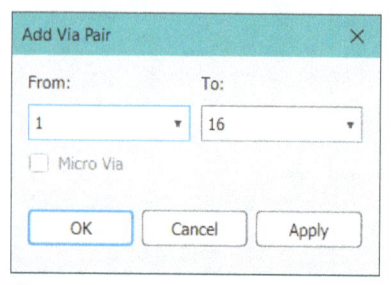

Figure-8. Add Via Pair dialog box

- Select desired options from the **From** and **To** drop-downs to define the two layers to be joined.
- After setting desired parameters, click on the **Apply** button to create links. The via will be added between layers; refer to Figure-9. Click on the **OK** button to create via and exit the tool.

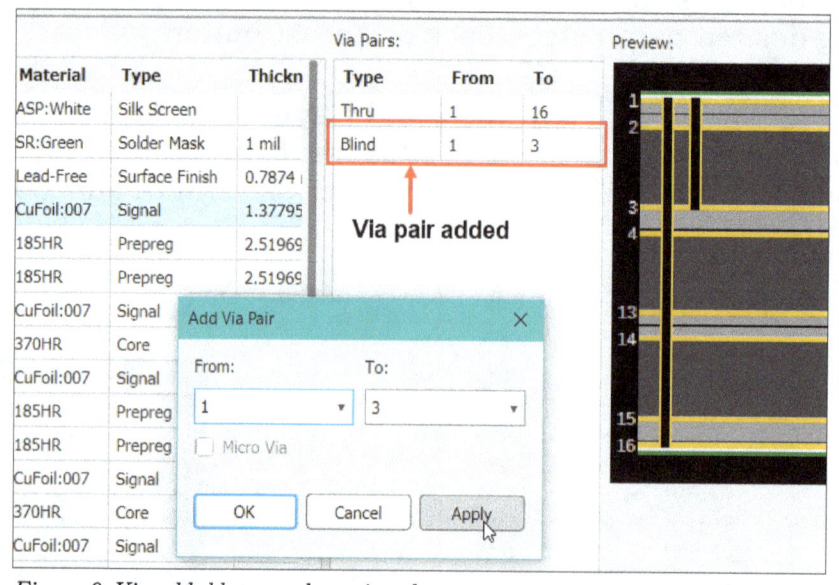

Figure-9. Via added between layers in pcb

Removing Selected Via Type

- Select desired via pair to be removed from the **Via Pairs:** section of the dialog box and click on the **Remove selected via type** tool from the toolbar.

Editing Material Properties

- Select the **Edit material properties** check box to modify various parameters of the PCB layers.
- After selecting check box, double-click in the **Layer Pairs** table to edit the parameters.

Clearance Parameters

- Select the **Clearance** tab from the dialog box to modify various clearance parameters like gap between wires, pads, and vias; refer to Figure-10. You can also set length of signal (wire) in the `Same Signals` section dialog box.

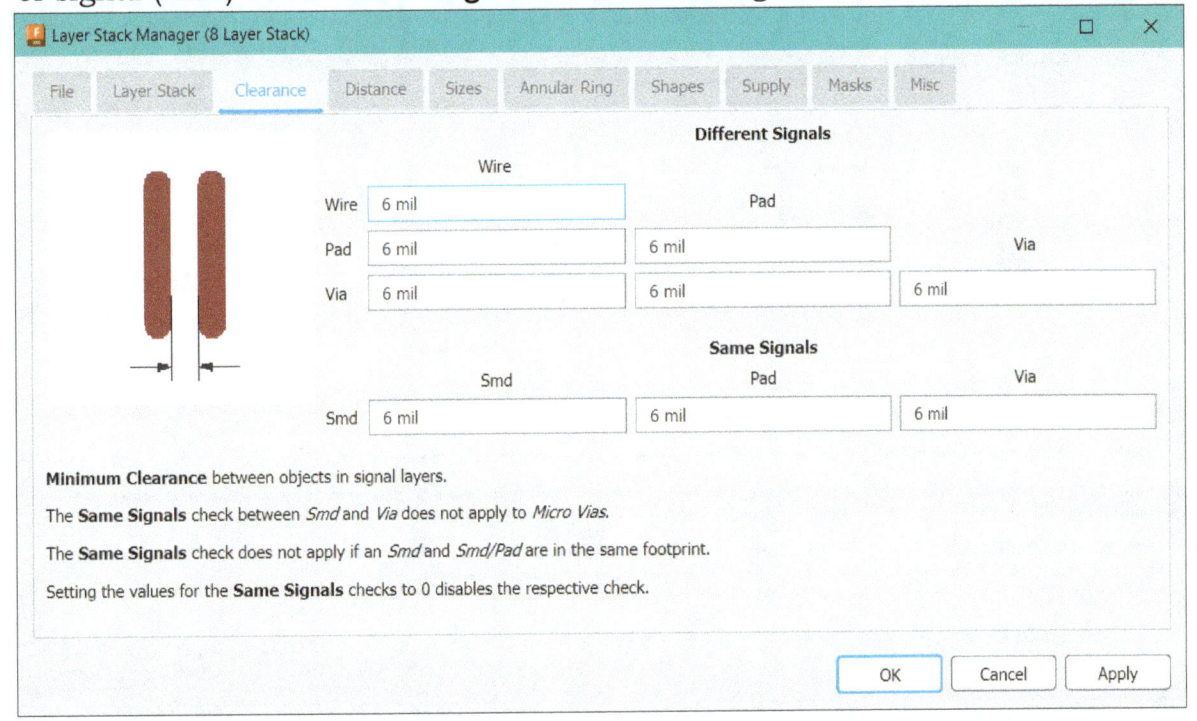

Figure-10. Clearance dialog box

Distance Parameters

- Select the **Distance** tab from the **Layer Stack Manager** dialog box. The options will be displayed as shown in Figure-11.

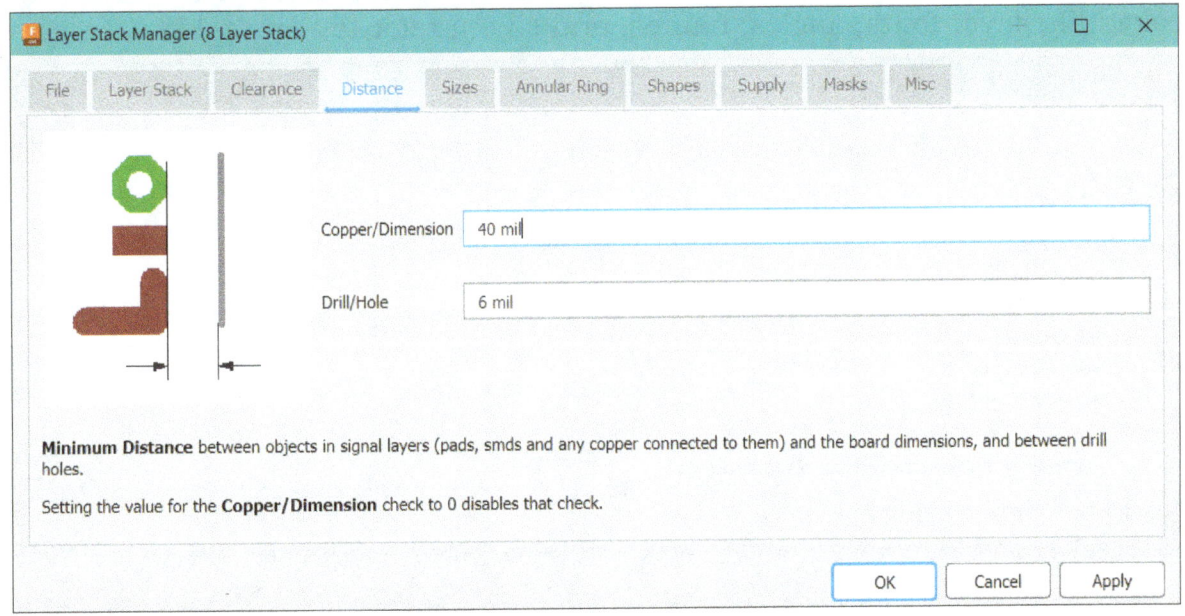

Figure-11. Distance tab of Layer Stack Manager

- Specify desired parameters in the **Copper/Dimension** and **Drill/Hole** edit boxes for defining minimum distance between two copper lines and two holes, respectively.

Size Parameters

Select the **Sizes** tab of the dialog box to define minimum sizes of various signal layers, drill holes, micro via, and blind via; refer to Figure-12.

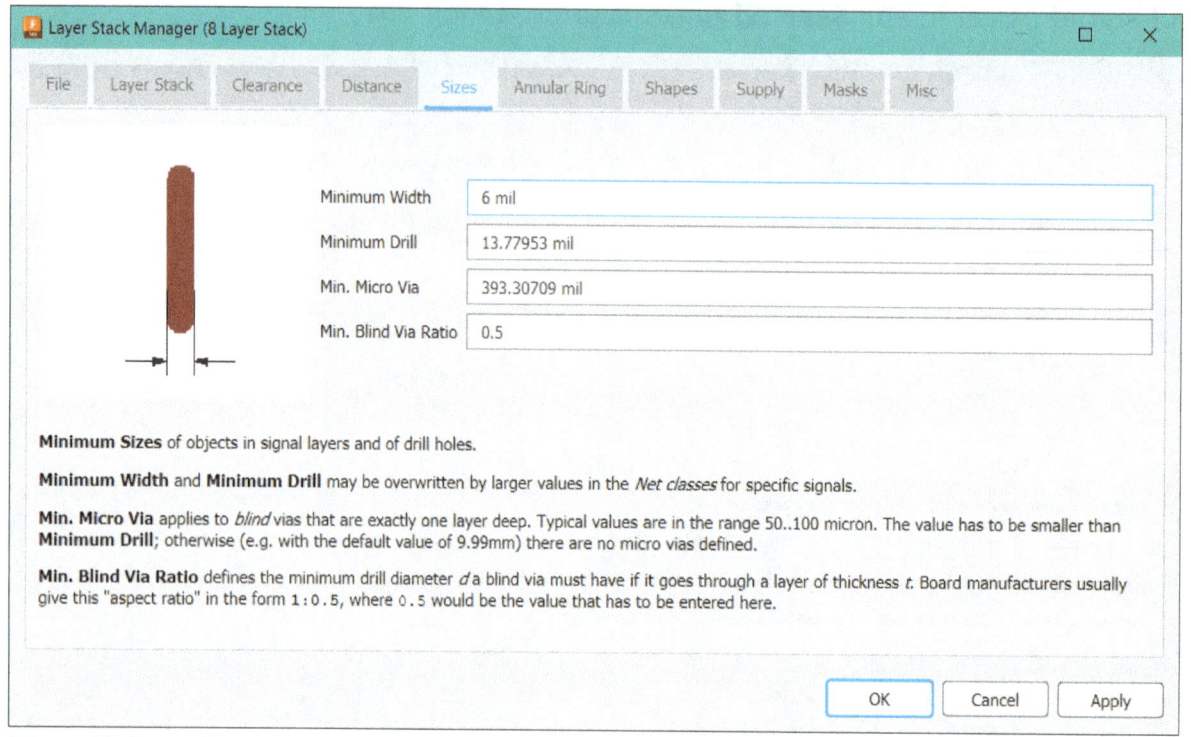

Figure-12. Sizes tab

Similarly, set the size and parameters in other tabs of the dialog box.

Miscellaneous Parameters

The options of **Misc** tab are used to define various parameters of the layers to be checked like layer fonts, angles, names, and so on; refer to Figure-13.

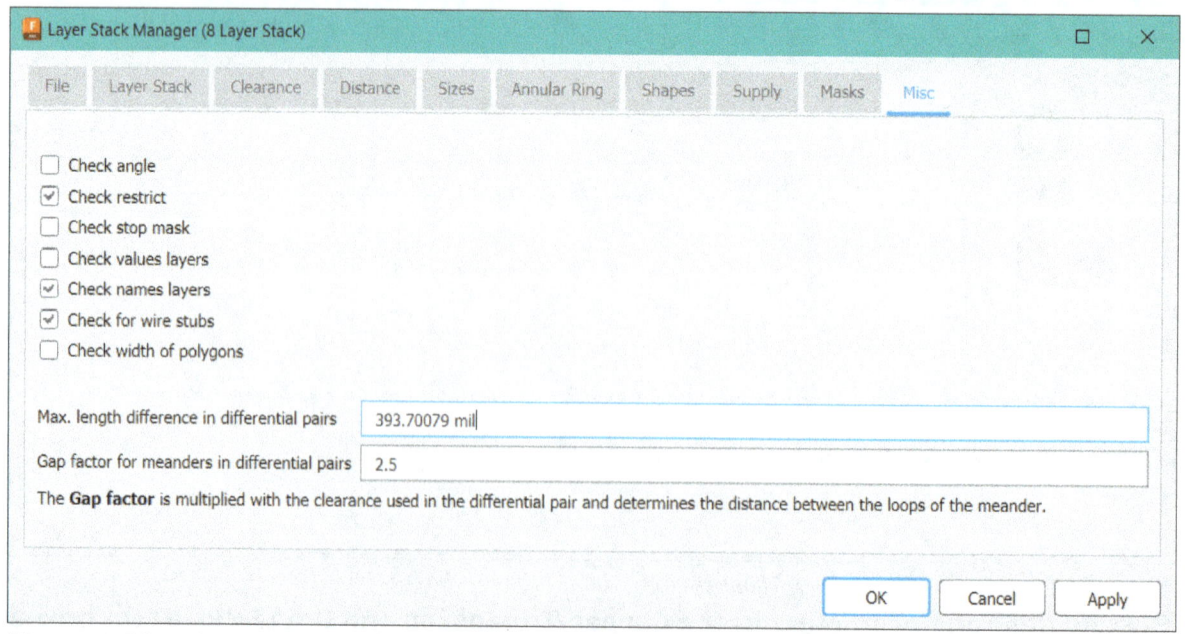

Figure-13. Misc tab

Select desired check boxes from the dialog box to define the parameters to be checked during analysis. Click on the **OK** button to set layer stack parameters.

CREATING BOARD SHAPES

The tools in the **BOARD SHAPE** drop-down are used to define the shape of PCB; refer to Figure-14. These tools are discussed next.

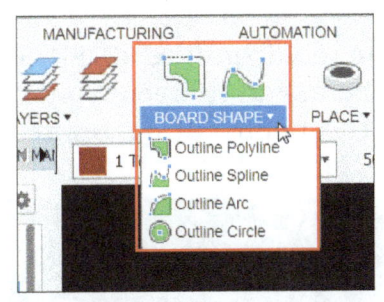

Figure-14. Board Shape drop-down

Creating Polyline Outline

The `Outline Polyline` tool is used to create outline of PCB using polyline shapes. The procedure to use this tool is given next.

- Click on the `Outline Polyline` tool from the **BOARD SHAPE** drop-down in the **DESIGN** tab of the **Ribbon**. The **OUTLINE POLYLINE MANAGER** will be displayed; refer to Figure-15.

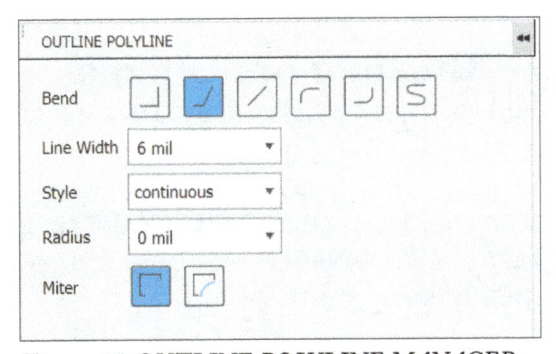

Figure-15. OUTLINE POLYLINE MANAGER

- Select desired button from the **Bend** section to define the line shape.
- Select desired option from the **Line Width** drop-down to specify width of line.
- Set the other parameters in **MANAGER** as discussed earlier and specify start point of polyline at desired location. You will be asked to specify next point.
- Click at desired locations to specify consecutive points and then close the outline at the end; refer to Figure-16. Press **ESC** to exit the tool.

Figure-16. Outline of PCB

Creating Spline Outline

The **Outline Spline** tool is used to create PCB outline using spline geometry. The procedure to use this tool is given next.

- Click on the **Outline Spline** tool from the **BOARD SHAPE** drop-down in the **DESIGN** tab of the **Ribbon**. You will be asked to specify start point of spline.
- Click at desired location to specify start point of the spline and then next consecutive points; refer to Figure-17.

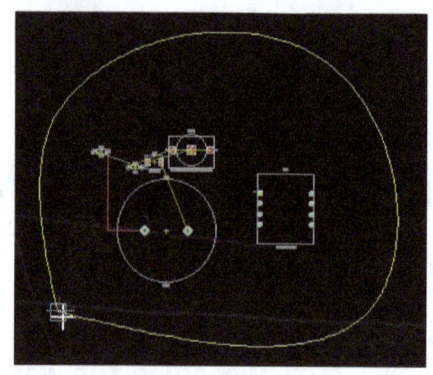

Figure-17. Spline outline of PCB

- After creating spline outline, press **ESC** to exit the tool.

Creating Arc Outline

The **Outline Arc** tool is used to create PCB outline using arc geometry. The procedure to use this tool is given next.

- Click on the **Outline Arc** tool from the **BOARD SHAPE** drop-down in the **DESIGN** tab of the **Ribbon**. The **OUTLINE ARC MANAGER** will be displayed; refer to Figure-18 and you will be asked to specify start point of arc.

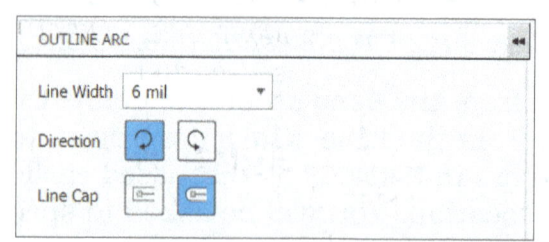

Figure-18. OUTLINE ARC MANAGER

- Set the parameters as discussed earlier in the **MANAGER**.
- Click in the drawing area to specify start point of arc, diameter of arc, and end point of arc; refer to Figure-19.

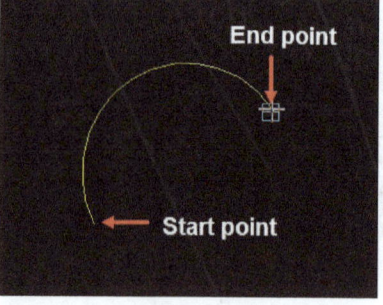

Figure-19. Creating arc outline

- Press **ESC** to exit the tool.

Creating Circle Outline

The **Outline Circle** tool is used to create PCB outline using circle geometry. The procedure to use this tool is given next.

- Click on the **Outline Circle** tool from the **BOARD SHAPE** drop-down in the **DESIGN** tab of the **Ribbon**. The **OUTLINE CIRCLE MANAGER** will be displayed.
- Specify the line width of outline circle in **MANAGER** and click in the drawing area to specify center of circle.
- You will be asked to specify diametrical point of the circle. Click at desired location to specify diameter of circle outline.
- Press **ESC** to exit the tool.

PLACING OBJECTS IN DRAWING

The tools in **PLACE** drop-down are used to place various components and PCB objects in the Design; refer to Figure-20. These tools are discussed next.

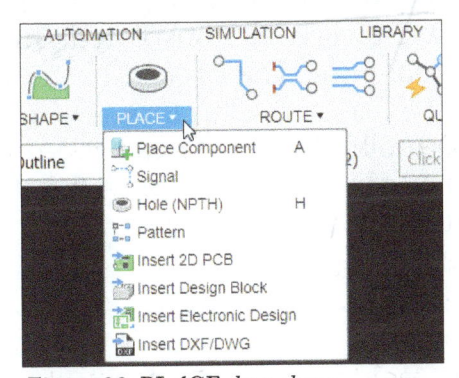

Figure-20. PLACE drop-down

Place Components

The **Place Component** tool is used to display **PLACE COMPONENTS MANAGER** so that you can place desired components as discussed earlier.

Creating Signal Wires

The **Signal** tool is used to create signal air wires in the individual PCB designs. The procedure to use this tool is given next.

- Click on the **Signal** tool from the **PLACE** drop-down in the **DESIGN** tab of the **Ribbon**. The **SIGNAL Dialog** will be displayed; refer to Figure-21 and you will be asked to specify the start point of air wire.

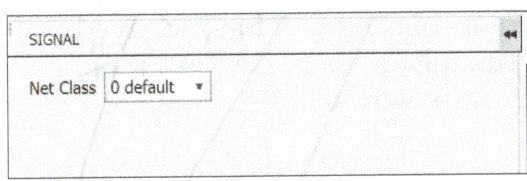

Figure-21. SIGNAL Dialog

- Select desired class option from the drop-down in the **Dialog** and click in the drawing area to specify the start point and end point of signal air wires; refer to Figure-22. Press **ESC** to exit the tool.

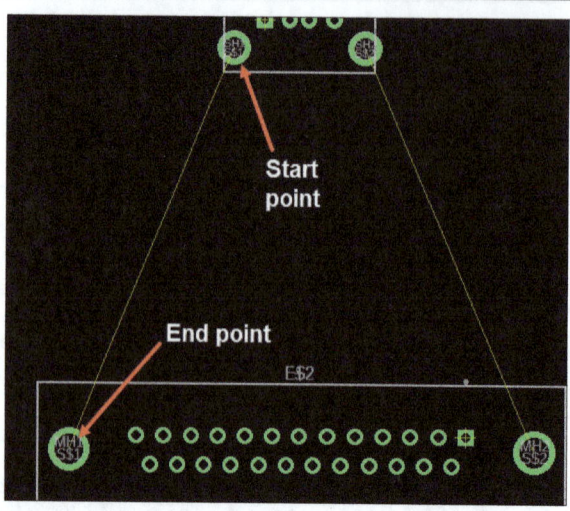

Figure-22. Creating signal air wire

Creating Hole (NPTH)

The **Hole (NPTH)** tool is used to create non-plated through holes in the PCB for connecting devices and components. The procedure to use this tool is given next.

- Click on the **Hole (NPTH)** tool from the **PLACE** drop-down in the **DESIGN** tab of the **Ribbon**. The **HOLE (NPTH) Dialog** will be displayed; refer to Figure-23 and you will be asked to specify location of hole.

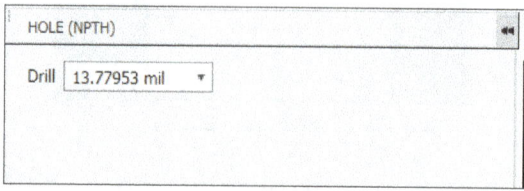

Figure-23. HOLE (NPTH) Dialog

- Select desired size of hole from the **Drill** drop-down.
- Click in the drawing area to specify location of drill hole. Press **ESC** to exit the tool.

The other tools of **PLACE** drop-down have been discussed earlier.

Creating Route Manually

The **Route Manual** tool is used to manually create wire route on air wires. The procedure to use this tool is given next.

- Click on the **Route Manual** tool from the **ROUTE** drop-down in the **DESIGN** tab of the **Ribbon**. The **ROUTE MANUAL Dialog** will be displayed; refer to Figure-24.
- Set the parameters in **Dialog** as discussed earlier and click on the air wire earlier created in design. The route wire will be attached to cursor; refer to Figure-25.

Autodesk Fusion 360 PCB Black Book 5-13

Figure-24. ROUTE MANUAL Dialog

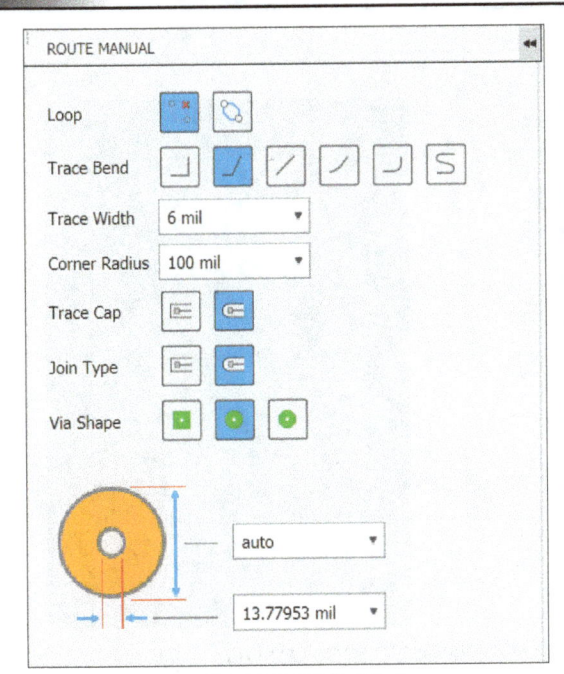

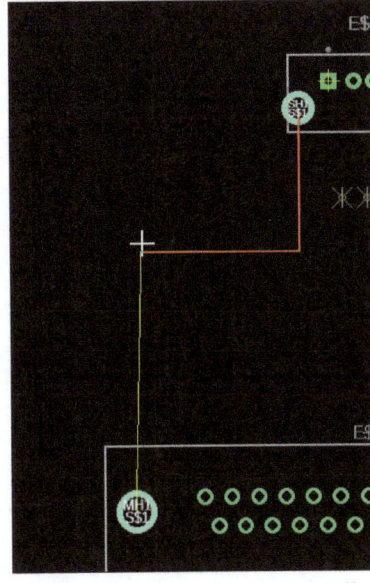

Figure-25. Creating route manually

- Create at desired locations to specify consecutive points of the route and then click at the last point of air wire to complete creating route.
- Press **ESC** to exit the tool.

Creating Differential Pair Routing

The Differential pair in routing is used for minimizing the signal quality loss due to surrounding noise signals coming from various near by devices like phones, Wi-Fi routers, and so on. In different pair, one wire carries + signal and other wire carries - signal value. Assume that due to surrounding noise, there is an addition of 0.1 V in signal strength then + signal will be added by 0.1 V and - signal will be reduced by 0.1 V hence keeping the net value of signal same. The **Route Differential Pair** tool is used to perform routing of differential pairs. Note that in Autodesk Fusion 360 PCB, the differential pairs are identified by their prefixes. For example, two net wires in schematic with **xxxx P** and **xxxx N** will be identified as differential pairs; refer to Figure-26. The procedure to perform differential pair routing is given next.

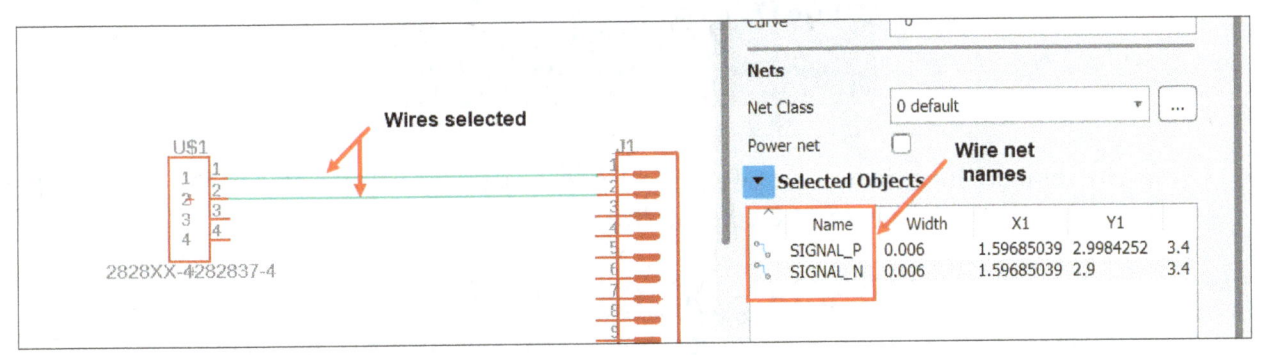

Figure-26. Wire nets named for differential in schematic

- Click on the **Route Differential Pair** tool from the **ROUTE** drop-down in the **DESIGN** tab of the **Ribbon**. The differential pairs will be highlighted in the drawing area and **ROUTE DIFFERENTIAL PAIR Dialog** will be displayed; refer to Figure-27.

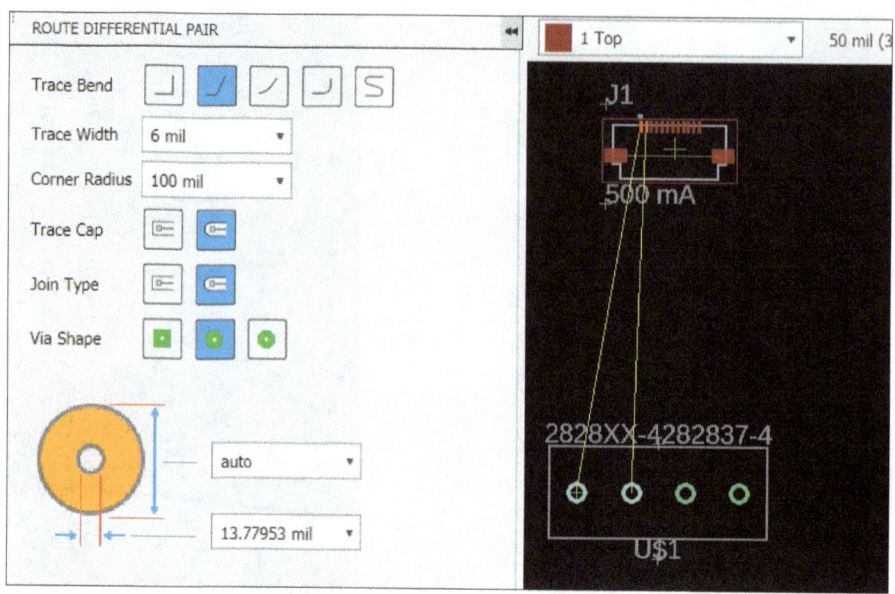

Figure-27. Differential pairs highlighted in drawing

- Set the parameters in **Dialog** as discussed earlier and click on the differential pair air wires. The routing wires will get attached to cursor; refer to Figure-28.

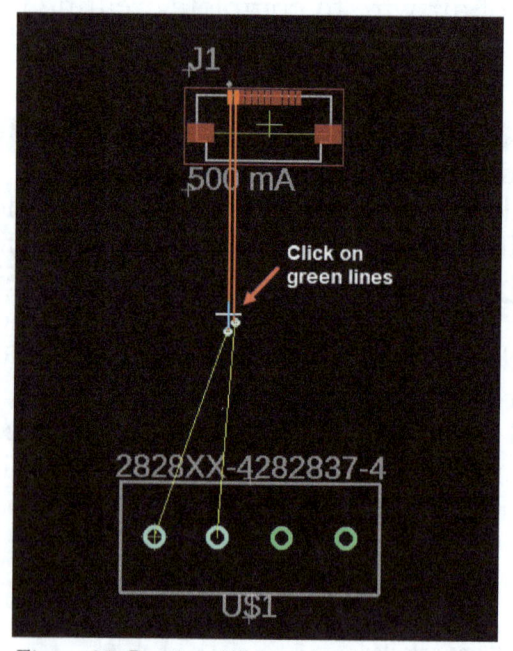

Figure-28. Routing differential pair

- Click at desired locations to specify routing points for the differential pair. After performing routing, press **ESC** to exit the tool.

Routing Multiple Airwires

The **Route Multiple** tool is used to convert multiple airwires into route wires. The procedure to use this tool is given next.

- Click on the **Route Multiple** tool from the **ROUTE** drop-down in the **DESIGN** tab of the **Ribbon**. The **ROUTE MULTIPLE Dialog** will be displayed; refer to Figure-29.

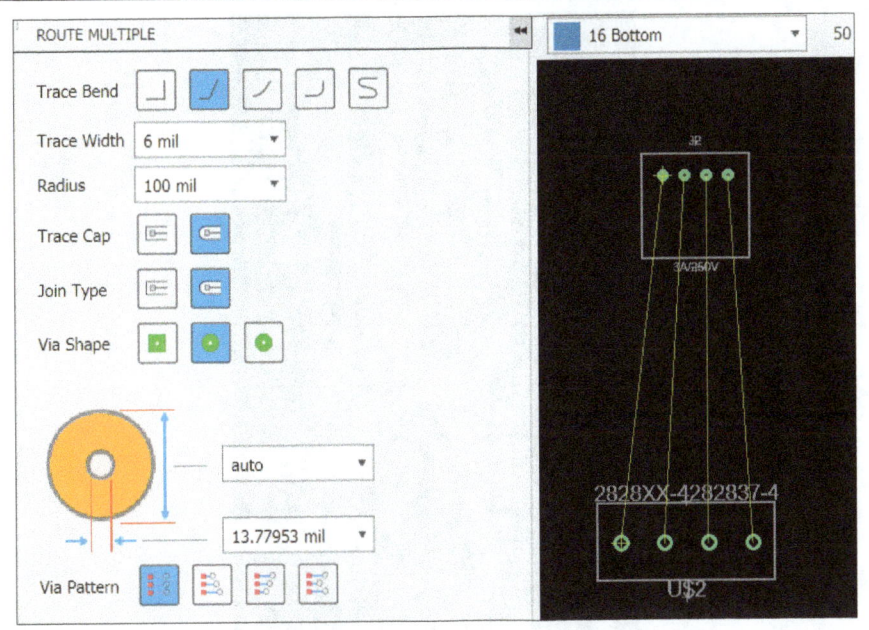

Figure-29. ROUTE MULTIPLE Dialog

- Set desired parameters in the **Dialog** to define shape and size of route wires.
- Click in the drawing area and create a line crossing all the pads of electronic component to be connected; refer to Figure-30.
- Once you have marked all the pads to be routed then right-click in the drawing area. The route wires will get attached to cursor; refer to Figure-31.

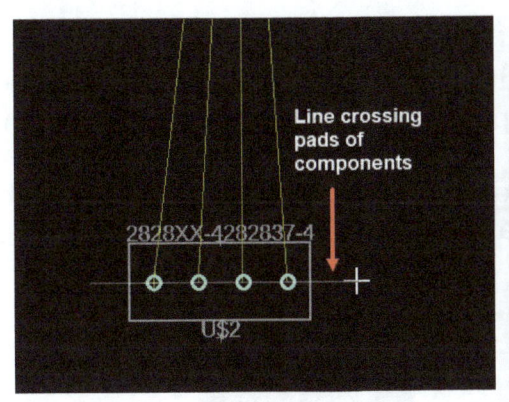

Figure-30. Selecting pads to be connected

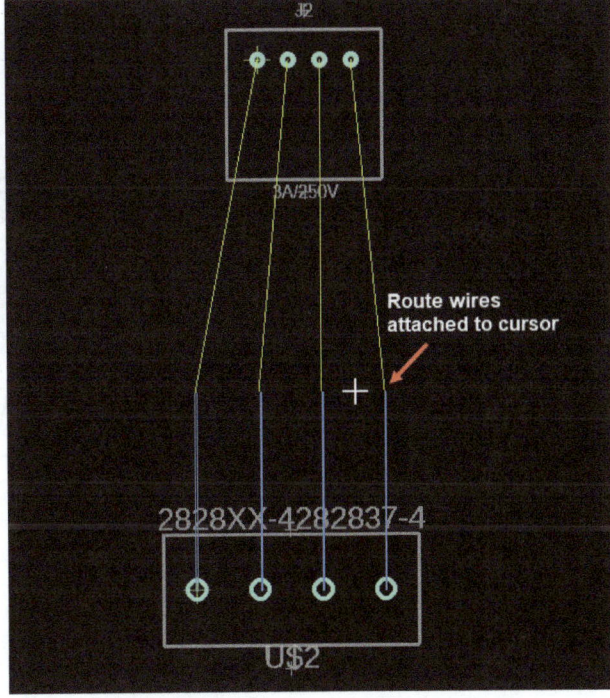

Figure-31. Route wires attached to cursor

- Click at desired locations to specify intermediate points of route and at the end click on one of the pads in target component to complete the route; refer to Figure-32.

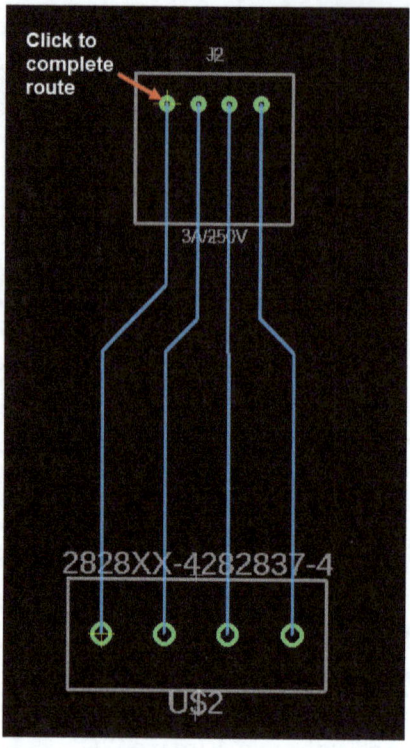

Figure-32. Completing the routing

- Press **ESC** to exit the tool.

Placing Vias in PCB

The **Via** tool is used to place via (current conducting small hole in PCB) connecting different layers of PCB at desired location. The procedure to use this tool is given next.

- Click on the **Via** tool from the **ROUTE** drop-down in the **DESIGN** tab of the **Ribbon**. The via symbol will get attached to cursor and **VIA Dialog** will be displayed; refer to Figure-33.

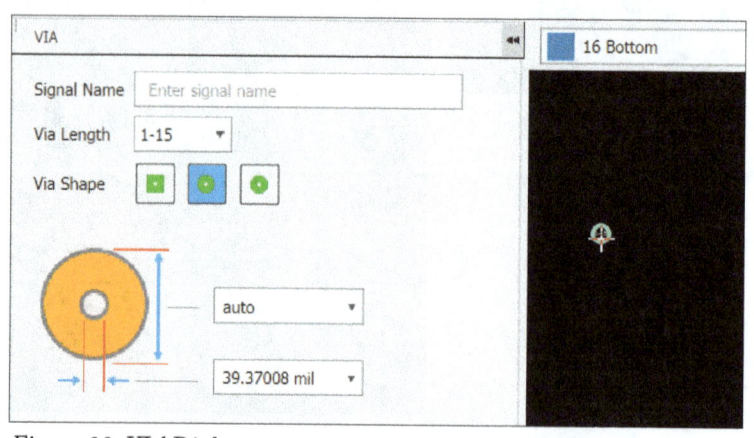

Figure-33. VIA Dialog

- Specify the name of the via in **Signal Name** edit box of Dialog.
- Select desired option from the **Via Length** drop-down to define the via to be used for creating respective section on the PCB.
- Set the other parameters in Dialog as discussed earlier and click at desired location on the PCB to place the symbol.
- After placing the symbols, press **ESC** to exit the tool.

Meander

The **Meander** tool is used to balance the length of individual wires in differential pairs to keep the losses of signals in sync. The procedure to use this tool is given next.

- Click on the **Meander** tool from the **ROUTE** drop-down in the **DESIGN** tab of the **Ribbon**. You will be asked to select the wire of differential pair to be balanced.
- Click at desired wire of pair. The selected wire will be highlighted and cursor will change to drag symbol; refer to Figure-34.

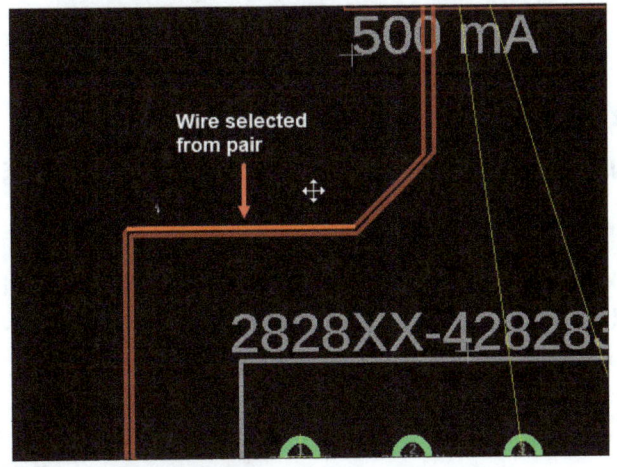

Figure-34. Selecting wire from differential pair

- Click at desired location on selected line from where you want to modify the line for balancing length and move cursor left/right. Preview of balanced pair will be displayed; refer to Figure-35.

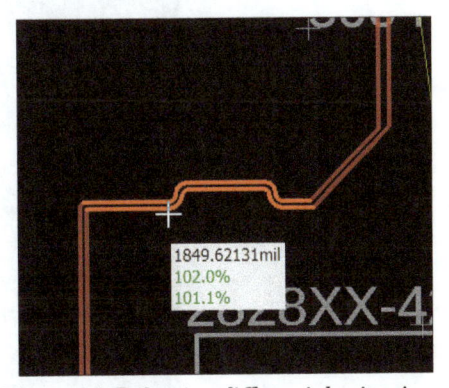

Figure-35. Balancing differential pair wires

- Click at desired location to apply changes. Press **ESC** to exit the tool.

QUICK ROUTE TOOLS

The tools in **QUICK ROUTE** drop-down are used to perform automatically route selected air wires; refer to Figure-36. The tools of this drop-down are discussed next.

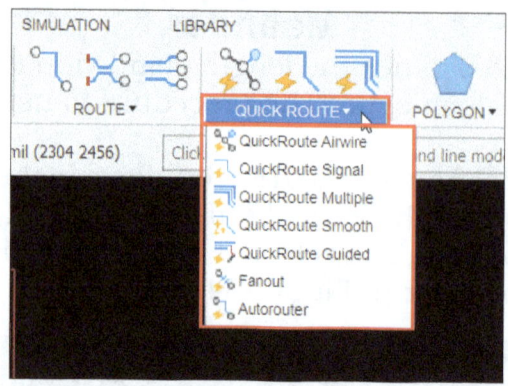

Figure-36. QUICK ROUTE drop-down

Quick Routing Air wires

The **QuickRoute Airwire** tool is used to automatically route selected air wires. The procedure to use this tool is given next.

- Click on the **QuickRoute Airwire** tool from the **QUICK ROUTE** drop-down in the **DESIGN** tab of the **Ribbon**. The **QUICKROUTE AIRWIRE Dialog** will be displayed; refer to Figure-37.

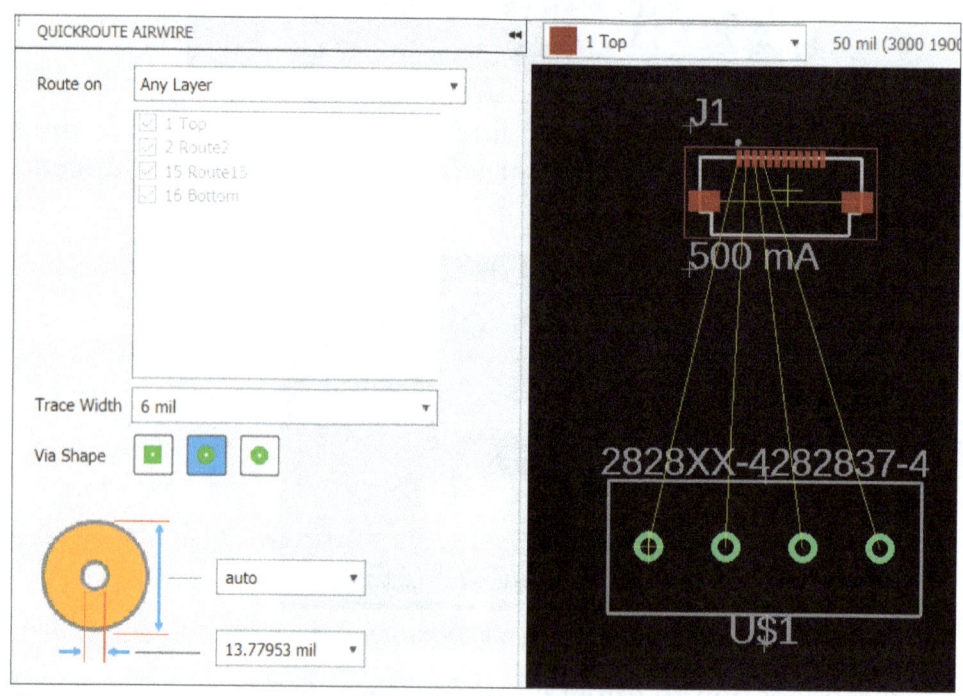

Figure-37. QUICKROUTE AIRWIRE Dialog

- Select desired option from the **Route on** drop-down to specify the layer on which route wires will be created.
- Specify the trace width and other parameters in the **Dialog** as discussed earlier.
- After setting desired parameters, click at desired air wires to convert them to route wires; refer to Figure-38.
- Press **ESC** to exit the tool.

Autodesk Fusion 360 PCB Black Book 5-19

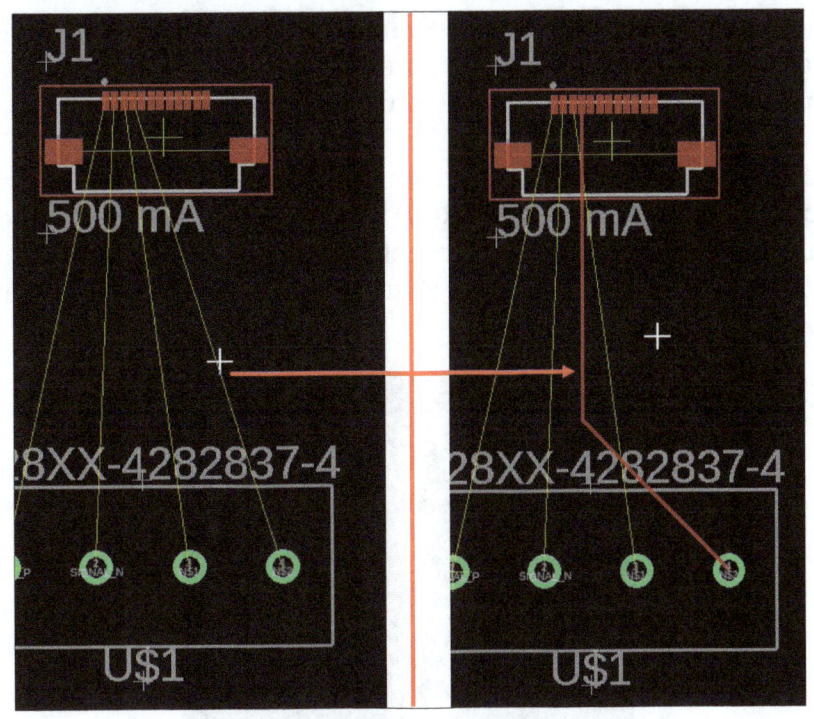

Figure-38. Air net to route wire

Quick Routing Signals

The **QuickRoute Signal** tool is used to automatically route group of signal wires with same names like differential pairs. The procedure to use this tool is given next.

- Click on the **QuickRoute Signal** tool from the **QUICK ROUTE** drop-down in the **DESIGN** tab of the **Ribbon**. The **QUICKROUTE SIGNAL Dialog** will be displayed similar to **Dialog** discussed earlier and you will be asked to select wire from the group.
- Select desired wire from the group. The other wires of group will also be routed automatically; refer to Figure-39.

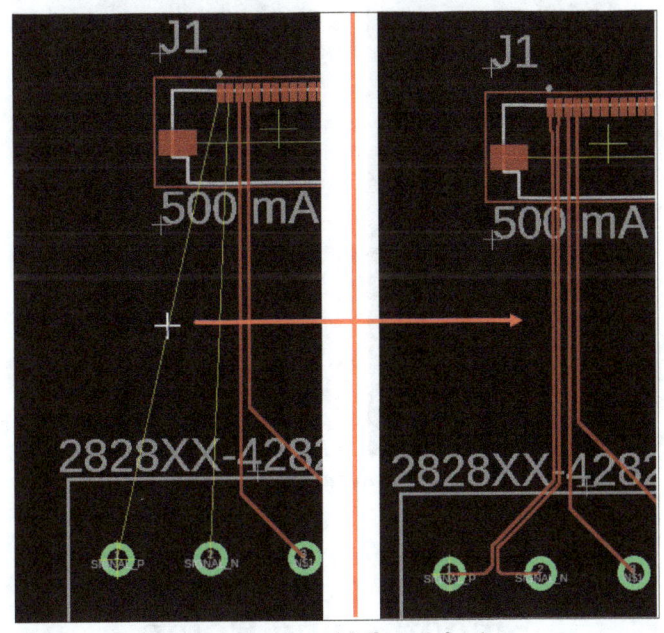

Figure-39. Automatic routing of differential pair

- Press **ESC** to exit the tool.

Quick Routing Multiple Wires

The **QuickRoute Multiple** tool is used to automatically generate route wires from multiple air wires. The procedure to use this tool is given next.

- Click on the **QuickRoute Multiple** tool from the **QUICK ROUTE** drop-down in the **DESIGN** tab of the **Ribbon**. The **QUICKROUTE MULTIPLE Dialog** will be displayed and you will be asked to create a line intersecting with pads of component to be routed.
- Create the reference line; refer to Figure-40. The connected airwires will be highlighted.

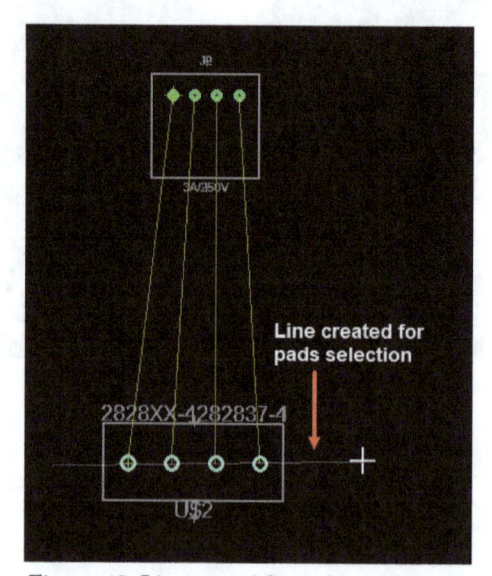

Figure-40. Line created for pads selection

- After the connected airwires have been highlighted then right-click in the drawing area. The route wires will be created automatically; refer to Figure-41.

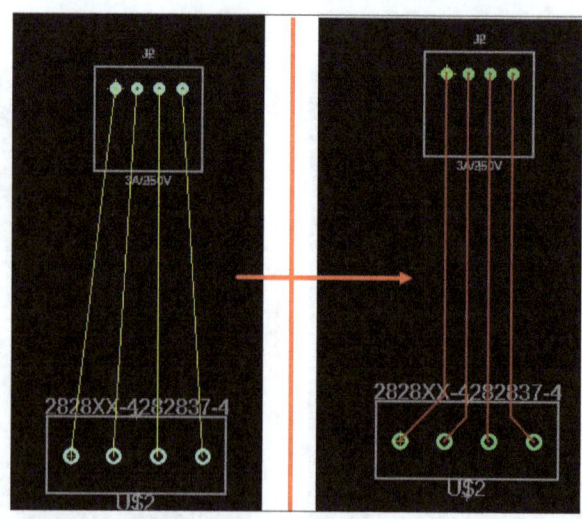

Figure-41. Routing multiple airwires

- Press **ESC** to exit the tool.

Smoothening Route

The **QuickRoute Smooth** tool is used to reduce extra bends in selected wire routes. The procedure to use this tool is given next.

- Click on the **QuickRoute Smooth** tool from the **QUICK ROUTE** drop-down in the **DESIGN** tab of the **Ribbon**. You will be asked to select the signal wire for smoothening.
- Select desired routed wire from the drawing area to smoothen.

Guided Quick Routing

The **QuickRoute Guided** tool is used to route multiple wires using specified guide points. The procedure to use this tool is given next.

- Click on the **QuickRoute Guided** tool from the **QUICK ROUTE** drop-down in the **DESIGN** tab of the **Ribbon**. The **QUICKROUTE GUIDED Dialog** will be displayed and you will be asked to draw a line for selecting pads of component.
- Draw the line intersecting pads of component to be routed; refer to Figure-42. The connected airwires will be highlighted along with the pads.

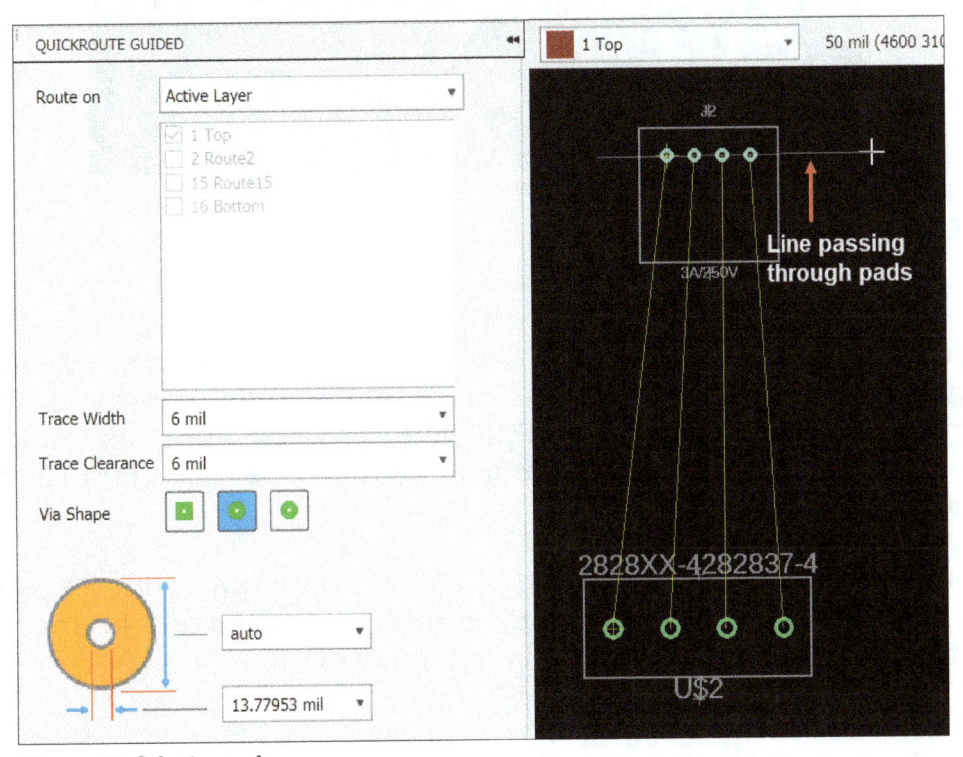

Figure-42. Selecting pads

- Right-click in the drawing area to complete selection of pads/airwires. The routing point of airwires will be attached to cursor.
- Click at desired locations to create routing guide line and then press **ENTER**. The route path will be created; refer to Figure-43.

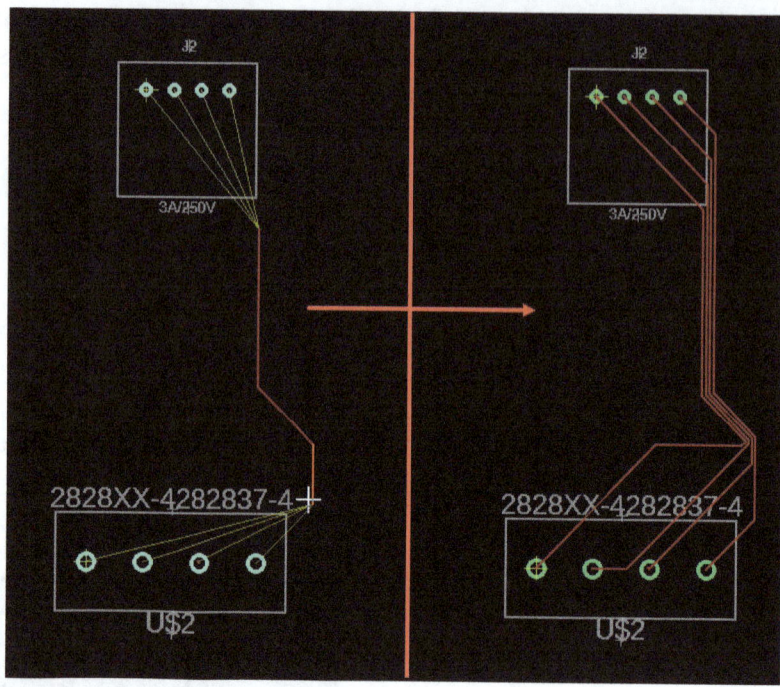

Figure-43. Creating guided route

- Press **ESC** to exit the tool.

Performing Fanout

The **Fanout** tool is used to create routed wire and connected via from selected onboard components. Note that some components are not directly soldered to PCB rather they have legs which are soldered to PCB, like USB port, jumpers, and so on. For such cases, you need to use fanout for generating connection vias. The procedure to use fanout is given next.

- Click on the **Fanout** tool from the **QUICK ROUTE** drop-down in the **DESIGN** tab of the **Ribbon**. The **FANOUT Dialog** will be displayed; refer to Figure-44 and you will be asked to select pad/wire/airwire/via to apply fanout.

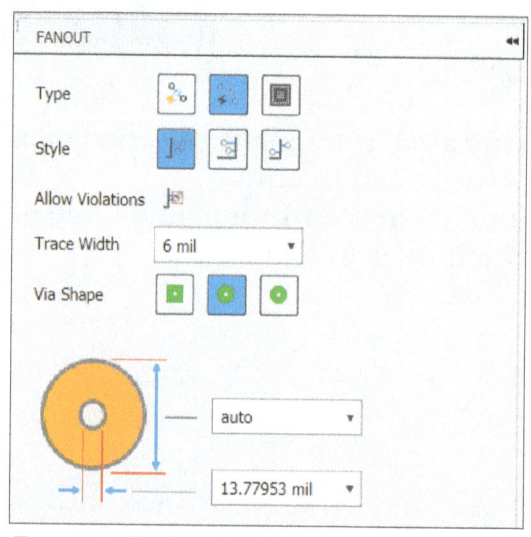

Figure-44. FANOUT Dialog

- Select the **Fanout Device** button from the **Type** section of **Dialog** to create via at shortest distance to selected pads of device.

Autodesk Fusion 360 PCB Black Book   5-23

- Select the **Fanout Signal** button from **Type** section to add via for each pad along the signal path.
- Select the **BGA Autorouter** button from the **Type** section to use Ball Grid Array autorouter for automatically generating via for pads of selected component. On selecting this button, the **BGA Settings** dialog box will be displayed; refer to Figure-45. Select desired element from the **Elements** section and click on the **>>** button. The selected element will be added in the list. Click on the **Edit BGA** button from the dialog box. The **Edit BGA** dialog box will be displayed; refer to Figure-46. Double-click on the signal to modify via connections. The **Layers** dialog box will be displayed. Select desired layers (while holding **CTRL** key) to be connected by via and click on the **OK** button from dialog box. Next, click on the **OK** button from **Edit BGA** dialog box and **BGA Settings** dialog box. The route will be created automatically.

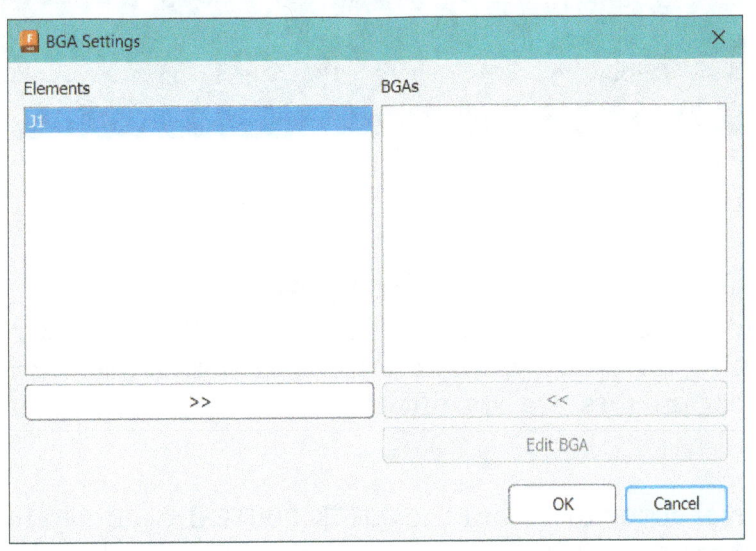

Figure-45. BGA Settings dialog box

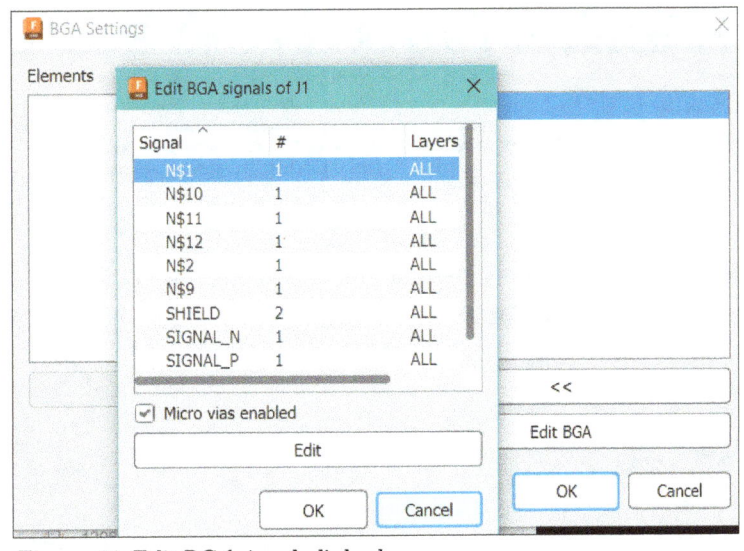

Figure-46. Edit BGA signals dialog box

- Select desired button from the **Style** section of **FANOUT Dialog** to apply fanout symbol on outside, inside, and alternate side of part.
- After setting desired parameters, click on the airwire of component. The pad vias will be created; refer to Figure-47.

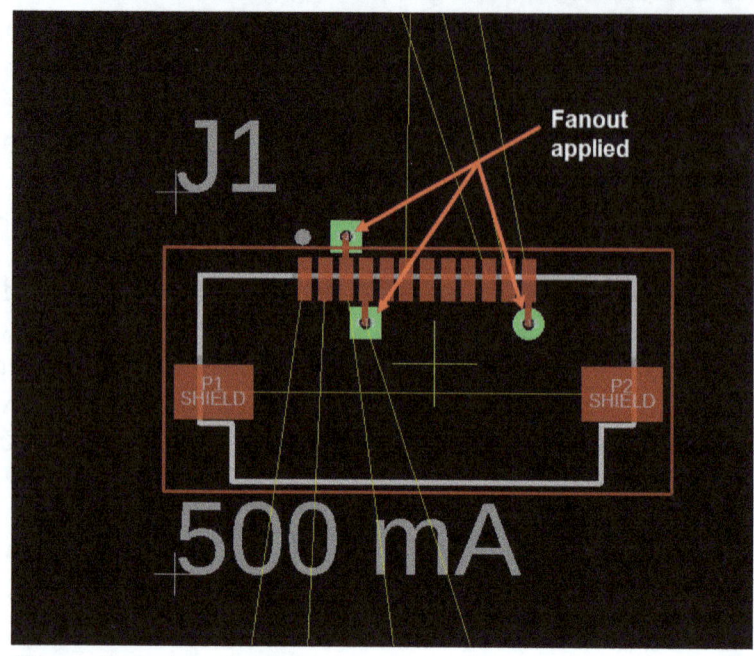

Figure-47. Applying fanouts

- Press **ESC** to exit the tool.

Autorouter

The **Autorouter** tool is used to automatically route all the air wires in the drawing area. Before running autorouter, make sure that net classes, grid sizes, drill hole sizes, and other parameters are specified correctly. The procedure to use this tool is given next.

- Click on the **Autorouter** tool from the **QUICK ROUTE** drop-down in the **DESIGN** tab of **Ribbon**. The **Autorouter Main Setup** dialog box will be displayed; refer to Figure-48.

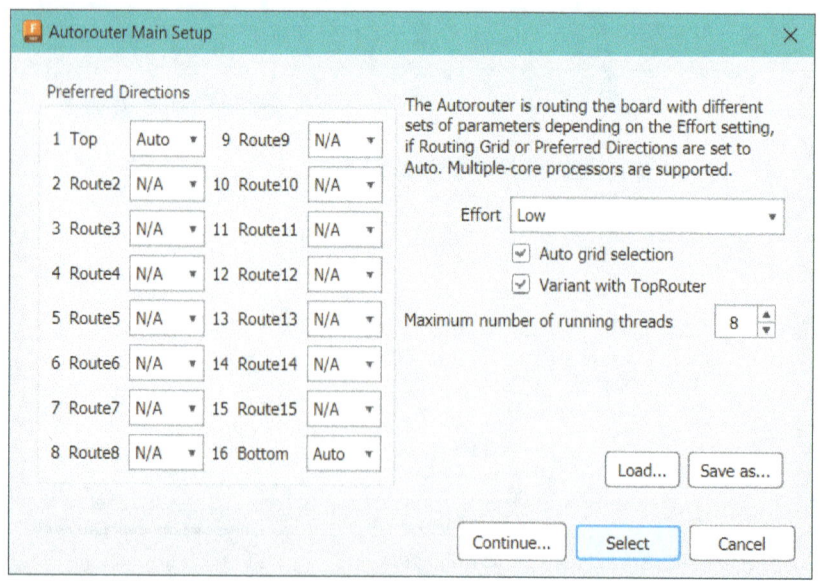

Figure-48. Autorouter Main Setup dialog box

- Set desired options in the drop-downs of **Preferred Directions** area to define directions of routing for various route layers.
- Select desired option from the **Effort** drop-down to define the level of processing to be performed for generating optimal routes.

- Select the **Auto grid selection** check box to automatically select suitable grid lines for routing.
- Select the **Variant with TopRouter** check box to generate routes with minimum vias.
- Set desired number in the **Maximum number of running threads** spinner to define the number of simultaneous computing jobs that can run in CPU for routing. Make sure that this parameter should be less than the number of threads available in CPU.
- Click on the **Load** button to load autorouter parameters from selected CTL file.
- Click on the **Save as** button to save current autorouter parameters in a CTL file.
- Click on the **Select** button from the dialog box to select signal airwires from the drawing area for routing. If you do not select specific airwires then all the airwires will be routed.
- Click on the **Continue** button from the dialog box. The **Routing Variants Dialog** will be displayed; refer to Figure-49.

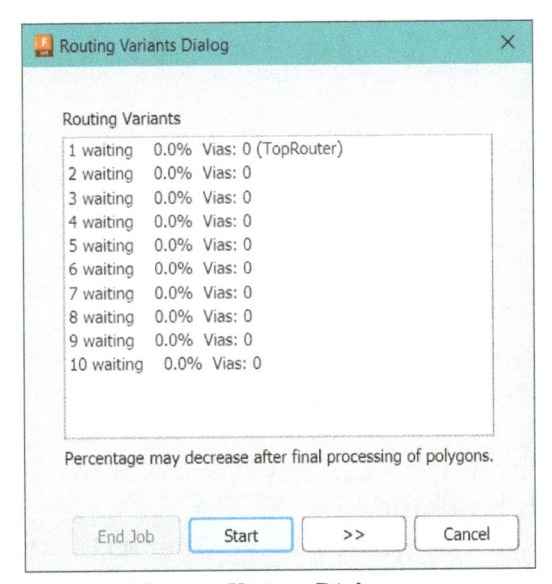

Figure-49. Routing Variants Dialog

- Click on the **Start** button from the dialog box to start routing. Once the process is complete, the route will be created.
- Select desired variant from the dialog and click on the **End Job** button to generate respective routing.

POLYGONS

The polygon is used to create a sheet of copper on PCB for signal. The tools to create and manage polygons are available in the **POLYGON** drop-down of **Ribbon**; refer to Figure-50. The tools of this drop-down are discussed next.

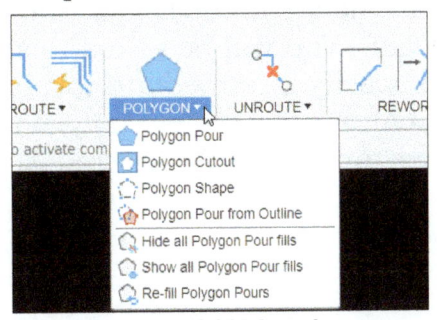

Figure-50. POLYGON drop-down

Creating Polygon Pour

The **Polygon Pour** tool is used to create a sheet of copper on routing layer for connecting multiple signals. The procedure to use this tool is given next.

- Click on the **Polygon Pour** tool from the **POLYGON** drop-down in the **DESIGN** tab of the **Ribbon**. The **POLYGON POUR Dialog** will be displayed; refer to Figure-51.

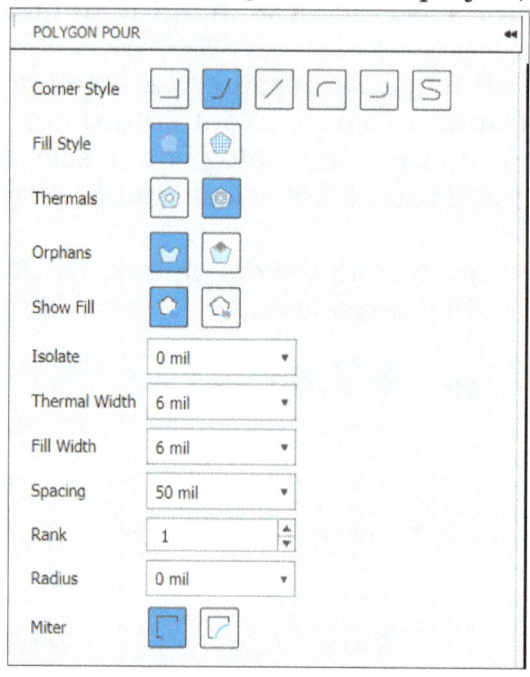

Figure-51. POLYGON POUR Dialog

- Select desired button from the **Corner Style** section to define boundary of polygon pour.
- Select desired button from the **Fill Style** section to completely solid fill polygon or fill the polygon with hatching.
- Select the **Thermals on** button from the **Thermals** section to reduce thermal conductivity of copper sheet while still allowing current conduction. This enables to allow easy soldering. Select the **Thermals off** button to allow free thermal conduction.
- Orphans are the small polygons which do not have connections to signals but they are generated due to specified design rules when creating polygon. Select the **Drop orphans** button if you want to delete orphan polygons and select the **Keep orphans** button if you want to keep orphan polygons.
- Select the **Show Fill** button from the **Show Fill** section to display polygon fill in drawing area. Select the **Hide Fill** button to hide polygon fill in drawing area.
- Specify desired value in the **Isolate** edit box to define the minimum gap to be maintained from nearby signals.
- Similarly, specify other parameters in the Dialog and then click in drawing area to specify start point of polygon. You will be asked to specify next point of polygon.
- Click at desired points to specify intermediate corner points of polygon and then double-click at desired location to complete polygon creation. The **Signal** dialog box will be displayed; refer to Figure-52.

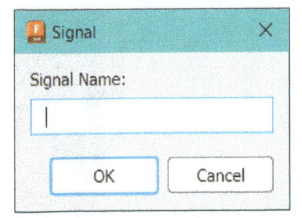

Figure-52. Signal dialog box

- Specify desired name of polygon in the edit box and click on the **OK** button. The polygon will be created; refer to Figure-53. Press **ESC** to exit the tool.

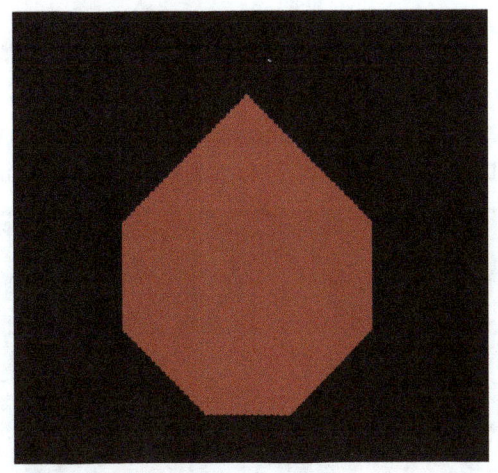

Figure-53. Polygon pour created

Creating Polygon Cutout

The **Polygon Cutout** tool is used to create a void area in polygon pour earlier created. The procedure to use this tool is given next.

- Click on the **Polygon Cutout** tool from the **POLYGON** drop-down in the **DESIGN** tab of the **Ribbon**. The **POLYGON CUTOUT Dialog** will be displayed; refer to Figure-54.

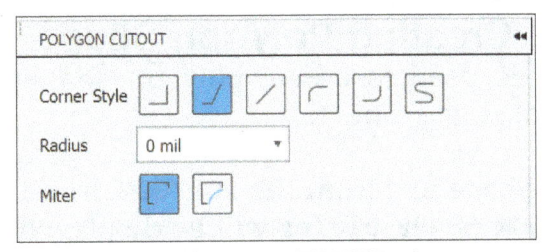

Figure-54. POLYGON CUTOUT Dialog

- Specify the parameters in Dialog as discussed earlier and click in the polygon pour at desired location to specify start point cutout; refer to Figure-55.

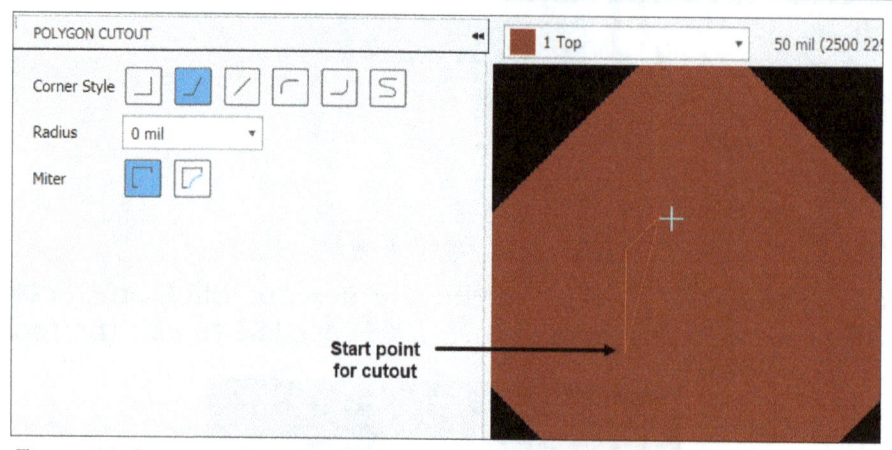

Figure-55. Creating polygon cutout

- Specify consecutive points of the cutout and then double-click to complete the cutout creation; refer to Figure-56. Press **ESC** to exit the tool.

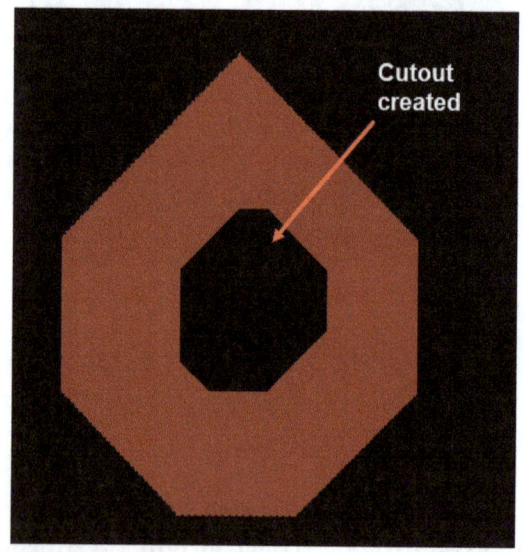

Figure-56. Cutout created

Creating Polygon Shape

The **Polygon Shape** tool is used to create polygon shaped non-signal solids. The procedure to use this tool is given next.

- Click on the **Polygon Shape** tool from the **POLYGON** drop-down in the **DESIGN** tab of the **Ribbon**. The **POLYGON SHAPE Dialog** will be displayed and you will be asked to specify start point of the polygon.
- Set the parameters in **Dialog** as discussed earlier and click in the drawing area to specify start point of the polygon; refer to Figure-57.

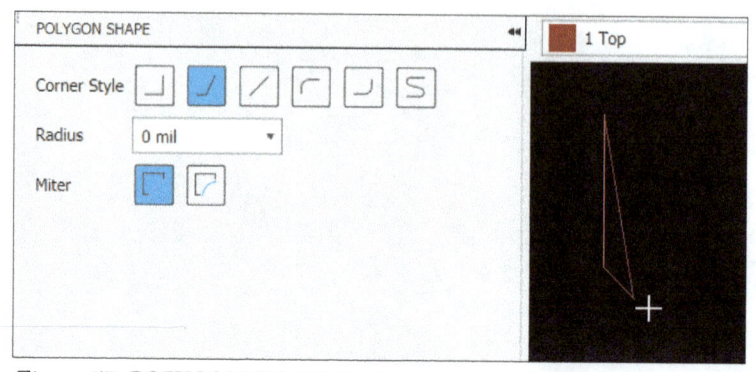

Figure-57. POLYGON SHAPE Dialog

- Click in the drawing area to specify consecutive points and then double-click at last point to complete creation of polygon. Press **ESC** twice to exit the tool.

Creating Polygon Pour from Outline

The `Polygon Pour from Outline` tool is used to convert selected board outlines to polygon pours. The procedure to use this tool is given next.

- Click on the `Polygon Pour from Outline` tool from the **POLYGON** drop-down in the **DESIGN** tab of the **Ribbon**. You will be asked to select outlines of board.
- Select desired outline of board from drawing area. The **Layer** dialog box will be displayed; refer to Figure-58.

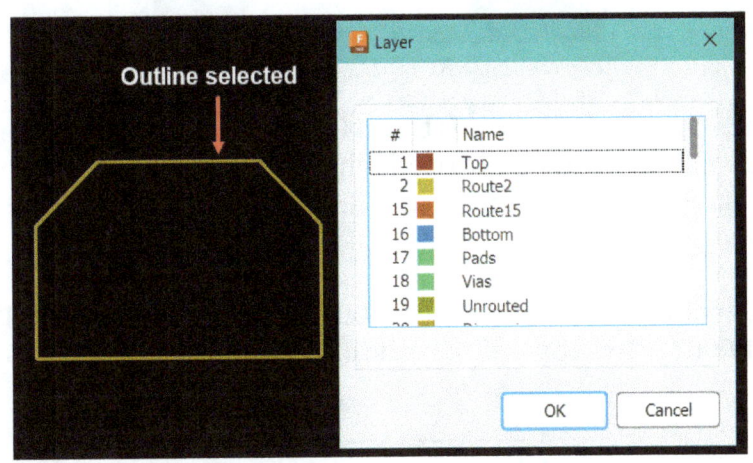

Figure-58. Selecting outline

- Select desired layer from the dialog box to define the layer on which polygon pour will be created and click on the **OK** button. The pour will be created; refer to Figure-59.

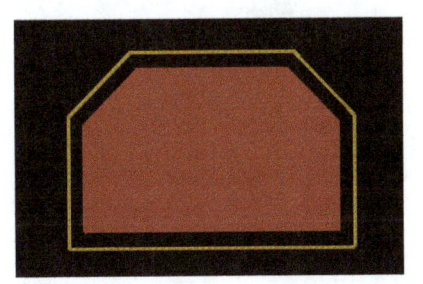

Figure-59. Polygon pour created using outline

Hiding All Polygon Pour Fills

Click on the `Hide all Polygon Pour fills` tool to hide all the polygon pour fills from the drawing area.

Showing All Polygon Pour Fills

Click on the `Show all Polygon Pour fills` tool to display all the polygon pour fills from the drawing area.

Repairing Polygon Pours

Click on the `Re-fill Polygon Pours` tool to repair the polygon pours after you have moved components and cutouts.

UNROUTING TOOLS

Unrouting is performed to convert routed wires & Vias to unrouted signals for performing rerouting later. This is useful when you have performed major changes in design and want to re-run Autorouter tool later. The tools to perform unrouting are available in the **UNROUTE** drop-down of **Ribbon**; refer to Figure-60. Various tools of this drop-down are discussed next.

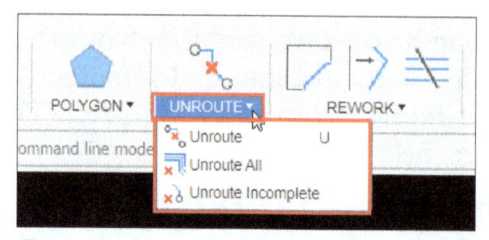

Figure-60. UNROUTE drop-down

Unrouting

The **Unroute** tool is used to convert routed wires, vias, and polygon pours to unrouted signals based on specified selection criteria. The procedure to use this tool is given next.

- Click on the **Unroute** tool from the **UNROUTE** drop-down in the **DESIGN** tab of the **Ribbon**. The **UNROUTE Dialog** will be displayed; refer to Figure-61.

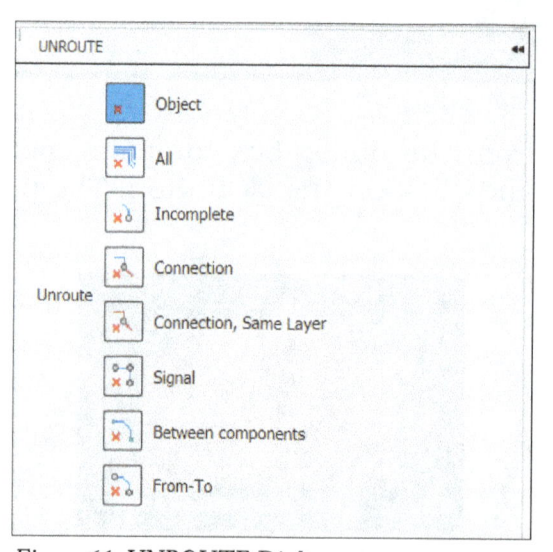

Figure-61. UNROUTE Dialog

- Select desired button from the **Dialog** to define the type of unroute operation to be performed. Select the **Object** button to convert selected routed segments to unroute signals. Select the **All** button to convert all the routed segments to unrouted signals. Select the **Incomplete** button to convert incomplete routed segments to unrouted signals. Select the **Connection** button to convert all copper connected to selected segment to unrouted signals. Select the **Connection, Same Layer** button to convert all copper connections of selected segment to unrouted signals on same layer. Select the **Signal** button to convert all segments of same name as selected segment to unrouted signals. Select the **Between components** button to convert all segments between two selected components to unrouted signals. Select the **From-To** button to convert all segments between two selected terminals to unrouted signals.

- After selecting desired button, perform respected selection in the drawing area.
- Click on the **Done** button from the **Dialog** to complete operation. A warning message box will be displayed asking whether you want to convert the routed segments to unrouted signals. Click on the **Yes** button to apply changes.

Unrouting All

The **Unroute All** tool is used to convert all the routed segments to unrouted signals. The procedure to use this tool is given next.

- Click on the **Unroute All** tool from the **UNROUTE** drop-down in the **DESIGN** tab of the **Ribbon**. A message box will be displayed; refer to Figure-62.

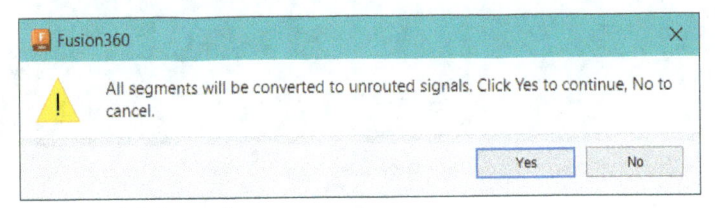

Figure-62. Message box

- Click on the **Yes** button to perform the operation.

Unroute Incomplete

The **Unroute Incomplete** tool is used to convert all the routed segment which are not complete to unrouted signals. A message box will be displayed for confirmation of performing operation. Click on the **Yes** button from message box to complete the operation.

REWORK TOOLS

The tools in the **REWORK** drop-down of **Ribbon** are used to modify routed segments in PCB; refer to Figure-63. Most of the tools of this drop-down have been discussed earlier in previous chapters. The procedure to use **Reroute** tool is discussed next.

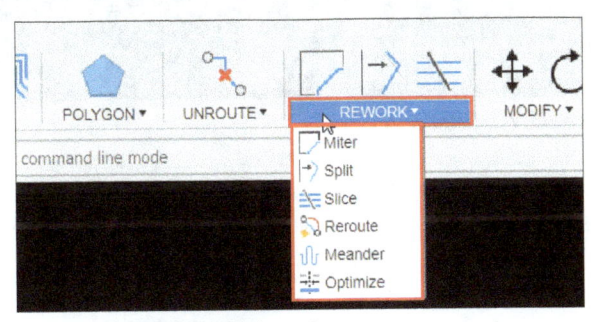

Figure-63. REWORK drop-down

Rerouting

The **Reroute** tool is used to reroute selected wires on same layer after modifying position of related components. The procedure to use this tool is given next.

- Once you have performed position changes of the components, you will find that the original route is not in optimum shape; refer to Figure-64. To reroute the wires, click on the **Reroute** tool from the **REWORK** drop-down in the **DESIGN** tab of the **Ribbon**. You will be asked to select the pads/devices/wires to be rerouted.

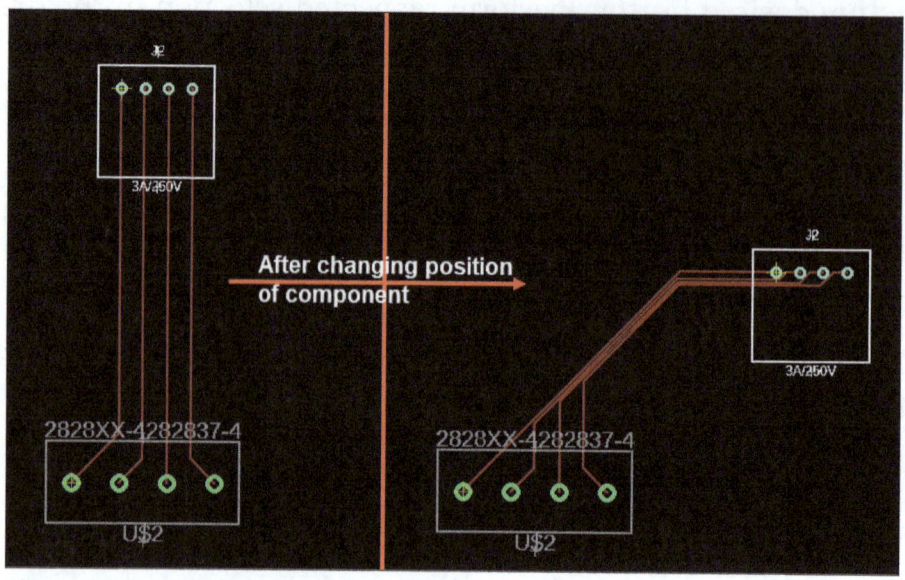

Figure-64. Changing position of component

- Click on the device/component to reroute all the wires connected to selected device; refer to Figure-65.

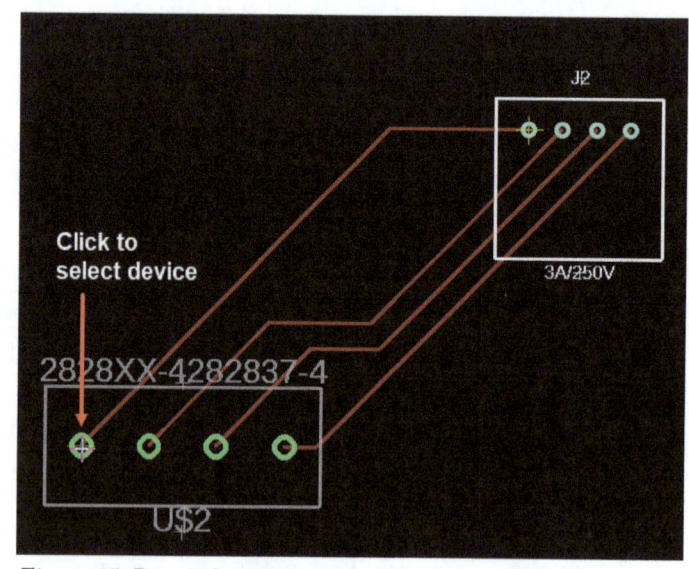

Figure-65. Rerouted wires

Self Assessment

Q1. What is the use of Flip Board tool?

Q2. What are the parameters to be specified in Clearance tab of Layer Stack Manager dialog box?

Q3. When components are placed on top layer of PCB then they are shown with green padding but when components are placed on bottom layer of PCB then they are displayed with red padding. (T/F)

Q4. The Signal tool is used to balance the length of individual wires in differential pairs to keep the losses of signals in sync. (T/F)

Q5. The Unroute Incomplete tool is used to convert all the routed segments to unrouted signals. (T/F)

Q6. The .... tool is used to create non-plated through holes in the PCB for connecting devices and components.

Q7. The .... in routing is used for minimizing the signal quality loss due to surrounding noise signals coming from various near by devices.

Q8. Which of the following tools is used to reduce extra bends in selected wire routes?

a) QuickRoute Guided
c) QuickRoute Signal
b) QuickRoute Airwire
d) QuickRoute Smooth

Q9. Which of the following tools is used to create a sheet of copper on routing layer for connecting multiple signals?

a) Polygon Pour
c) Fanout
b) Autorouter
d) Meander

Answers: **Ans 3.** T **Ans 4.** F **Ans 5.** F **Ans 6.** Hole (NPTH) **Ans 7.** Differential pair **Ans 8.** d **Ans 9.** a

FOR STUDENT NOTES

Chapter 6

3D PCB Manufacturing and Design Library

Topics Covered

The major topics covered in this chapter are:

- *Performing DRC*
- *Generating 3D PCB and Editing*
- *Manufacturing Operations*
- *Design Library*

INTRODUCTION

In previous chapter, you have learned to create and route schematics as well as PCBs. In this chapter, you will learn to push those designs to a 3D PCB and analyze your design in 3D. You will also learn to prepare your design for manufacturing via CAM. But before performing these operations, it is important to verify that the PCB passes all the design check. The process of performing this check is discussed next.

PERFORMING DRC

The DRC (Design Rule Check) is performed to ensure that all the route wires are connected properly and connections are maintained at specified design rule gaps. The procedure to perform DRC is given next.

- After creating design, click on the **DRC** tool from the **DRC** drop-down in the **RULES DRC/ERC** tab of **Ribbon**. The **DRC** dialog box will be displayed; refer to Figure-1.

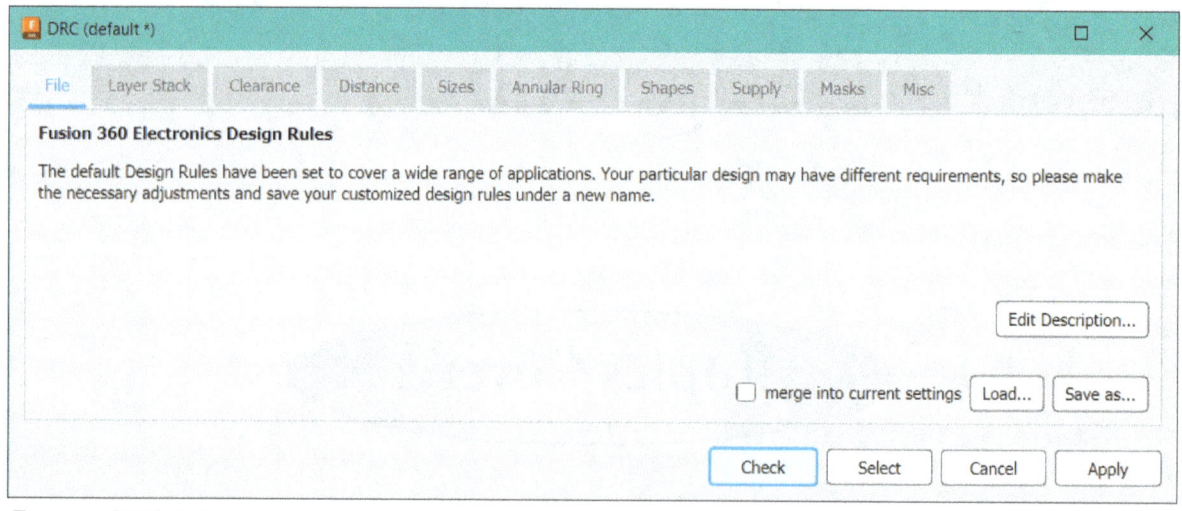

Figure-1. DRC dialog box

- Set the parameters in dialog box as discussed earlier and click on the **Check** button. If the design passes DRC then **DRC: No Errors** message box will be displayed in status bar.
- If there are errors in the design then **DRC Errors** dialog box will be displayed; refer to Figure-2.

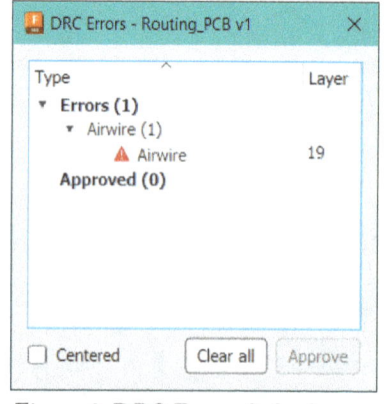

Figure-2. DRC Errors dialog box

- Select the error from the dialog box to check it in drawing area; refer to Figure-3.

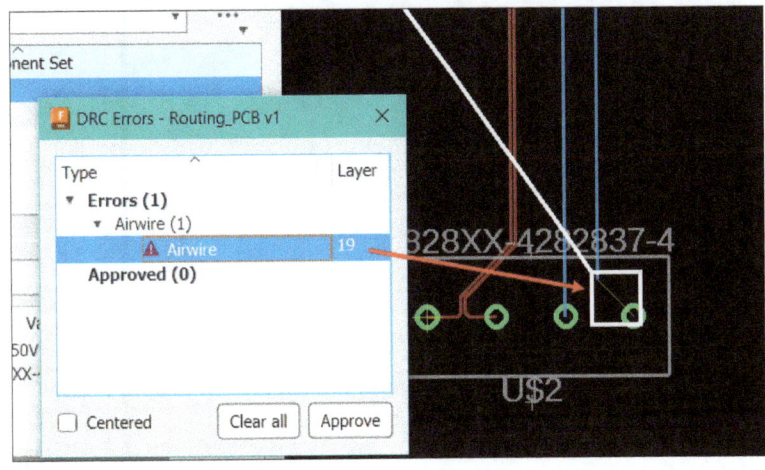

Figure-3. Error in drawing area

- If you think that the shown error is intentional in design then click on the **Approve** button otherwise close dialog box and modify the design accordingly. Note that after closing the dialog box, you can anytime check the errors by using the **Errors** tool from the **Ribbon** without re-running the DRC.

PUSHING PCB TO 3D PCB

The **Push to 3D PCB** tool is used to generate a 3D PCB using the existing PCB design. The procedure to use this tool is given next.

- Click on the **Push to 3D PCB** tool from the **SWITCH** drop-down in the **DESIGN** tab of the **Ribbon**. The **PUSH TO 3D PCB Dialog** will be displayed; refer to Figure-4.

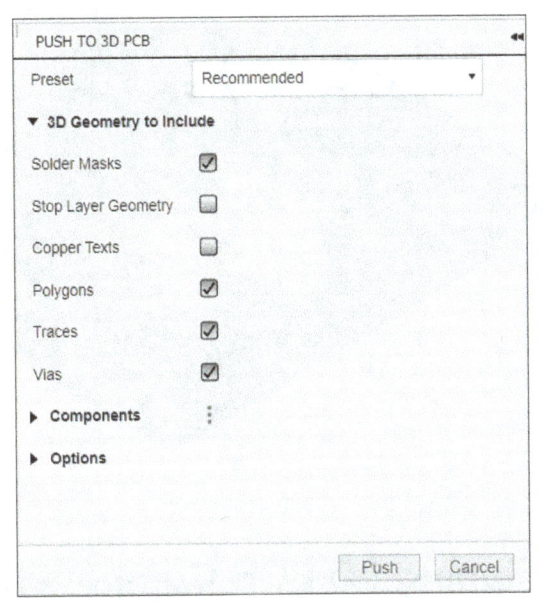

Figure-4. PUSH TO 3D PCB Dialog

- Select desired option from the **Preset** drop-down of **Dialog** to define the quality level for generating 3D model. Alternatively, you can select/clear desired check boxes from the **Dialog** to define the quality of 3D PCB.
- Expand the **Components** node of **Dialog** and select check boxes for components to be included in the 3D PCB; refer to Figure-5.

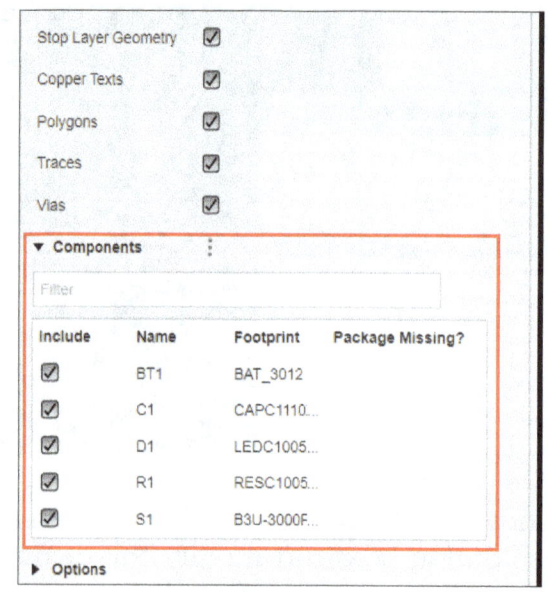

Figure-5. Components section

- Expand the **Options** node from the **Dialog** and select desired **Silkscreen Quality** button to define quality of silk screen.
- After setting desired parameters, click on the **Push** button. The 3D PCB will be created and displayed in 3D PCB environment; refer to Figure-6.

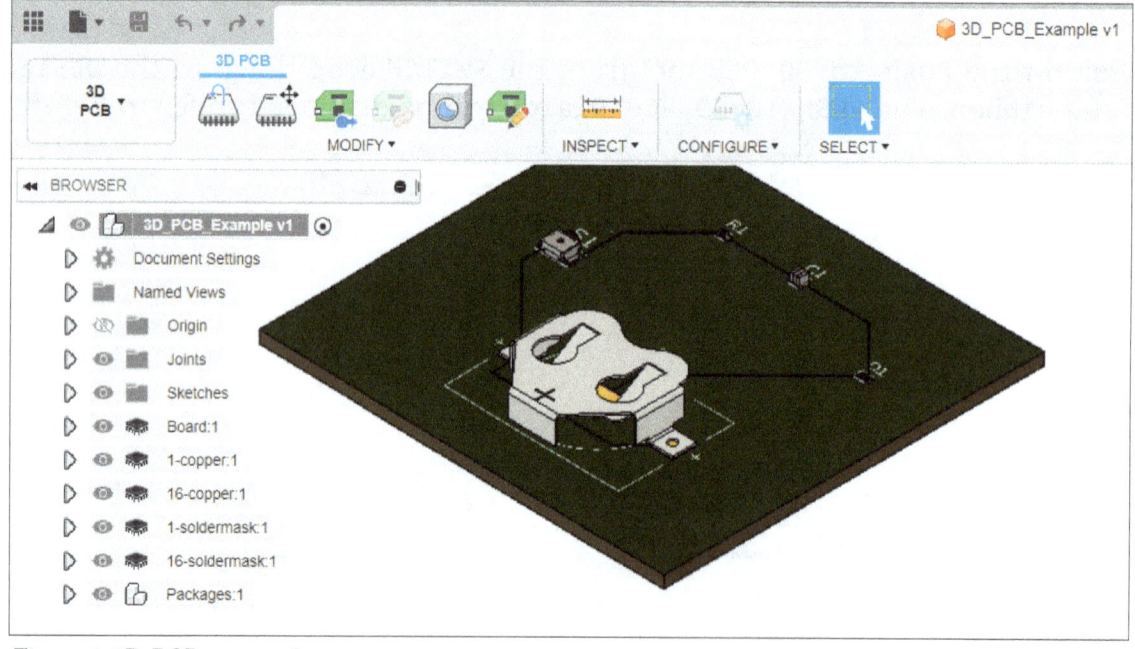

Figure-6. 3D PCB generated

Flipping 3D PCB Component

The **Flip 3D PCB Component** tool is used to flip the 3D Component on PCB. The procedure to use this tool is given next.

- Click on the **Flip 3D PCB Component** tool from the **MODIFY** drop-down in the **3D PCB** tab of the **Ribbon**. The **FLIP 3D PCB COMPONENT** dialog box will be displayed; refer to Figure-7.

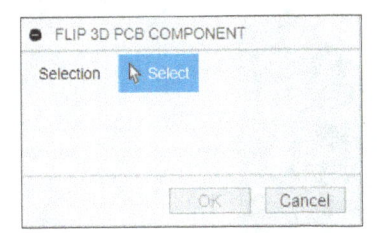

Figure-7. FLIP 3D PCB COMPONENT dialog box

- Select the component to be flipped from the drawing area. Preview of flipped component will be displayed; refer to Figure-8.

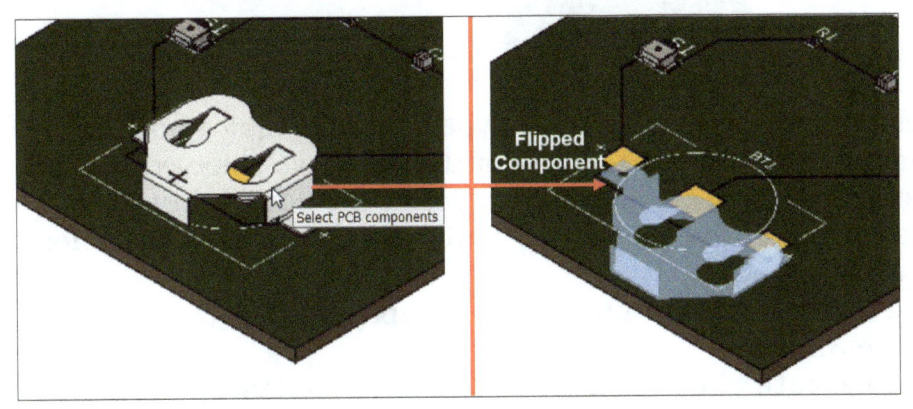

Figure-8. Flipped component

- Click on the **OK** button from the dialog box to flip the component.

Moving 3D PCB Components

The **Move 3D PCB Component** tool is used to move selected component at desired location on the 3D PCB. The procedure to use this tool is given next.

- Click on the **Move 3D PCB Component** tool from the **MODIFY** drop-down in the **3D PCB** tab of **Ribbon**. The **MOVE 3D PCB COMPONENT** dialog box will be displayed; refer to Figure-9.

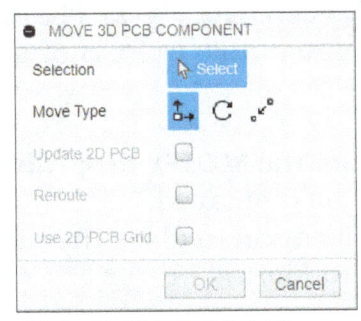

Figure-9. MOVE 3D PCB COMPONENT dialog box

- Select desired component to be moved from the PCB. The move handles will be displayed on component and the **MOVE 3D PCB COMPONENT** dialog box will be displayed; refer to Figure-10.

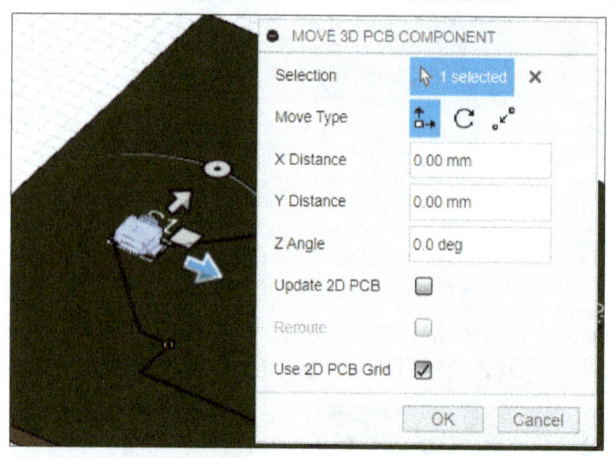

Figure-10. Updated dialog box

- Select desired button from the **Move Type** section of dialog box to define the method of movement. Select the **Free Move** button to move selected component in X and Y directions, and rotate about Z axis. Select the **Rotate** button to the component about Z axis. Select the **Point to Point** button to move component between two selected points.
- Select the **Update 2D PCB** check box to update the 2D PCB based on changes made in the 3D PCB.
- Select the **Reroute** check box to update wire routes in schematic and 2D PCB based on changes in 3D PCB.
- Move the component using drag handles displayed over component in drawing area or specify the parameters in edit boxes of dialog box.
- After setting desired parameters, click on the **OK** button to apply changes.

Push to 2D PCB

The **Push to 2D PCB** tool is used to push all the desired changes performed in 3D PCB to 2D PCB. Click on the **Push to 2D PCB** tool from the **MODIFY** drop-down in the **3D PCB** tab of the **Ribbon** to apply changes.

Creating PCB Hole

The **PCBHole** tool is used to create a hole in PCB of specified size. The procedure to use this tool is given next.

- Click on the **PCBHole** tool from the **MODIFY** drop-down of **3D PCB** tab in the **Ribbon**. The **PCBHOLE** dialog box will be displayed.
- Click on the PCB at desired location to place the hole. The updated **PCBHOLE** dialog box will be displayed; refer to Figure-11.

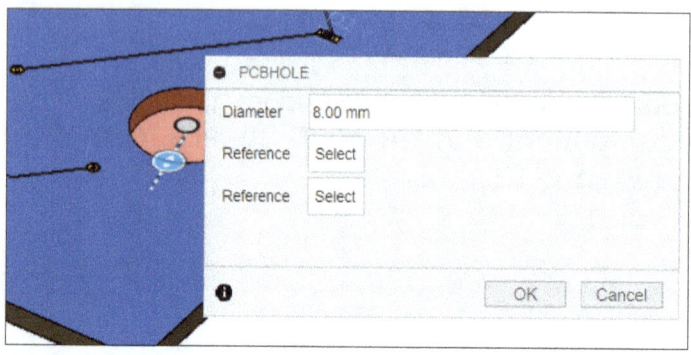

Figure-11. PCBHOLE dialog box

- Select two perpendicular edges of the PCB to specify distance reference parameters for placement of hole; refer to Figure-12.

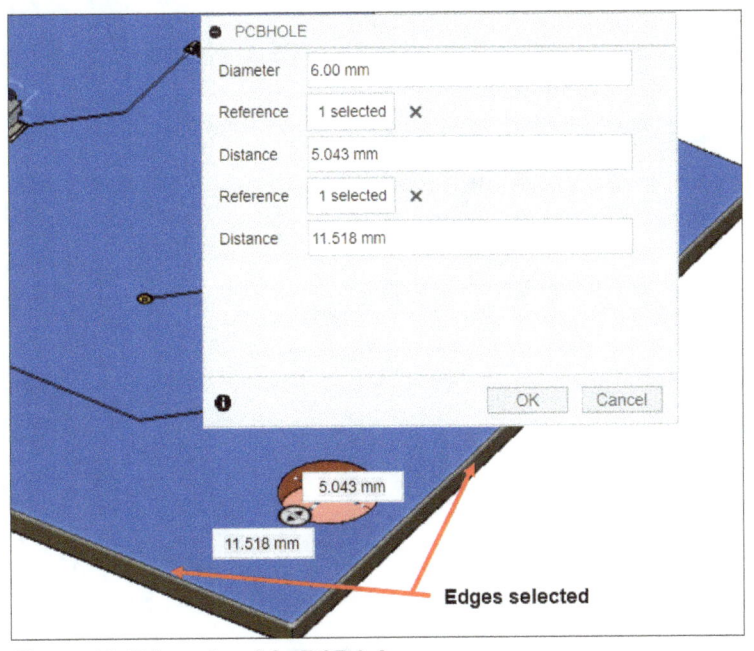

Figure-12. Edges selected for PCB hole

- After specifying desired parameters, click on the **OK** button to create the hole.

Similarly, you can use **Edit Board** tool from **Ribbon** to modify the PCB board boundary.

INSERTING 3D PCB IN ASSEMBLY

After creating 3D PCB, save it in the **Data Panel** at desired location and then you need to create two separate Fusion 360 model files; one for the enclosure of PCB and another for assembly of enclosure with 3D PCB. The procedure to do so is given next.

Creating Enclosure for PCB

- Start a new mechanical design in Autodesk Fusion 360 by using the **New Design** tool of **File** menu or press **CTRL+N**. Design workspace of Fusion 360 will be displayed.
- Click on the **Insert Derive** tool from the **INSERT** drop-down of **SOLID** tab in the **Ribbon**. An information box will be displayed asking you to save the file first; refer to Figure-13.

Figure-13. Fusion 360 information box

- Click on the **Save and Continue** button from the information box. The **Save** dialog box will be displayed as discussed earlier.
- Specify the name as pcb enclosure and save in desired cloud location. On clicking the **Save** button, the **Select Source** dialog box will be displayed; refer to Figure-14.

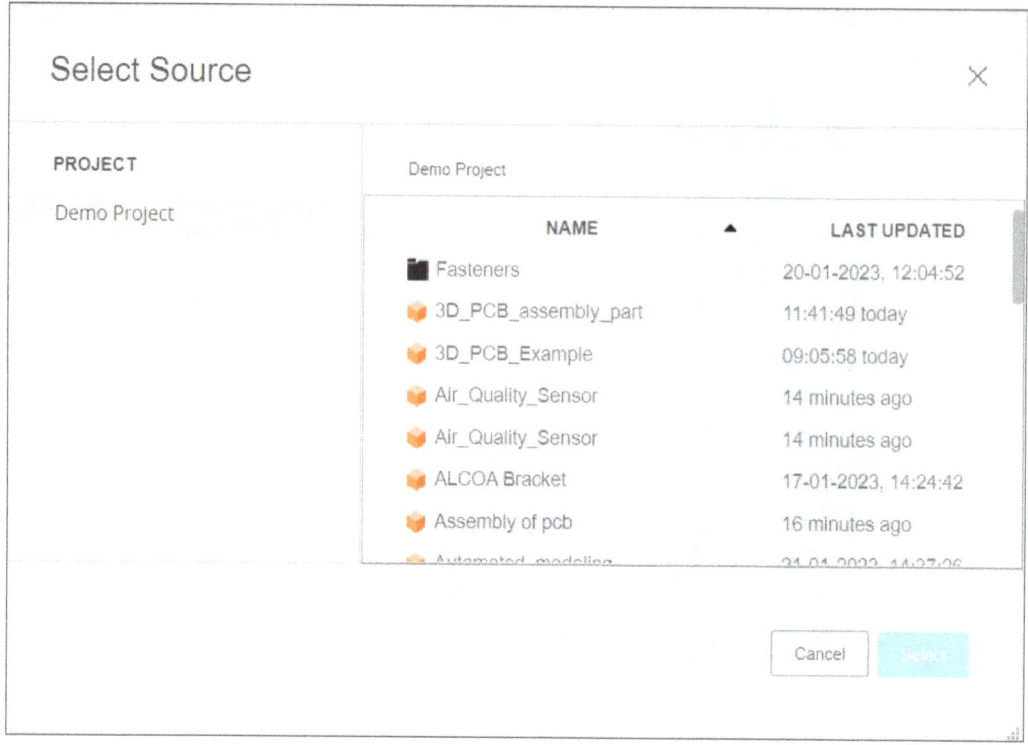

Figure-14. Select Source dialog box

- Select the 3D PCB model of desired electronics design (Air Quality Sensor in our case) from the dialog box and click on the **Select** button. The selected model file will open and **DERIVE** dialog box will be displayed; refer to Figure-15.

Figure-15. DERIVE dialog box

- Expand the **Sketches** node of **BROWSER** and select the **Outline** sketch; refer to Figure-15. The selected sketch will be added in the **Objects** area of **DERIVE** dialog box.
- Click on the **OK** button from the dialog box. The selected sketch will be added in the enclosure file earlier created; refer to Figure-16.

Autodesk Fusion 360 PCB Black Book   6-9

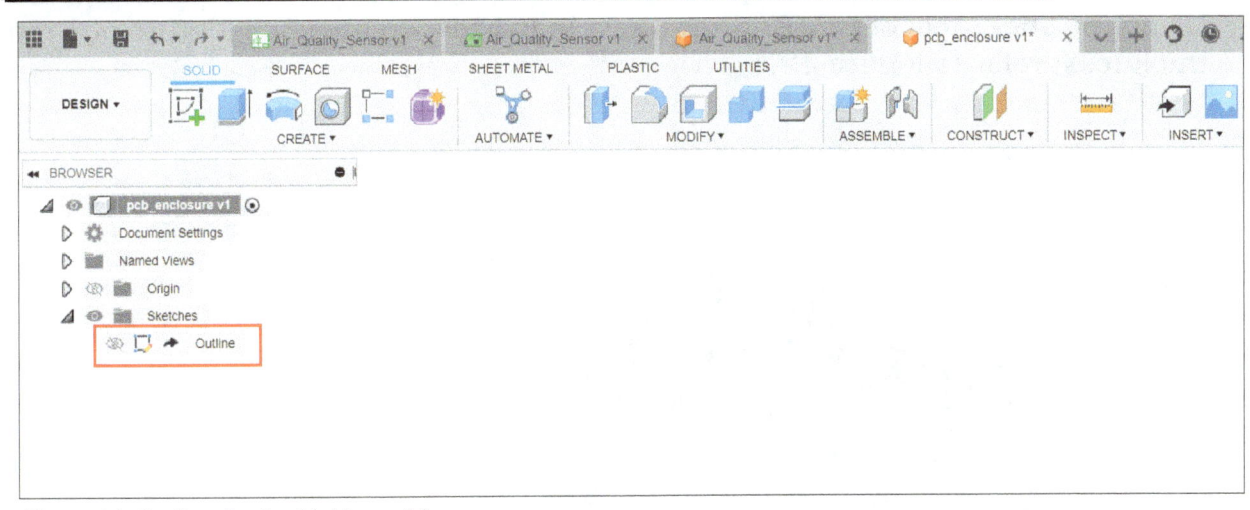

Figure-16. Outline sketch added in model

- Click on the visibility icon before Outline sketch in the **BROWSER** to set it to display; refer to Figure-17.

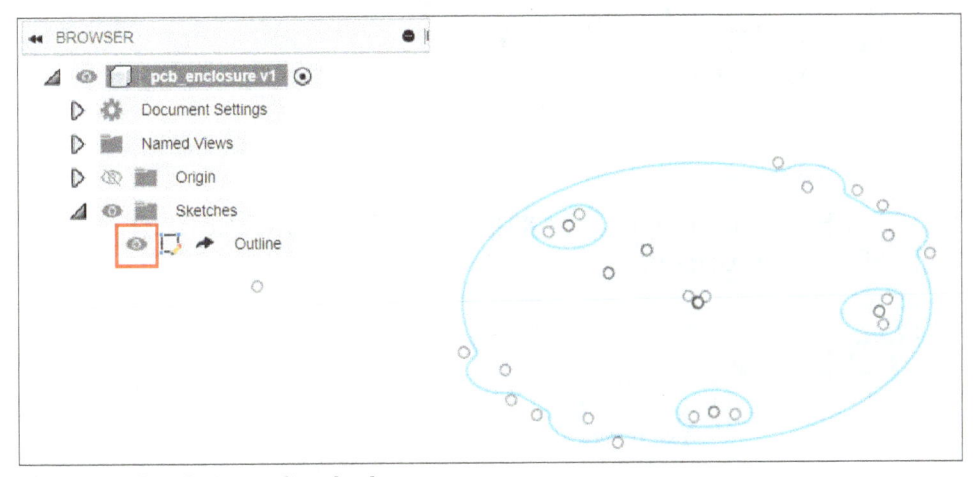

Figure-17. Displaying outline sketch

- Using the **Extrude** tool, create solid body from outline to generate base feature; refer to Figure-18.

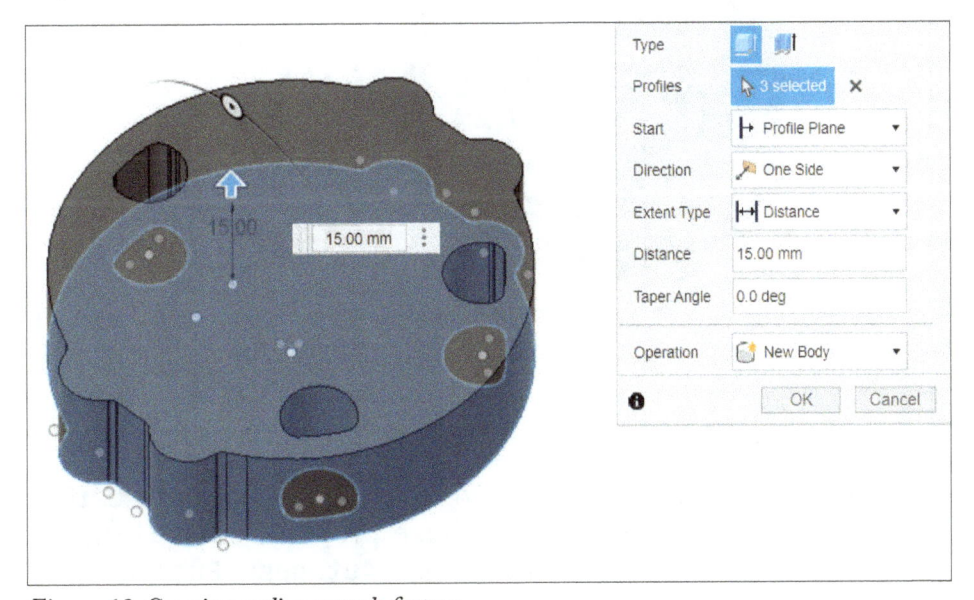

Figure-18. Creating outline extrude feature

- Using the **Shell** tool of **MODIFY** drop-down in **Ribbon**, create shell feature of 1 mm thickness; refer to Figure-19.

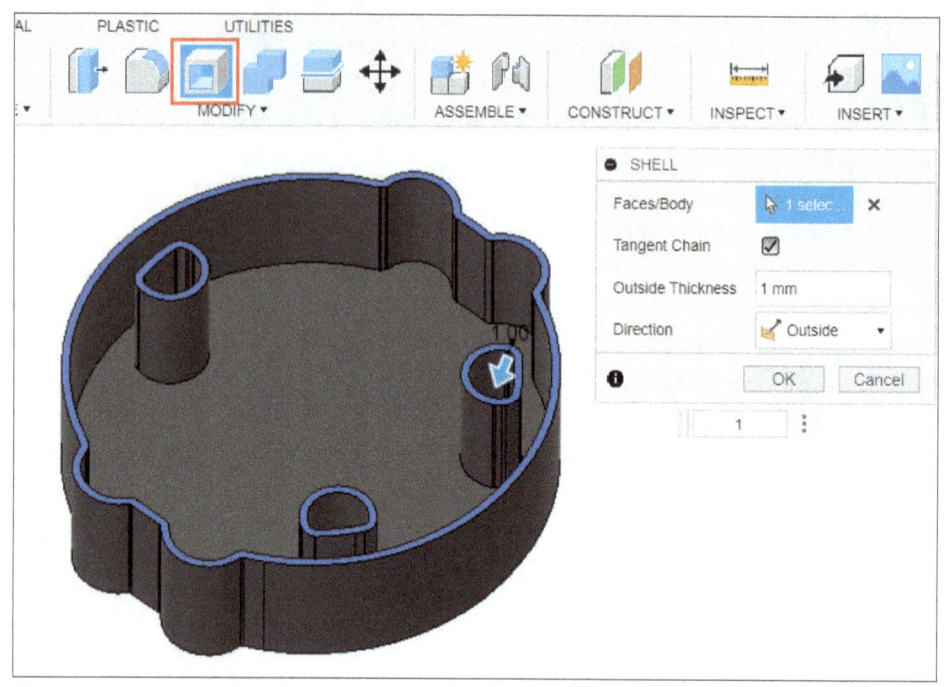

Figure-19. Applying shell tool

- Using the Thin Extrude feature of **EXTRUDE** tool, create the seat surface of PCB in enclosure; refer to Figure-20.

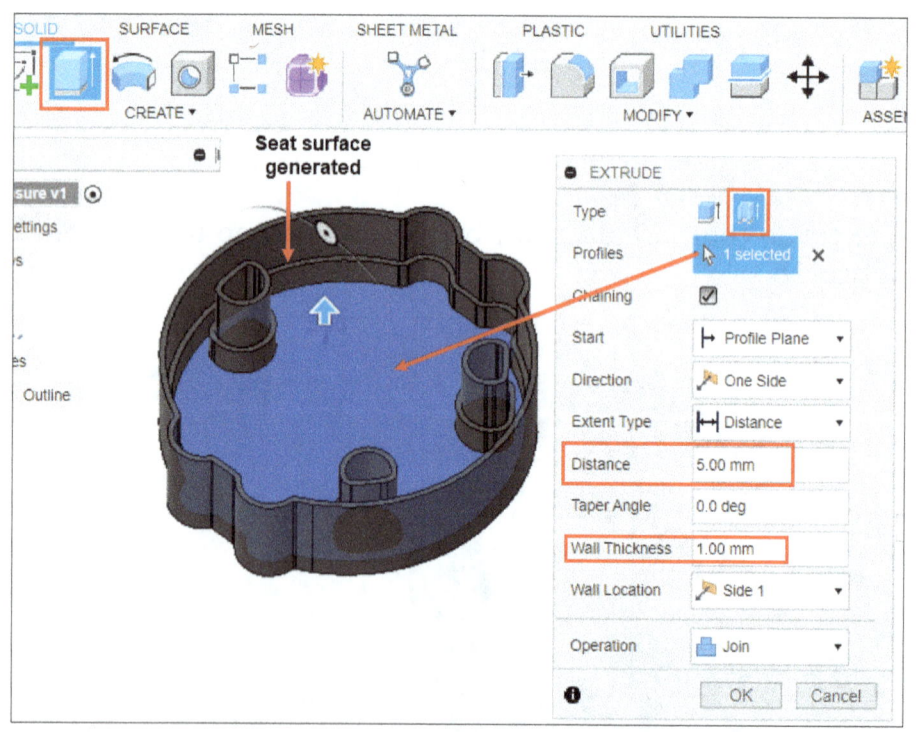

Figure-20. Creating thin extrude feature

- Save the file using **CTRL+S** key from keyboard. Although, you can perform a lot of modeling changes in the design but these procedures are out of the scope of this book. You can refer to our another book **Autodesk Fusion 360 Black Book (V2.0.15293)** as reference for mechanical design.

Autodesk Fusion 360 PCB Black Book 6-11

Assembly of PCB with Enclosure

- Start a new design file by pressing **CTRL+N** key and Save it by desired name.
- Expand the **Data Panel** by clicking on the **Show Data Panel** button at top left corner of application window. The contents of **Data Panel** will be displayed; refer to Figure-21.

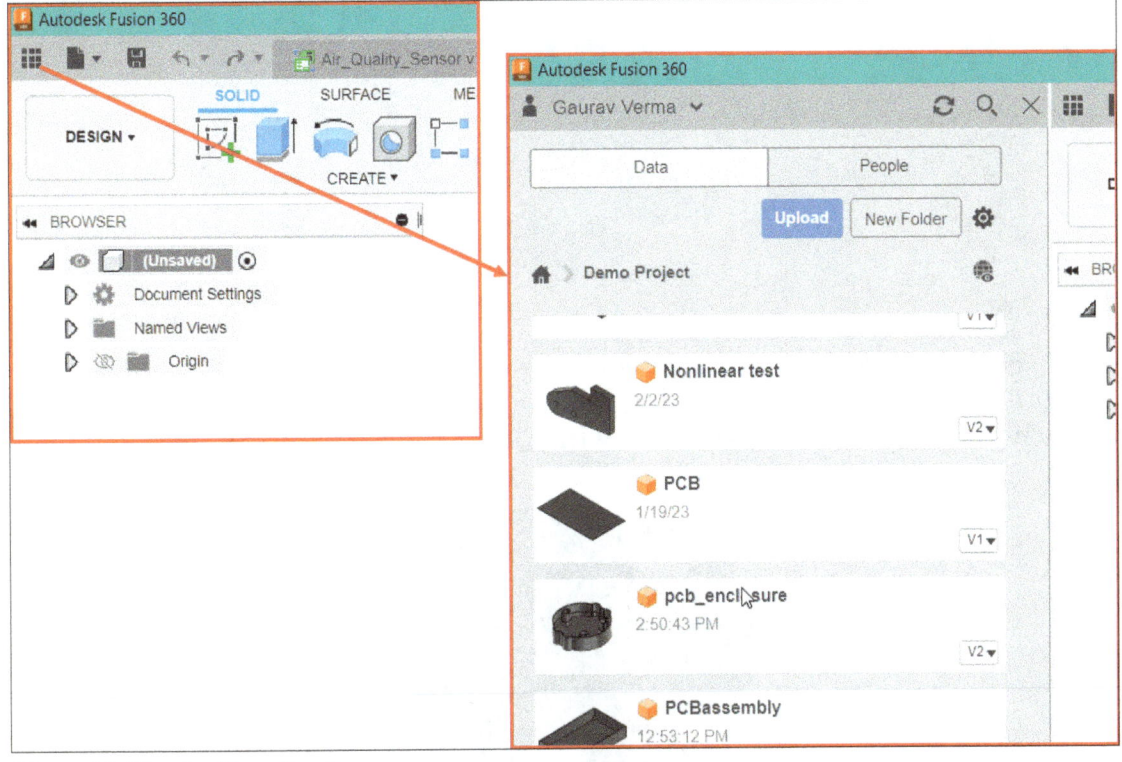

Figure-21. Showing Data Panel

- Right-click on the pcb enclosure model file created earlier, in the **Data Panel** and select the **Insert into Current Design** option from the shortcut menu; refer to Figure-22. The model will be inserted in drawing area and move handles will be displayed along with **MOVE/COPY** dialog box; refer to Figure-23.

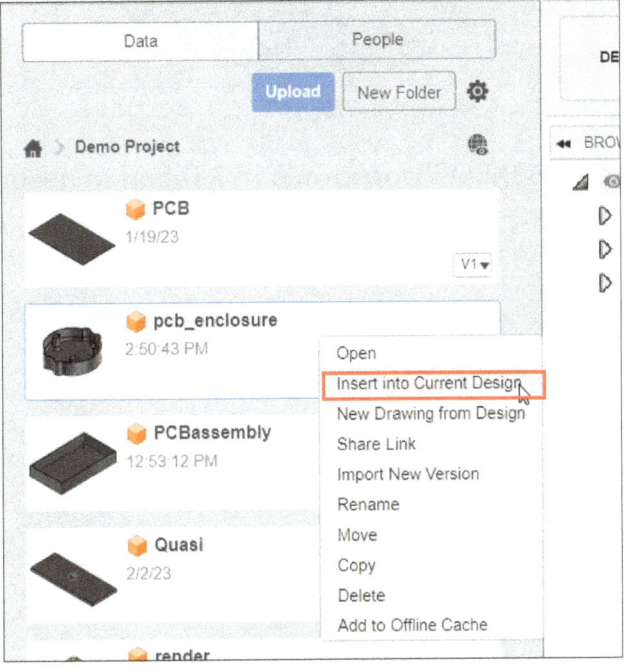

Figure-22. Insert into Current Design option

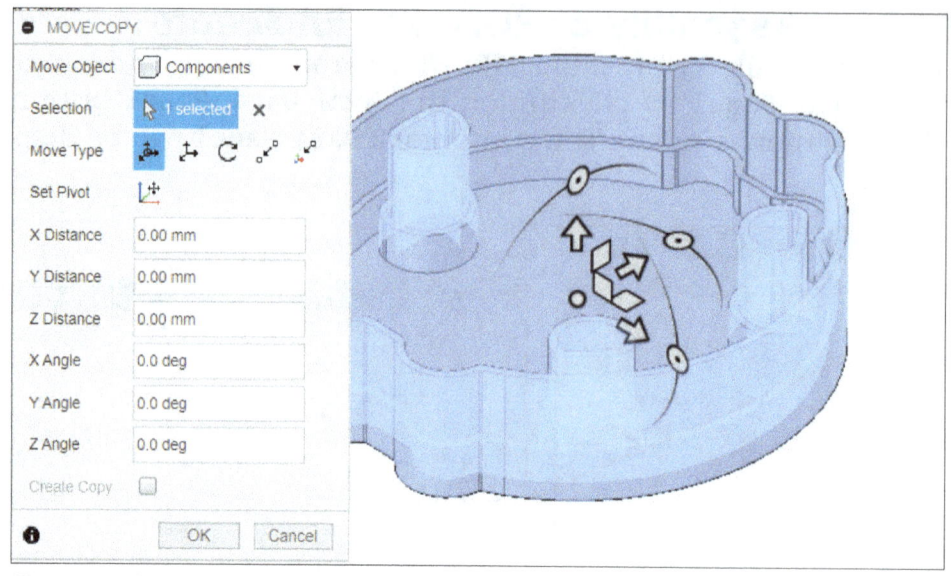

Figure-23. Component inserted in assembly

- Click on the **OK** button from dialog box to place the component.
- Similarly, insert the 3D PCB in drawing area; refer to Figure-24.

Figure-24. After inserting 3D PCB

- Use the **Joint** tool of **ASSEMBLE** drop-down in **Ribbon** to assemble the components; refer to Figure-25.

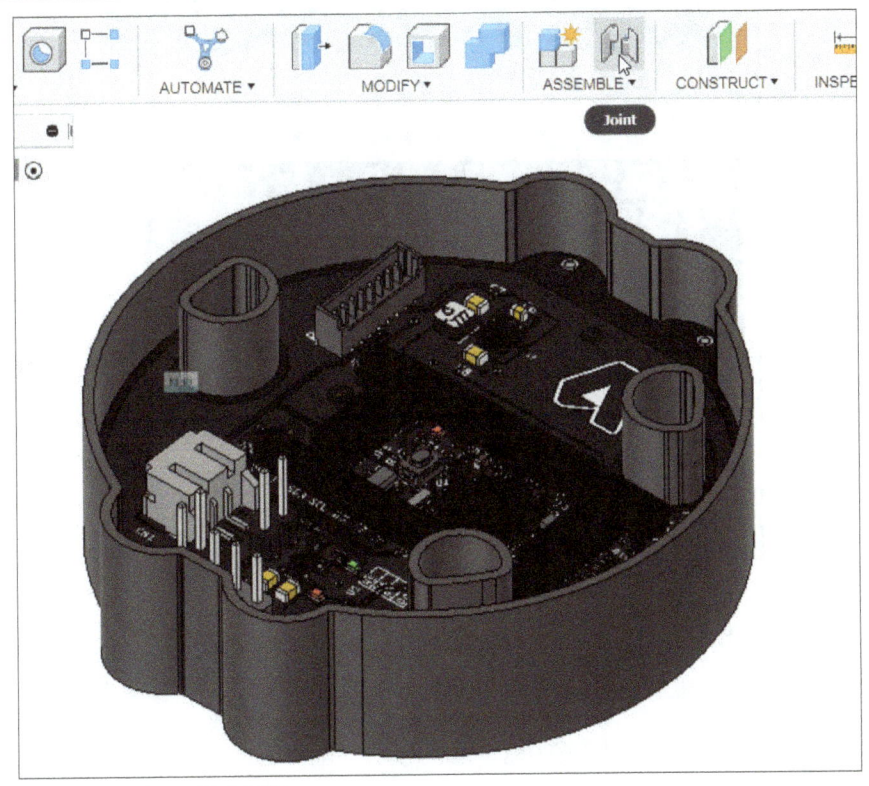

Figure-25. Assembled PCB

- Save the file by pressing **CTRL+S**.

MANUFACTURING (CAM)

The tools to generate CAM data of PCB are available in the **MANUFACTURING** tab of the **Ribbon** in 2D PCB environment; refer to Figure-26. These tools are discussed next.

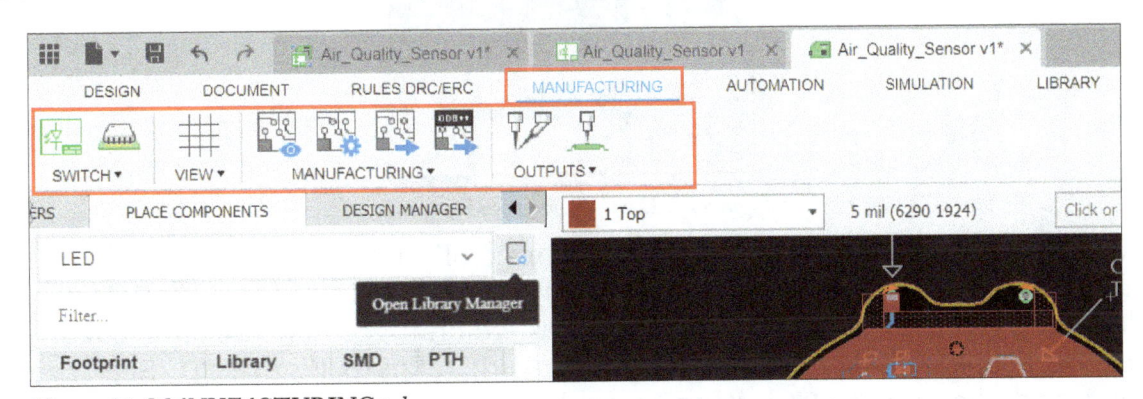

Figure-26. MANUFACTURING tab

CAM Preview

The **CAM Preview** tool is used to check the preview of CAM data to be generated for the PCB model. The procedure to use this tool is given next.

- Click on the **CAM Preview** tool from the **MANUFACTURING** drop-down in the **MANUFACTURING** tab of the **Ribbon**. The **Manufacturing Preview** dialog box will be displayed; refer to Figure-27.

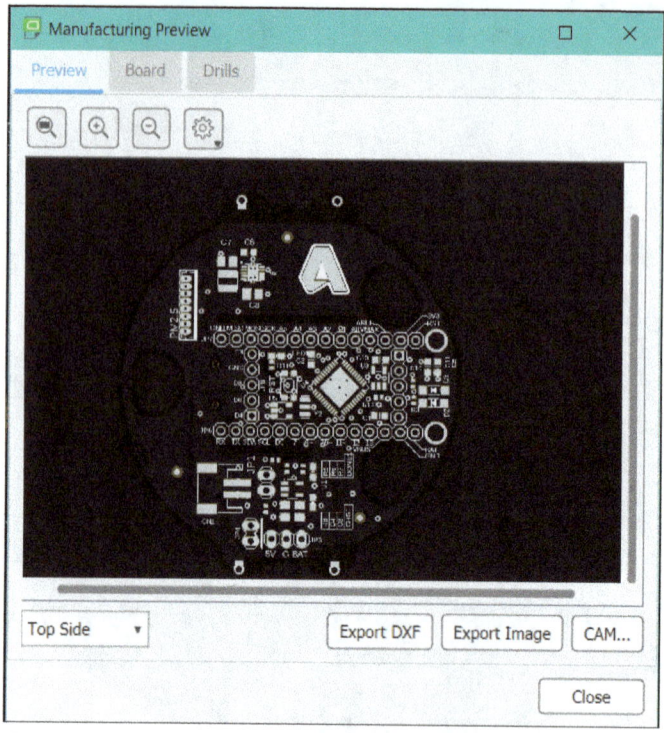

Figure-27. Manufacturing Preview dialog box

- Select desired option from the drop-down in bottom left corner of dialog box to specify the side/feature of PCB to be checked for manufacturing. By default, **Top Side** option is selected in the drop-down, so top of PCB is displayed in preview; refer to Figure-28.

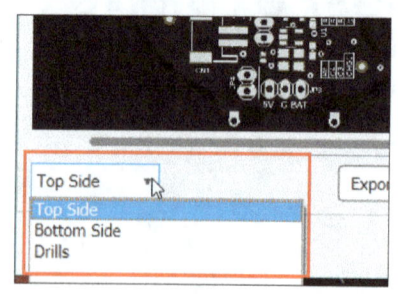

Figure-28. Drop-down for selecting sides

- Click on the **Configure Preview** button from toolbar at the top in the dialog box to modify colors of various entities in the preview; refer to Figure-29. Select desired element from drop-down whose color is to be changed. The **Select Color** dialog box for respective element will be displayed; refer to Figure-30. Select desired color and click on the **OK** button to apply the color.

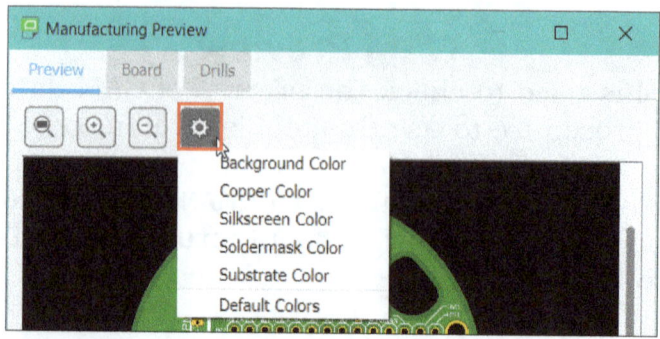

Figure-29. Configure Preview button

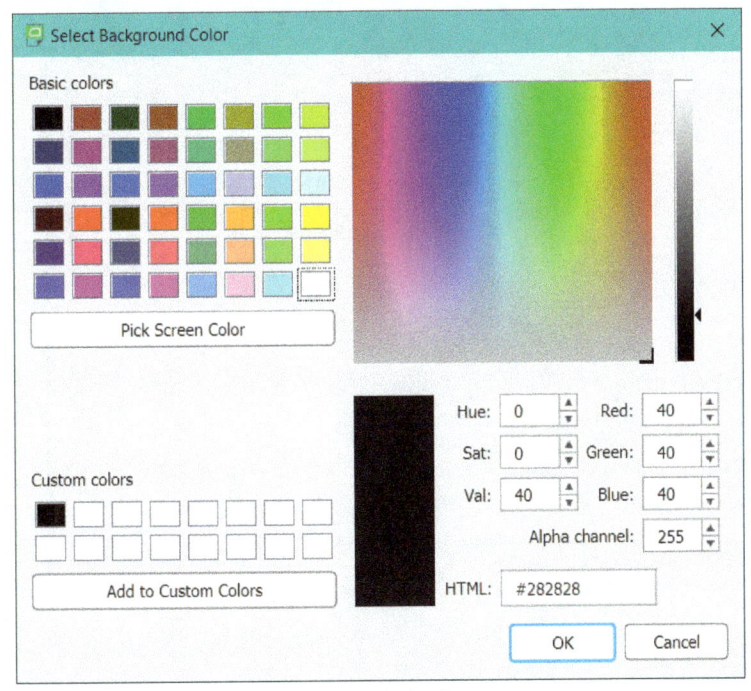

Figure-30. Select Background Color dialog box

- Click on the **Export DXF** button from the dialog box to export the PCB in DXF format file. On clicking this button, the **Save DXF To File** dialog box will be displayed; refer to Figure-31. Specify desired name of file and click on the **Save** button to save the file.

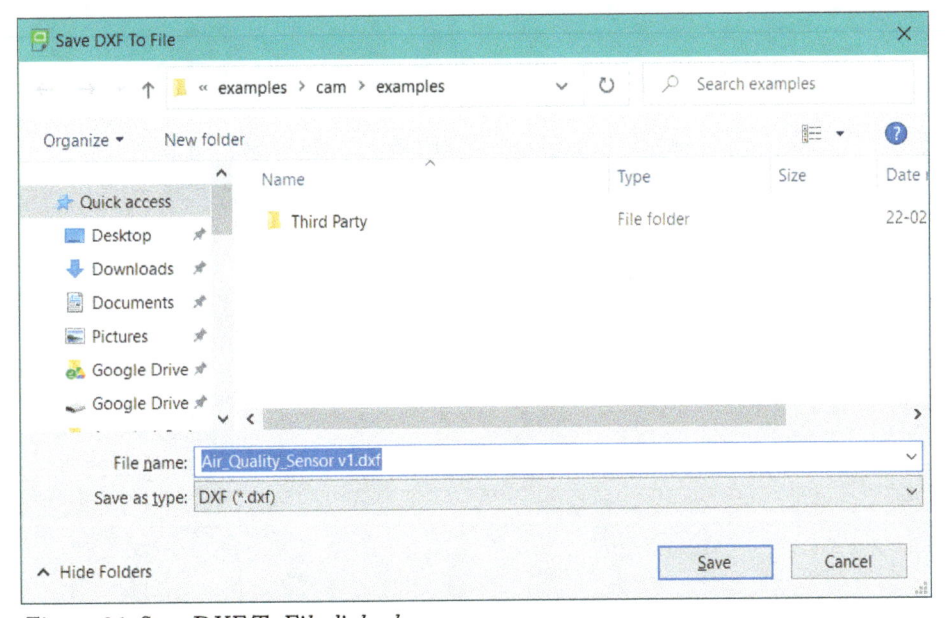

Figure-31. Save DXF To File dialog box

- Click on the **Export Image** button from the dialog box to export the current displayed model as an image file.
- Click on the **CAM** button of dialog box to activate CAM Processor. The options of **CAM Processor** dialog box are discussed in next topic.
- Click on the **Board** tab in the dialog box to check PCB parameters to be used by CAM; refer to Figure-32.

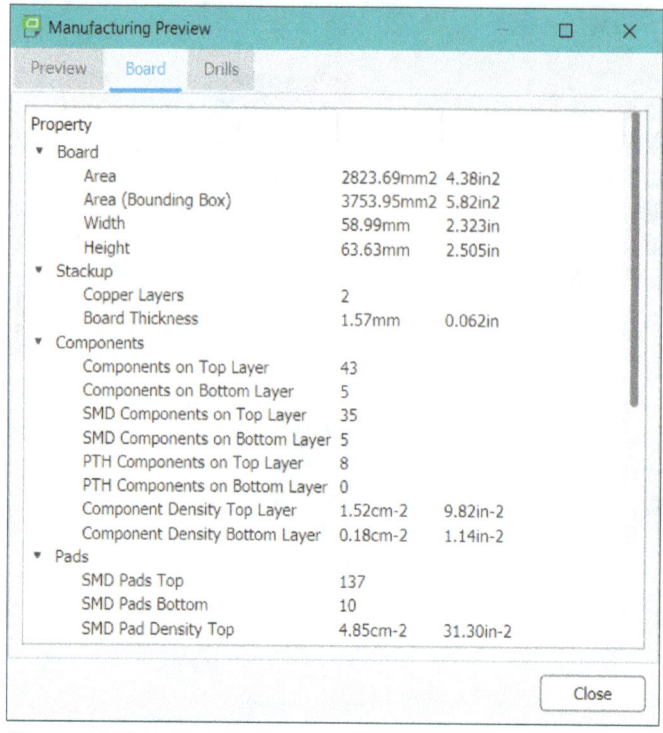

Figure-32. Board tab

- Similarly, click on the **Drills** tab in the dialog box to check the holes to be created in PCB and their related parameters; refer to Figure-33. You can export the drill data in CSV or JSON format by using respective button in this tab.

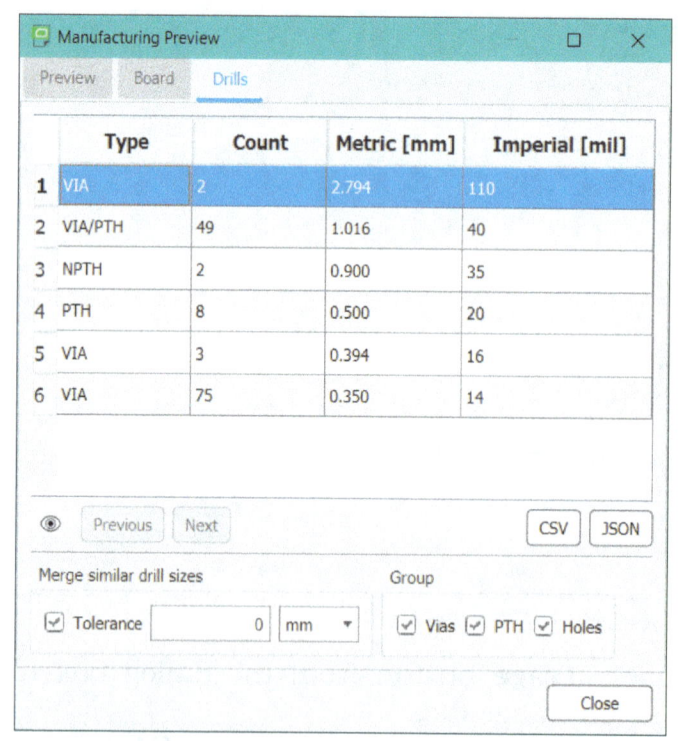

Figure-33. Drills tab

- Click on the **Close** button to exit the dialog box.

CAM Processor

The `CAM Processor` tool is used to generate various output files to be used by manufacturer for manufacturing PCB. The procedure to use this tool is given next.

- Click on the `CAM Processor` tool from the `MANUFACTURING` drop-down of `MANUFACTURING` tab in the `Ribbon`. The `CAM Processor` dialog box will be displayed; refer to Figure-34.

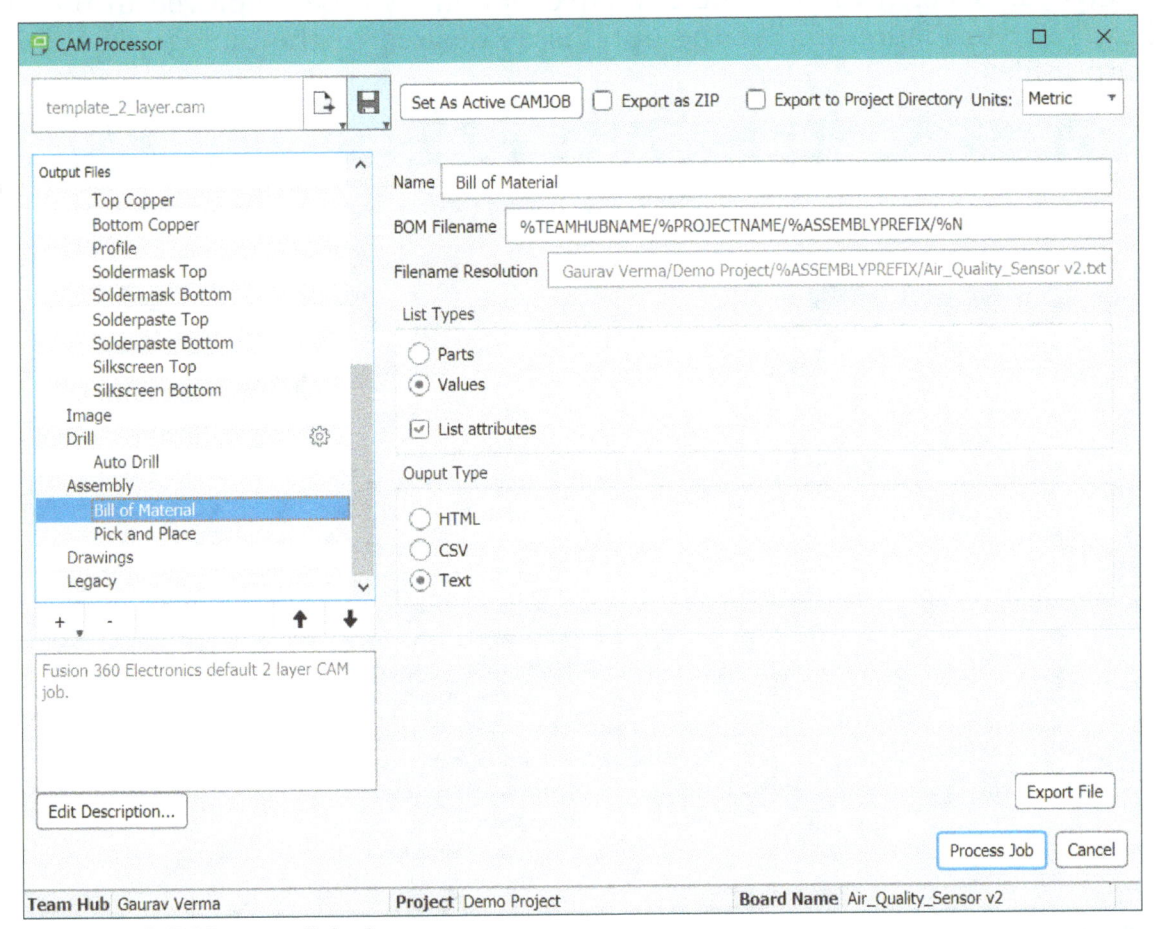

Figure-34. CAM Processor dialog box

- Select desired option from the `Load Job File` drop-down to define the template to be used for generating CAM data; refer to Figure-35 and click on the `Set As Active CAMJOB` button to make it active for CAM processing.

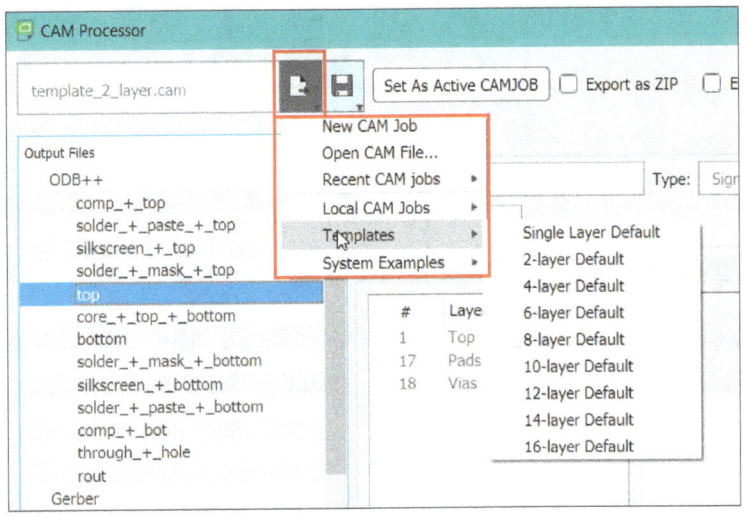

Figure-35. Load Job File drop-down

- Select the **Export as ZIP** check box to export all the CAM files in a compressed ZIP folder.
- Select the **Export to Project Directory** check box to save all CAM files in current project folder.
- Select desired option from the **Units** drop-down to define the unit of measurement to be used in exported files.
- Select the output file from left list box in the dialog box to modify the parameters related to selected CAM output. Related options will be displayed in the dialog box; refer to Figure-36. Set the options as desired in the dialog box to modify output files.

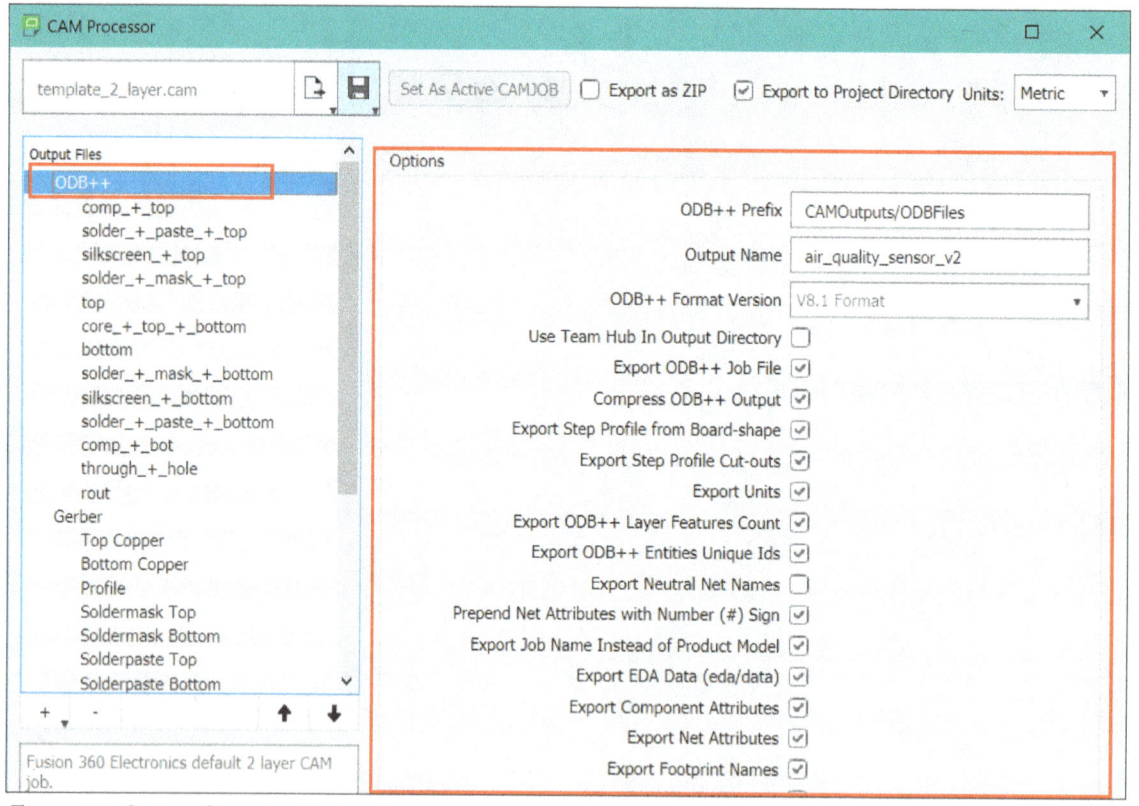

Figure-36. Output file and related options

- After performing desired changes, click on the **Process Job** button from the dialog box. Once the process is complete, a message box will be displayed; refer to Figure-37.

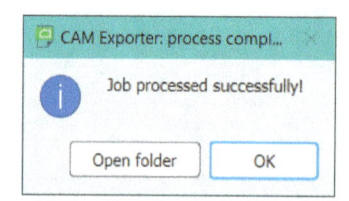

Figure-37. CAM Exporter message box

- Click on the **Open folder** button from the message box to check the folder of output files.
- Click on the **Save Job File As** button from **Save Job** drop-down to save current job parameters to reuse later; refer to Figure-38.

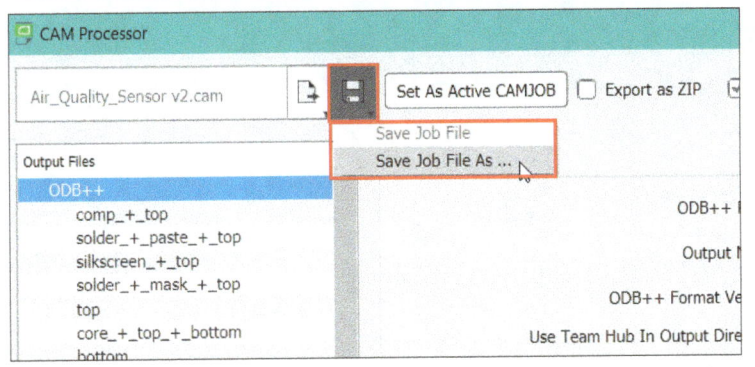

Figure-38. Save Job File As button

- Close the **CAM Processor** dialog box by using **X** button at the top right corner.

Exporting Gerber, NC Drill, Assembly, and Drawing Outputs

The **Export Gerber, NC Drill, Assembly and Drawing Outputs** tool is used to export respective files in a Zip folder at desired location. The procedure to use this tool is given next.

- Click on the **Export Gerber, NC Drill, Assembly and Drawing Outputs** tool from the **MANUFACTURING** drop-down in the **MANUFACTURING** tab of **Ribbon**. The **CAM Export File List** dialog box will be displayed; refer to Figure-39.

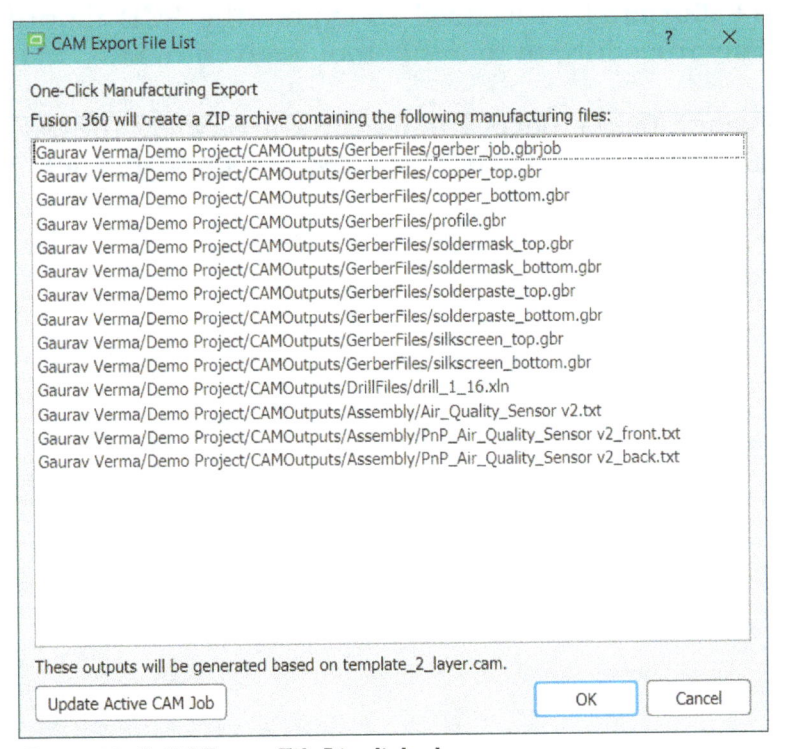

Figure-39. CAM Export File List dialog box

- By default, the current active CAM job will be used as template for exporting files. To change the active job, click on the **Update Active CAM Job** button from the dialog box. The **CAM Processor** dialog box will be displayed which has been discussed earlier.
- Click on the **OK** button from the **CAM Export File List** dialog box to export the files in a Zip folder. The **Create Zip CAM archive** dialog box will be displayed; refer to Figure-40.

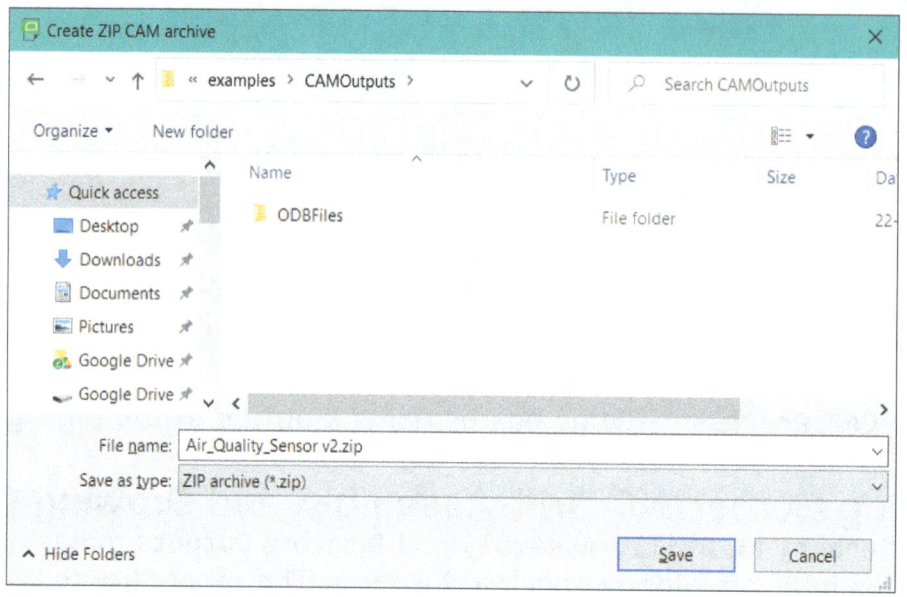

Figure-40. Create ZIP CAM archive dialog box

- Specify desired name of the zip file in the **File name** edit box and click on the **Save** button to save the file.

Exporting ODB++ Output

The **Export ODB++ Output** tool is used to export the PCB manufacturing data in ODB++ format. The procedure to use this tool is given next.

- Click on the **Export ODB++ Output** tool from the **MANUFACTURING** drop-down in the **MANUFACTURING** tab of the **Ribbon**. The **Export ODB++ Output** dialog box will be displayed; refer to Figure-41.

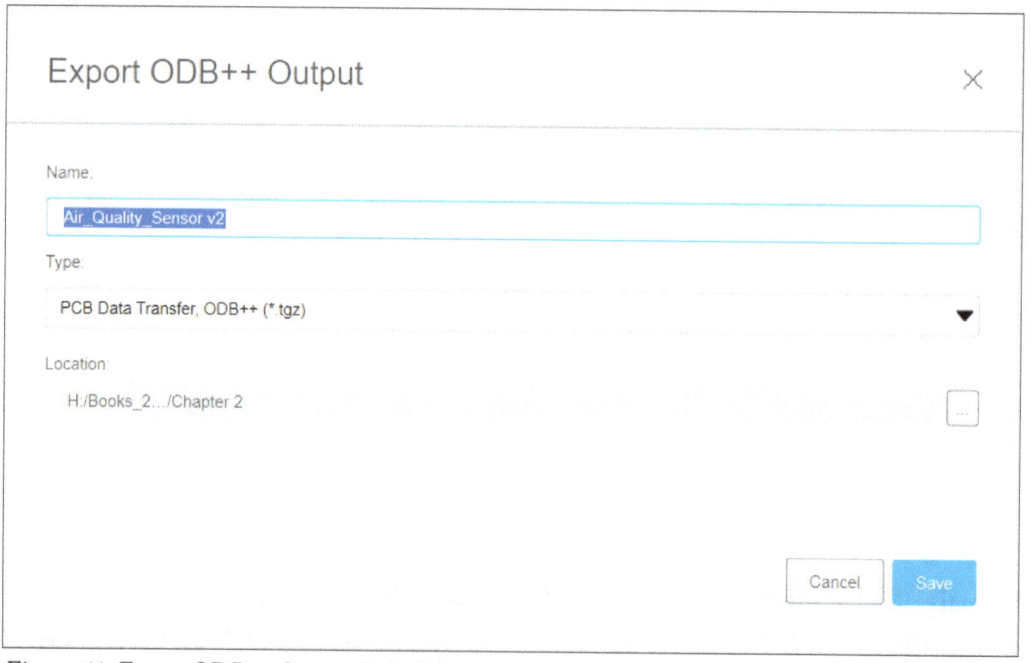

Figure-41. Export ODB++ Output dialog box

- Specify desired name of output file in the **Name** edit box, select the format of output file in the **Type** drop-down, and set the location of output file in the **Location** field.
- Click on the **Save** button from the dialog box to save the file.

Saving IPC Netlist

The **IPC-D-356 Netlist** tool is used to generate netlist data file to be used by manufacturers. The procedure to use this tool is given next.

- Click on the **IPC-D-356 Netlist** tool from the **OUTPUTS** drop-down in the **MANUFACTURING** tab of the **Ribbon**. The **Save IPC-D-356 File** dialog box will be displayed; refer to Figure-42.

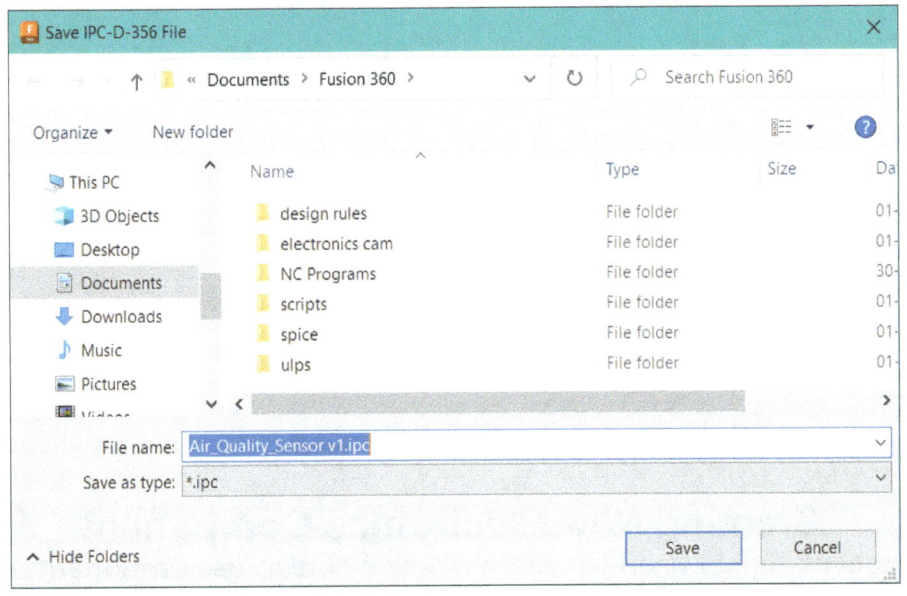

Figure-42. Save IPC-D-356 File dialog box

- Specify desired name of the file in the **File name** edit box and then click on the **Save** button to save the netlist data.

Similarly, you can use the **PickAndPlace** tool to save the file in **MNT** format.

NEW ELECTRONICS LIBRARY

The **New Electronics Library** tool is used to create a new library of electronics component. Click on the **New Electronics Library** tool from the **File** menu to activate electronics library; refer to Figure-43. The Electronics library environment will be displayed; refer to Figure-44. The tools of this environment are used to create new components, symbols, footprints, and packages. These tools are discussed next.

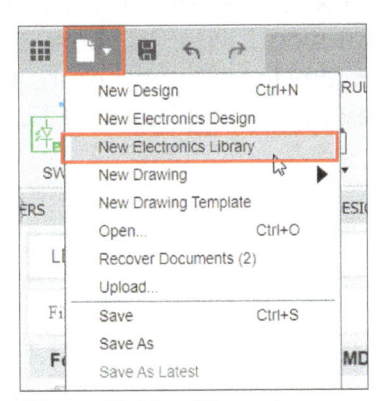

Figure-43. New Electronics Library dialog box

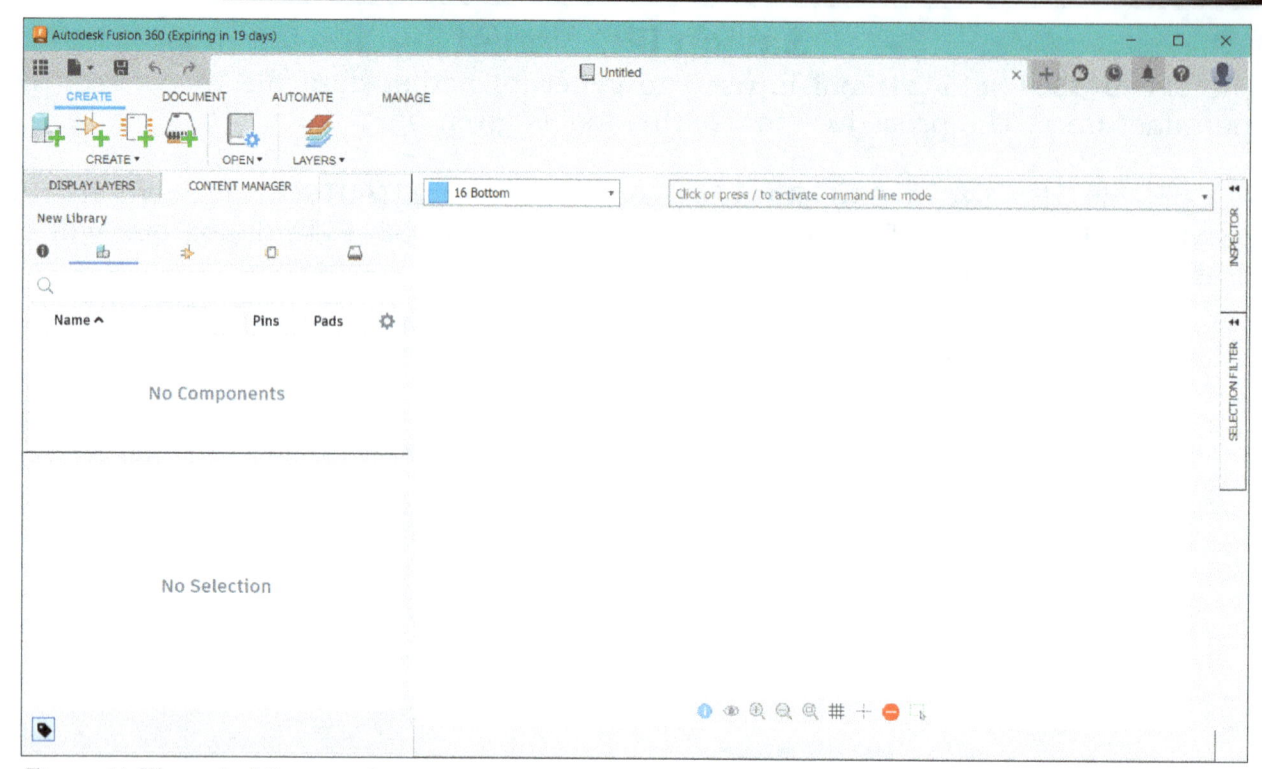

Figure-44. Electronics Library environment

Creating New Electronics Component

The **New Component** tool is used to create a new electronics component in the library. The procedure to use this tool is given next.

- Click on the **New Component** tool from the **CREATE** drop-down of **CREATE** tab in the **Ribbon**. The **Add Device** dialog box will be displayed; refer to Figure-45.

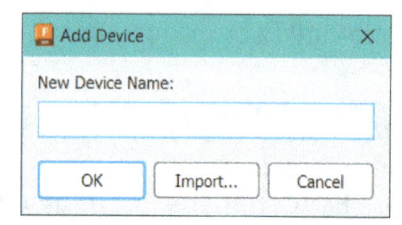

Figure-45. Add Device dialog box

- Specify the name of new device in the **New Device Name** edit box and click on the **OK** button from the dialog box to create a blank component. The tools to create new component will be displayed in the **Ribbon**; refer to Figure-46. Click on the **Import** button from the **Add Device** dialog box to use earlier created component as template. The **Import Device** dialog box will be displayed; refer to Figure-47.

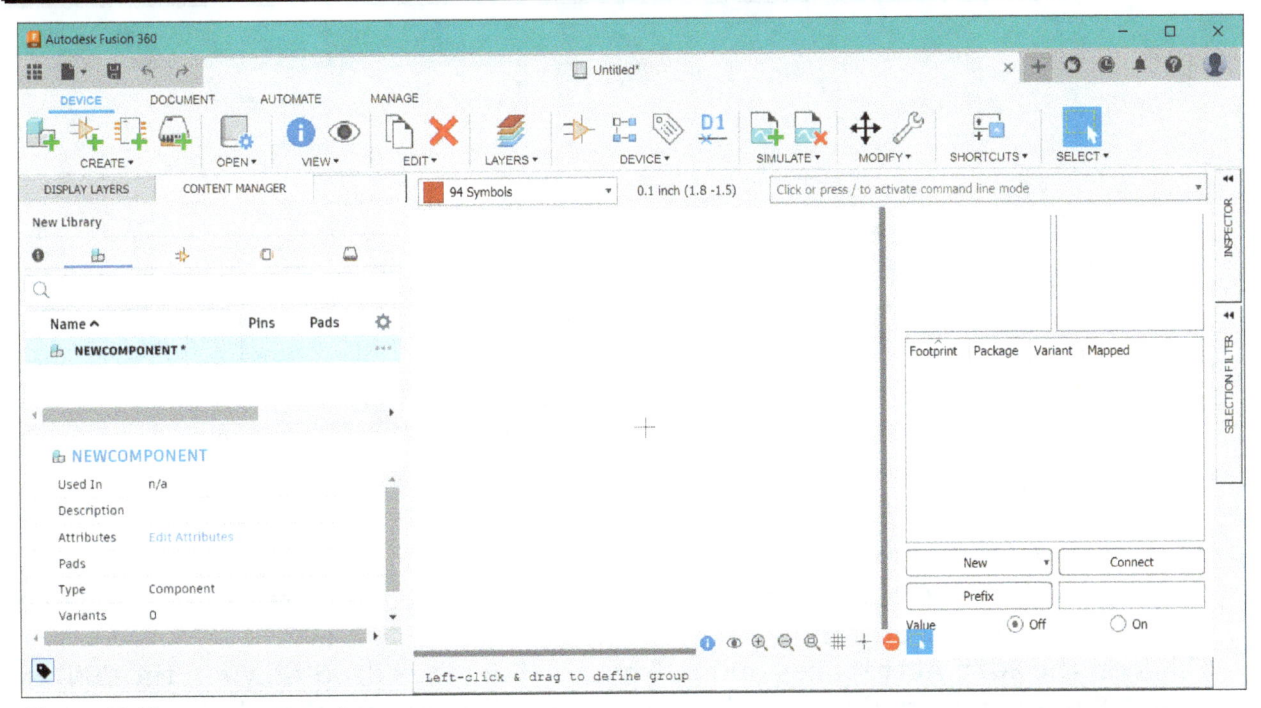

Figure-46. New component environment

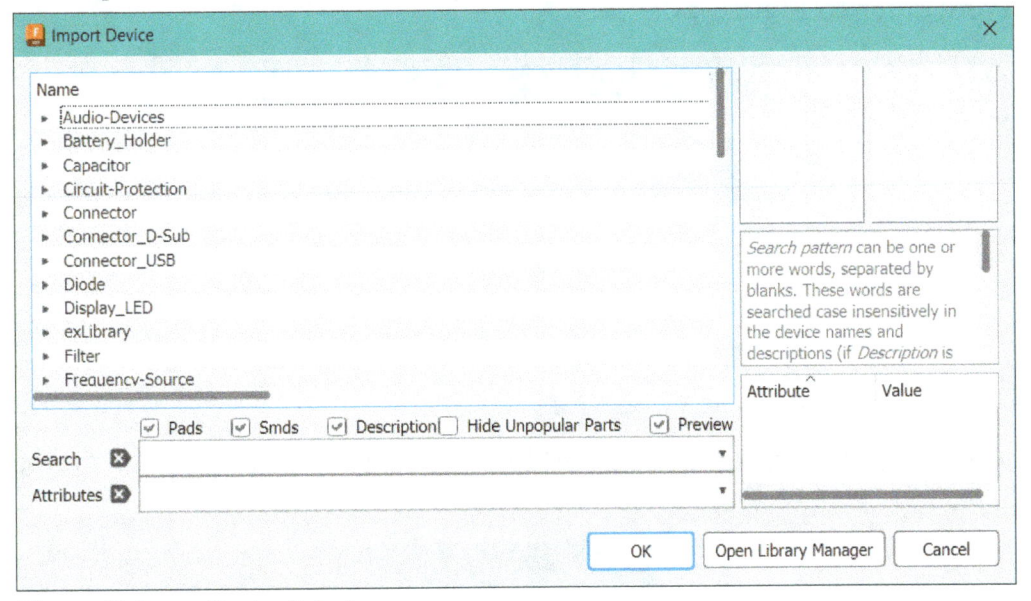

Figure-47. Import Device dialog box

- Select desired device(s) from the list and click on the **OK** button. The devices will be added in the **CONTENT MANAGER**.
- Click on the **Edit Description** option from the **Settings** drop-down in **CONTENT MANAGER** to modify description of the component; refer to Figure-48. The **Description** dialog box will be displayed; refer to Figure-49.

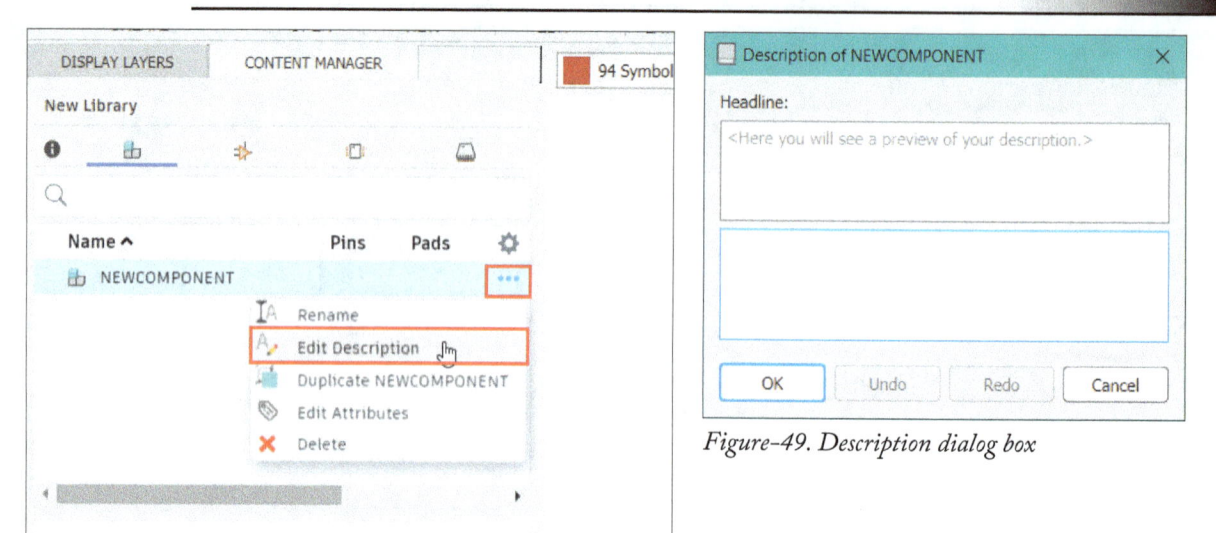

Figure-48. Edit Description option

Figure-49. Description dialog box

- Specify desired description text and headline in the edit boxes of the dialog box and click on the **OK** button.
- Click on the **Edit Attributes** option from the **Settings** drop-down in the **CONTENT MANAGER** to modify attributes of the component. The **Attributes** dialog box will be displayed; refer to Figure-50.

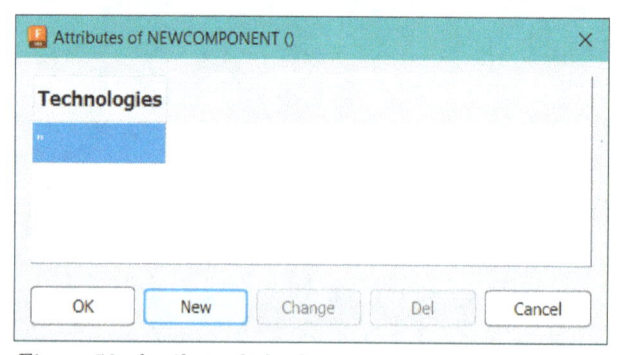

Figure-50. Attributes dialog box

- Click on the **New** button from the dialog box. The **New attribute** dialog box will be displayed; refer to Figure-51.

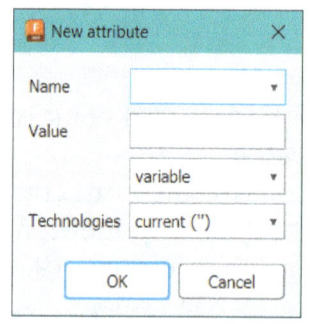

Figure-51. New attribute dialog box

- Specify desired name and value of the attributes in the **Name** and **Value** edit boxes of the dialog box.
- Select desired option from the drop-down below **Value** edit box to define whether the value is constant or varying.
- Select desired option from the **Technologies** drop-down and click on the **OK** button to apply the attributes.

- Click on the **Add** tool from the **DEVICE** drop-down in the **DEVICE** tab of the **Ribbon** to add symbols from imported packages in the current component. The **Add** dialog box will be displayed; refer to Figure-52.

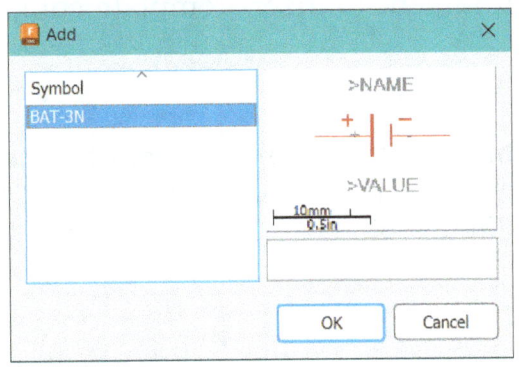

Figure-52. Add dialog box

- Select desired symbol from the list and click on the **OK** button. The symbol will get attached to cursor; refer to Figure-53.

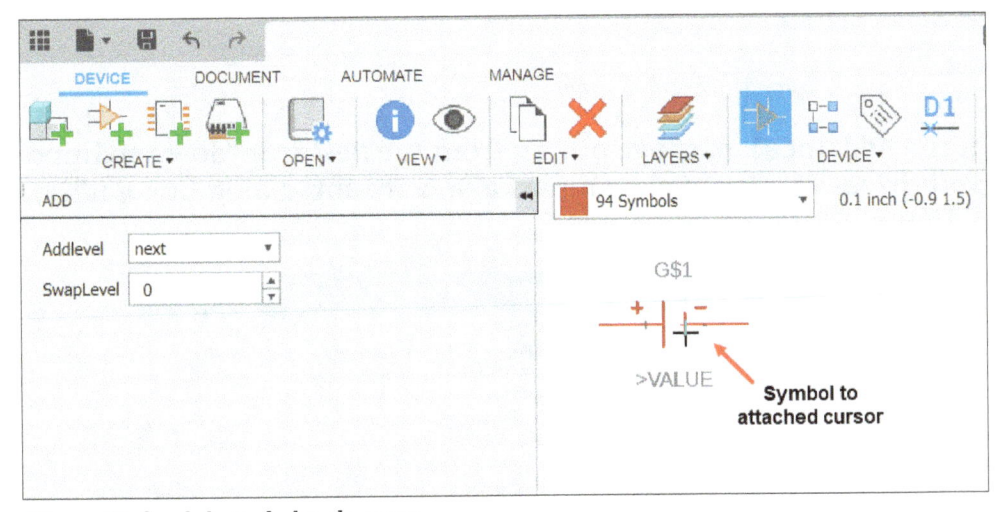

Figure-53. Symbol attached to the cursor

- Select desired option from the **Addlevel** drop-down of **ADD Dialog** to define the manipulator for swap level of component and specify desired swap level in the **SwapLevel** edit box. The value **0** in the **SwapLevel** edit box defines that pins of component cannot be swapped. Specify value other than **0** in edit box to make component pins swappable. After setting desired parameters in the **Dialog**, click in the drawing area to place the symbol(s).
- To add SPICE model information to the component, click on the **Add Spice Model** tool from the **SIMULATE** drop-down in the **DEVICE** tab of the **Ribbon**. The **Add Spice Model** dialog box will be displayed; refer to Figure-54.

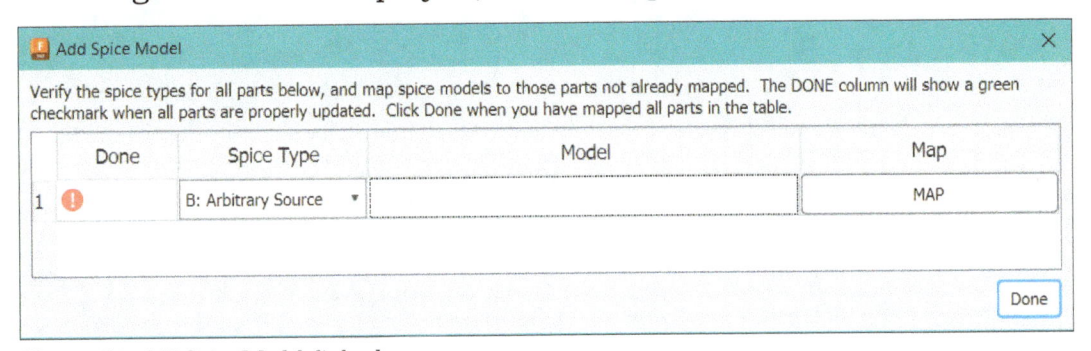

Figure-54. Add Spice Model dialog box

- Set the parameters for SPICE model as discussed earlier.

When creating a new component, you can also use the **New** drop-down in the component creation environment to insert template components. The procedure to do so is given next.

- Click on the **New** button at bottom-right corner of the application window to use component package available in the local drive or from the web. A drop-down will be displayed; refer to Figure-55.

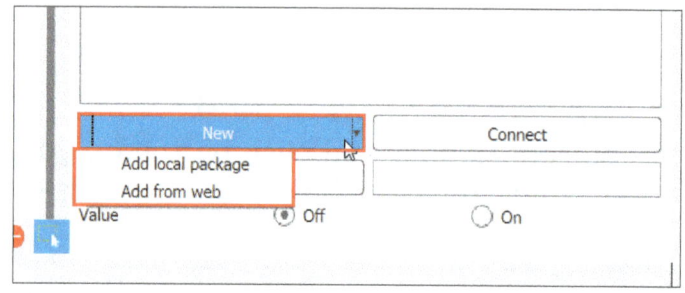

Figure-55. New drop-down

- Click on the **Add local package** option from the **New** drop-down to import package from local drive. The **Create new package variant** dialog box will be displayed; refer to Figure-56.

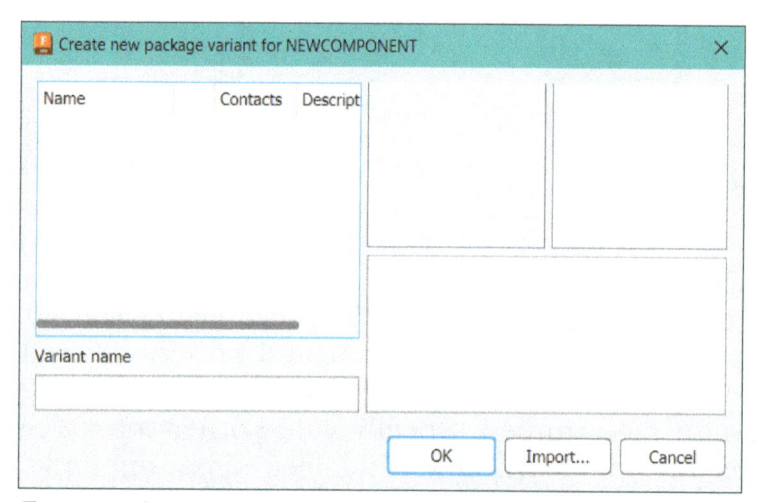

Figure-56. Create new package variant dialog box

- Click on the **Import** button from the dialog box. The **Import Package** dialog box will be displayed; refer to Figure-57.

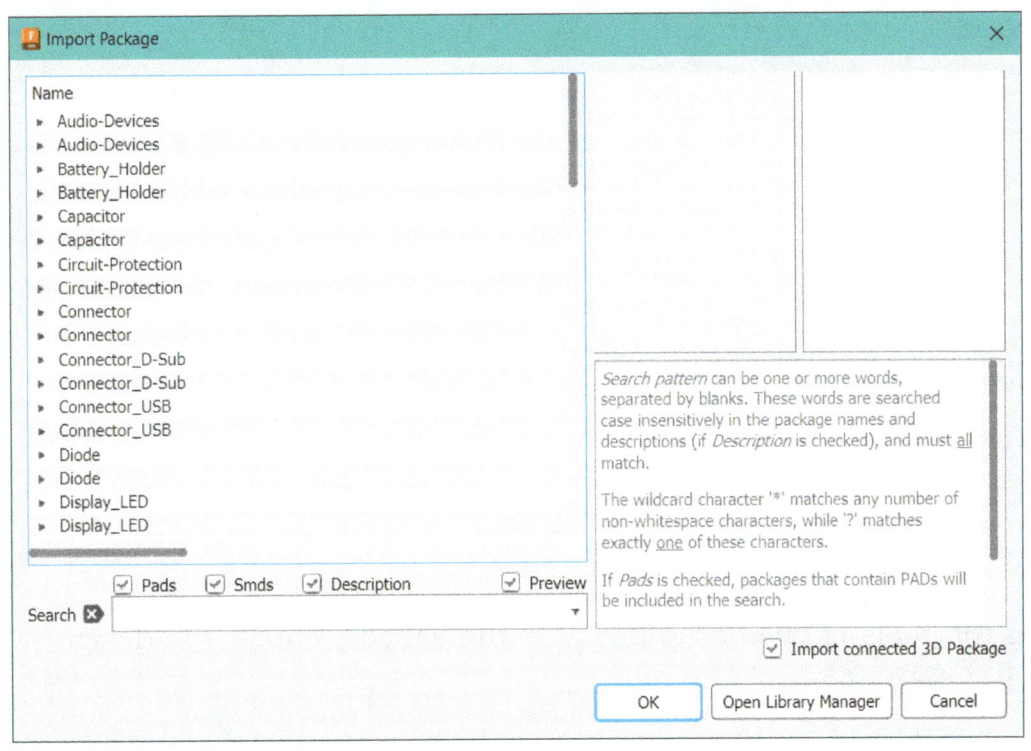

Figure-57. Import Package dialog box

- Select desired component package from the list box at the left in the dialog box to import it.
- Select the **Import connected 3D Package** check box to import connected 3D package as well. You can also use the **Library Manager** to import package using the **Open Library Manager** button.
- After setting desired parameters in the dialog box, click on the **OK** button.

Creating New Symbol

The **New Symbol** tool is used to create new electronics symbol in the current library. The procedure to use this tool is given next.

- Click on the **New Symbol** tool from the **CREATE** drop-down in the **DEVICE** tab of the **Ribbon**. The **Add Symbol** dialog box will be displayed; refer to Figure-58.

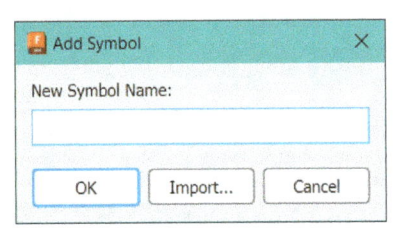

Figure-58. Add Symbol dialog box

- Specify desired name of the symbol in edit box and click on the **OK** button. The environment to create new symbol design will be displayed; refer to Figure-59.

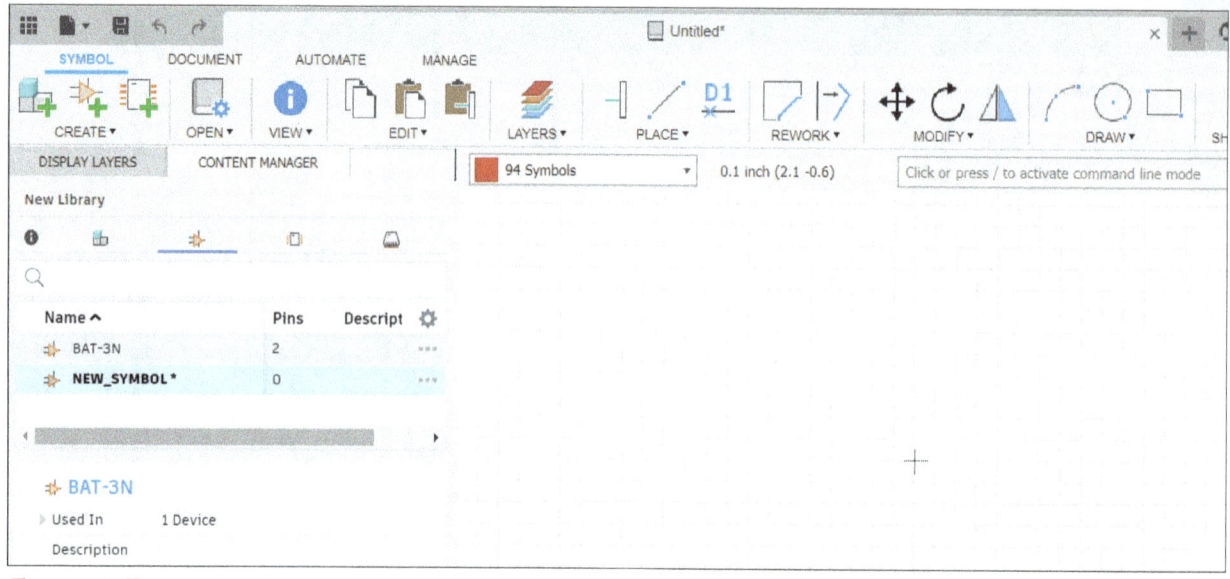

Figure-59. Environment for creating symbol

- Using the tools of **DRAW** drop-down in the **Ribbon**, create the design of symbol; refer to Figure-60.

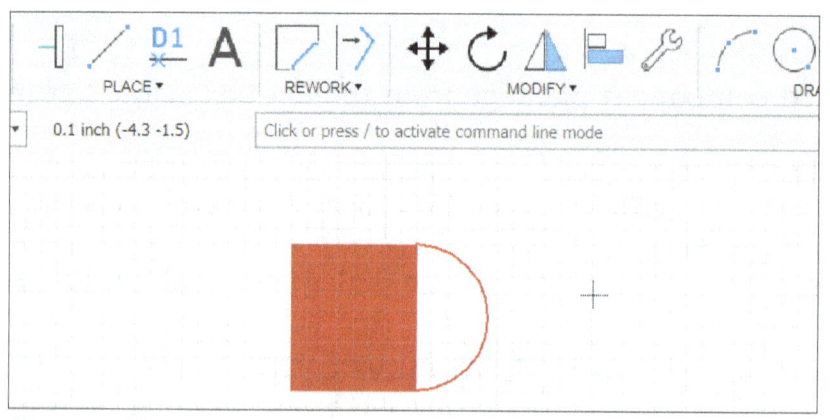

Figure-60. New symbol design created

- Click on the **Pin** tool from the **PLACE** drop-down in the **SYMBOL** tab of the **Ribbon**. The **PIN Dialog** will be displayed; refer to Figure-61.

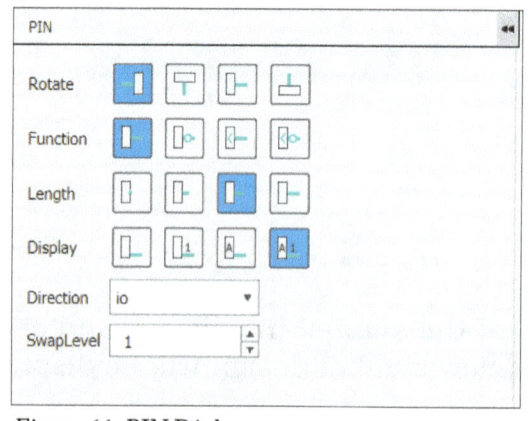

Figure-61. PIN Dialog

- Select desired button from the **Rotate** section to define angle of pin.
- Select desired button from the **Function** section to define the shape of pin connection.

- Select desired button from the **Length** section to define length of pin.
- Select desired button from the **Display** section to define the parameters to be displayed on the symbol.
- Select desired option from the **Direction** drop-down to define direction of pin. The options available in this drop-down are nc (not connected), in, out, io, oc (open collector), pwr, pas (passive), hiz (high impedance), and sup (general supply pin).
- Set desired value in **SwapLevel** drop-down as discussed earlier.
- Click in the drawing area at desired location to place the pin; refer to Figure-62.

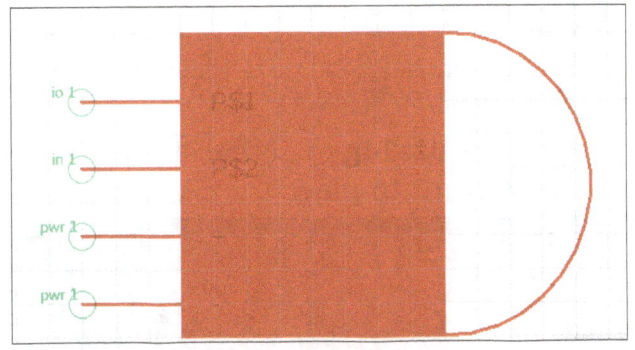

Figure-62. Pins placed in drawing area

- After placing pins, press **ESC** to exit the tool.

Similarly, you can use the **Pin Array** tool of **PLACE** drop-down in the **Ribbon** to create multiple pins.

Creating New Footprint

The **New Footprint** tool is used to create a new PCB footprint in the drawing. The procedure to use this tool is given next.

- Click on the **New Footprint** tool from the **CREATE** drop-down in the **SYMBOL** tab in the **Ribbon**. The **Add Footprint** dialog box will be displayed.
- Specify desired name of footprint in the dialog box and click on the **OK** button. The environment to create footprint will be displayed; refer to Figure-63.

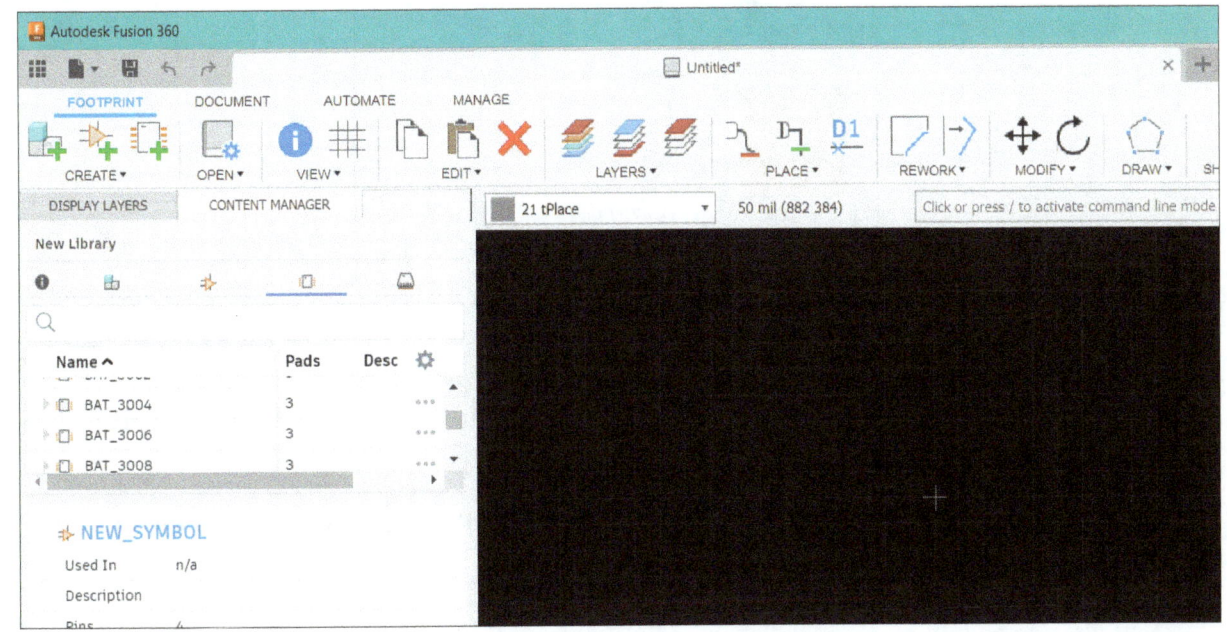

Figure-63. Environment for creating Footprint

- Create the shape of footprint using the tools in **DRAW** drop-down of **Ribbon**.
- Click on the **SMD Pad** tool from the **PLACE** drop-down in the **FOOTPRINT** tab of **Ribbon** to create surface-mounted device pad. The **SMD PAD Dialog** will be displayed; refer to Figure-64.

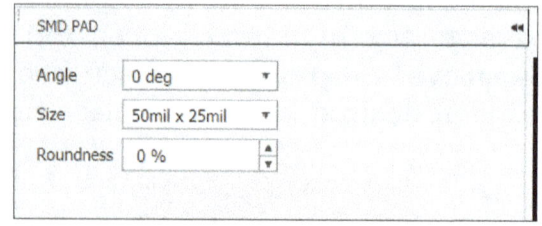

Figure-64. SMD PAD Dialog

- Set desired parameters in the **Dialog** to define shape of the SMD Pad and click at desired location on the footprint to place the SMD pad; refer to Figure-65.

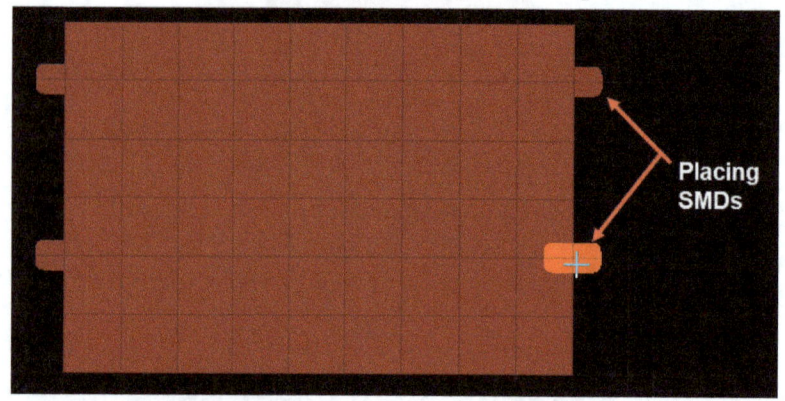

Figure-65. Placing SMD pads

- Click on the **PTH Pad** tool from the **PLACE** drop-down in the **FOOTPRINT** tab of the **Ribbon** to create plated through hole pad in the component. The **PTH PAD Dialog** will be displayed; refer to Figure-66.

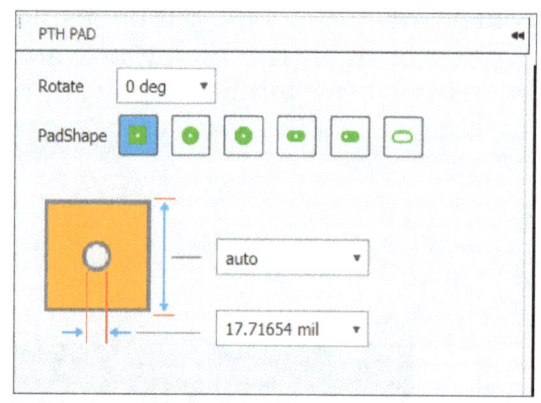

Figure-66. PTH PAD Dialog

- Select desired button from the **PadShape** section to define shape of the pad.
- Specify the size and other parameters of the pad, and click on the component at desired locations to place the pads. Press **ESC** to exit the tool.

Similarly, you can use the **SMD Pad Array** and **PTH Pad Array** tools to create arrays of pads.

Creating New Package

The **New Package** tool is used to generate 3D package of the component which can be used in 3D PCB. The procedure to use this tool is given next.

- Click on the **New Package** tool from the **CREATE** drop-down in the **FOOTPRINT** tab of the **Ribbon**. The environment to create package will be displayed with **PACKAGE GENERATOR**; refer to Figure-67.

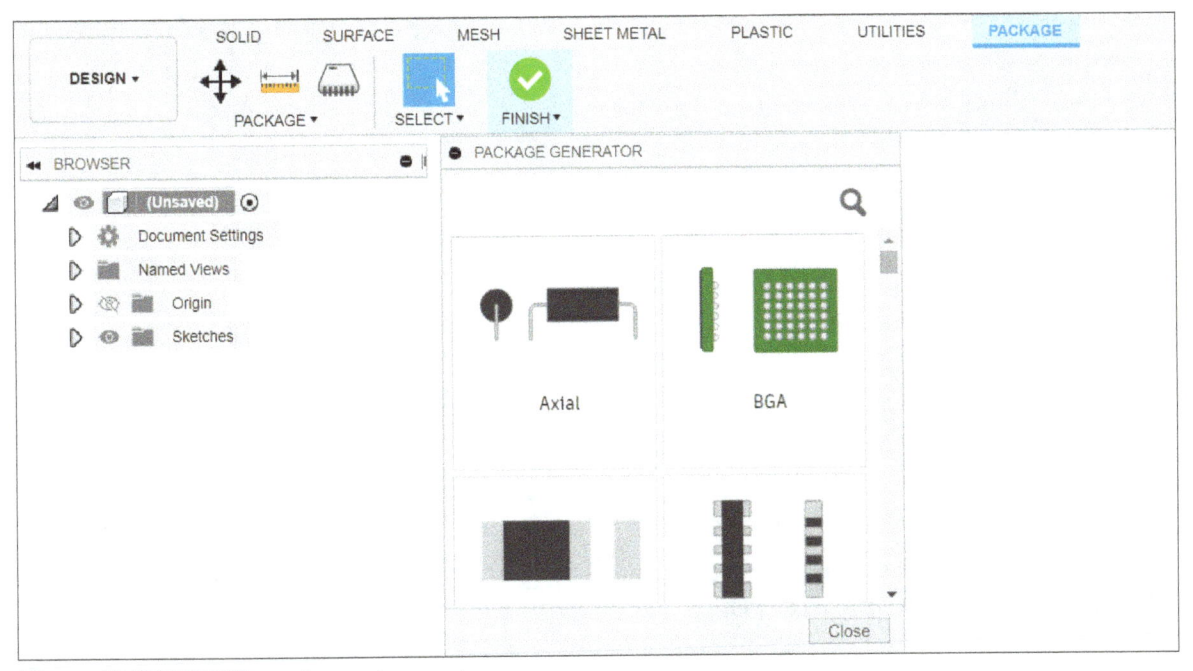

Figure-67. PACKAGE creation environment

- Select desired template from the **PACKAGE GENERATOR** to be used for generating package. The related dialog box will be displayed; refer to Figure-68.

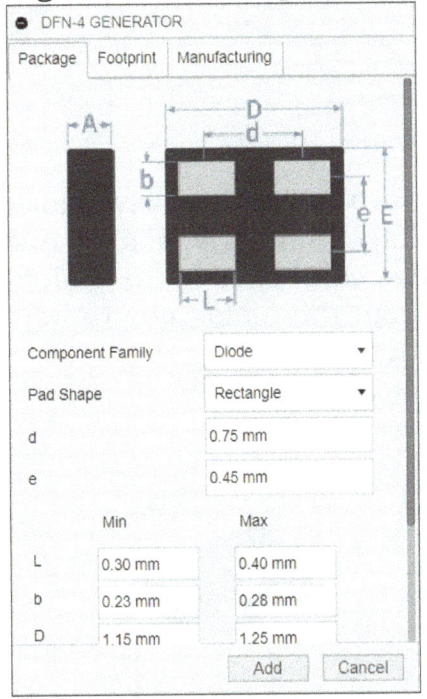

Figure-68. DFN-4 GENERATOR dialog box

- Set desired parameters in the dialog box to define size of various elements of package according to preview in the dialog box.

- Click on the **Footprint** tab of dialog box to modify parameters of package footprint. The options in dialog box will be displayed as shown in Figure-69.
- Select the **Custom Footprint** check box to modify size parameters footprint of package.

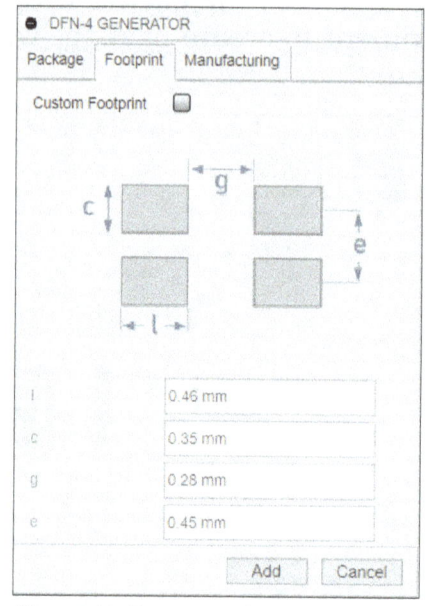

Figure-69. Footprint tab

- Select the **Manufacturing** tab from the dialog box to set manufacturing parameters of package; refer to Figure-70.

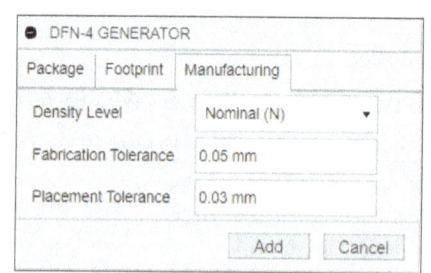

Figure-70. Manufacturing tab of dialog box

- Select desired option from the **Density Level** drop-down to define the manufacturing quality of component package. Similarly, set the tolerance values for fabrication and placement in respective edit boxes of the dialog box.
- Click on the **Add** button from the dialog box to add the custom package in the drawing area; refer to Figure-71.

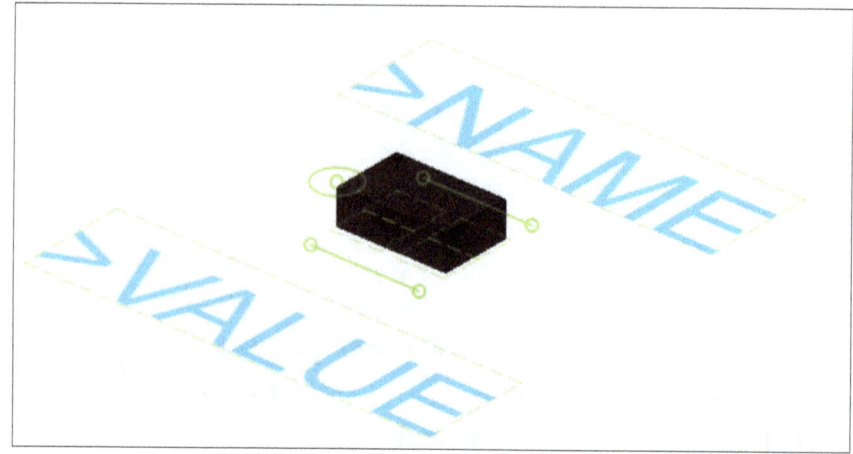

Figure-71. Custom package added in drawing area

- You can also use 3D Modeling tools to modify the package. Click on the **Finish** tool from the **FINISH** drop-down in the **PACKAGE** tab of **Ribbon**. The **Save** dialog box will be displayed.
- Specify desired name of the package and save the design at desired location on the cloud. Note that full library of created component will be saved on doing so.

SELF ASSESSMENT

Q1. Why the Design Rule Check is performed?

Q2. What is the use of IPC-D-356 Netlist tool?

Q3. The Push to 3D PCB tool is used to check the preview of CAM data to be generated for the PCB model. (T/F)

Q4. Select the Export to Project Directory check box in the CAM Processor to save all CAM files in current project folder. (T/F)

Q5. The CAM Processor tool is used to generate various output files to be used by manufacturer for manufacturing PCB. (T/F)

Q6. The Export ODB++ Output tool is used to export the PCB manufacturing data in .... format.

Q7. The .... tool is used to create a new PCB footprint in the drawing.

Q8. The .... tool is used to generate 3D package of the component which can be used in 3D PCB.

Answers: **Ans 3.** F **Ans 4.** T **Ans 5.** T **Ans 6.** ODB++ **Ans 7.** New Footprint **Ans 8.** New Package

For Student Notes

Chapter 7

Practical and Practice

Topics Covered

The major topics covered in this chapter are:

- ***Practical***
- ***Practice***

INTRODUCTION

In the chapter, you will work on various PCB drawings and create manufacturing model for the design. The practical on PCB is discussed next.

PRACTICAL

Create the schematic and 3D PCB design for the schematic drawing shown in Figure-1.

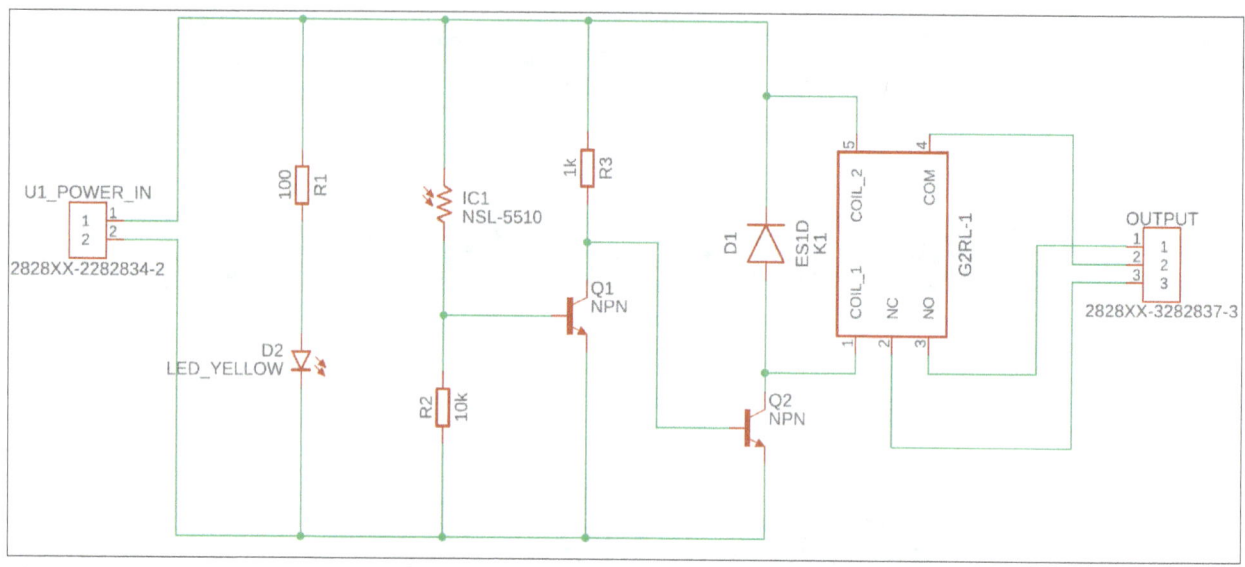

Figure-1. Practical 1 schematic

The steps to perform this practical are given next.

Creating Schematic Design

- Start Autodesk Fusion 360 software if not started yet.
- Click on the **New Electronics Design** tool from the **File** menu of Application window. A new electronics design document will open.
- Click on the **New Schematic** tool from the **CREATE** drop-down of **COMMON** tab in the **Ribbon**. The environment to create schematic design will be displayed.
- Now, we will insert the components as per the schematic drawing and then apply net wires to the components. Click on the **Place Component** tool from the **PLACE** drop-down in the **DESIGN** tab of the **Ribbon**. The **PLACE COMPONENTS MANAGER** will be displayed.
- Select the **Connector** option from the top drop-down in the **PLACE COMPONENTS MANAGER** and select 2828XX-2 component from the list; refer to Figure-2. Double-click on this component to insert it in the drawing. The component will get attached to cursor.

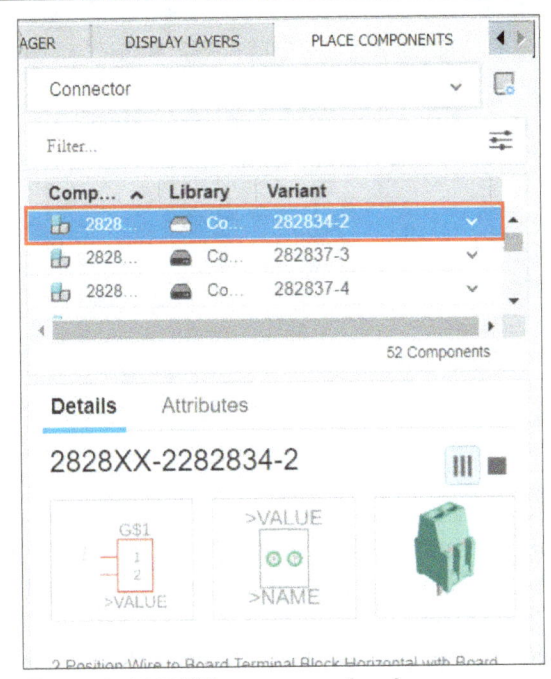

Figure-2. 2828XX component selected

- Click at desired location to place the component and press **ESC** to exit the tool.
- Similarly, select other components from the **Manager** and place them as per the schematic drawing; refer to Figure-3. Various components placed in the drawing are:

U$1 and U$2 are the components available in **Connector** category. Resistors with name R1, R2, and R3 are from **Resistor** category. D1 is from **LED** category and D2 is from **Diode** category. Q1 and Q2 are NPN transistors from **Transistor** category. K1 is component from **Relay** category. IC1 is light sensor component from **Sensor** category.

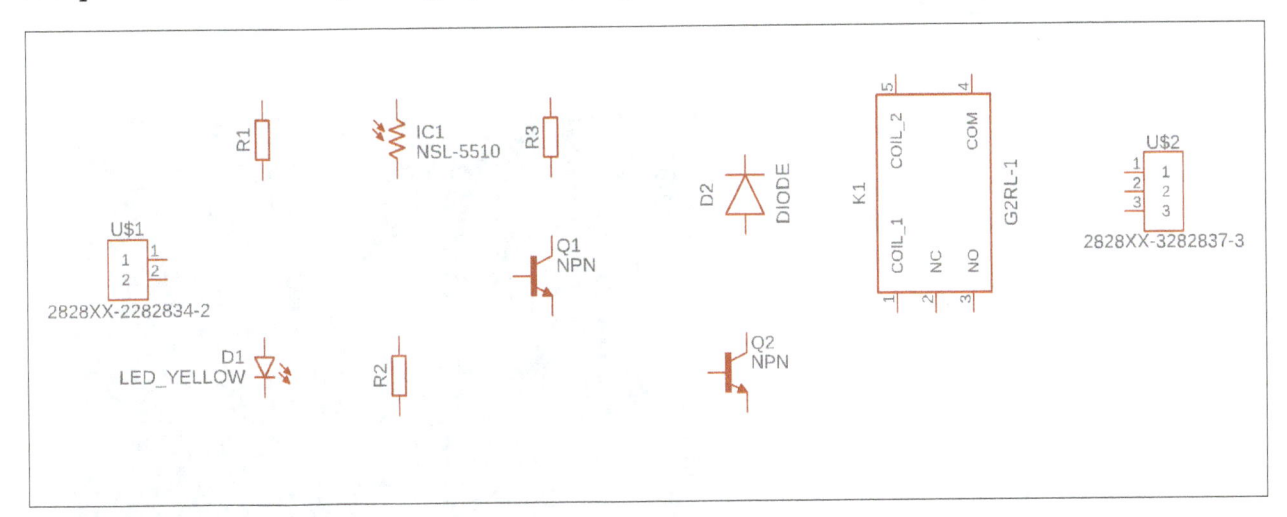

Figure-3. After placing components

- Click on the **Net** tool from the **CONNECT** drop-down in the **DESIGN** tab of the **Ribbon**. You will be asked to specify start point of the wire.
- Create the network of wire as per the schematic drawing; refer to Figure-4.

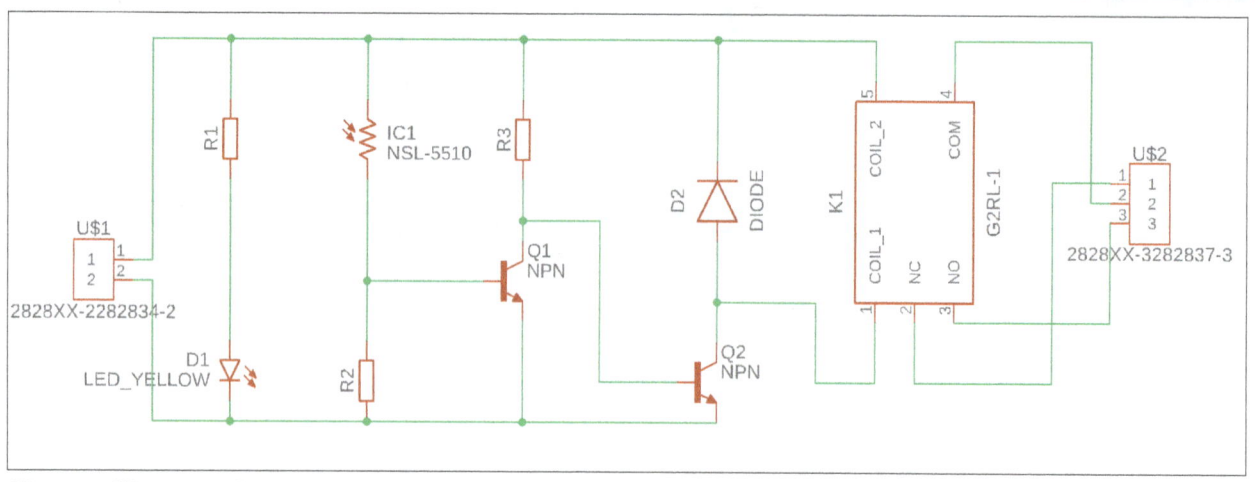

Figure-4. Wire network created in schematic

- Click on the **Value** tool from the **MODIFY** drop-down in the **DESIGN** tab of the **Ribbon**. You will be asked to select the component for changing its value.
- Change the values of resistors as per the schematic drawing.
- Click on the **Name** tool from the **CONNECT** drop-down in **Ribbon** and change the names of connectors as per the drawing.

Creating 2D PCB Design

- After creating the schematic of circuit, click on the `Switch to PCB document` tool from the **SWITCH** drop-down in **DESIGN** tab of the **Ribbon**. The environment to design 2D PCB will be displayed with all the electronics components connected by airwires; refer to Figure-5.

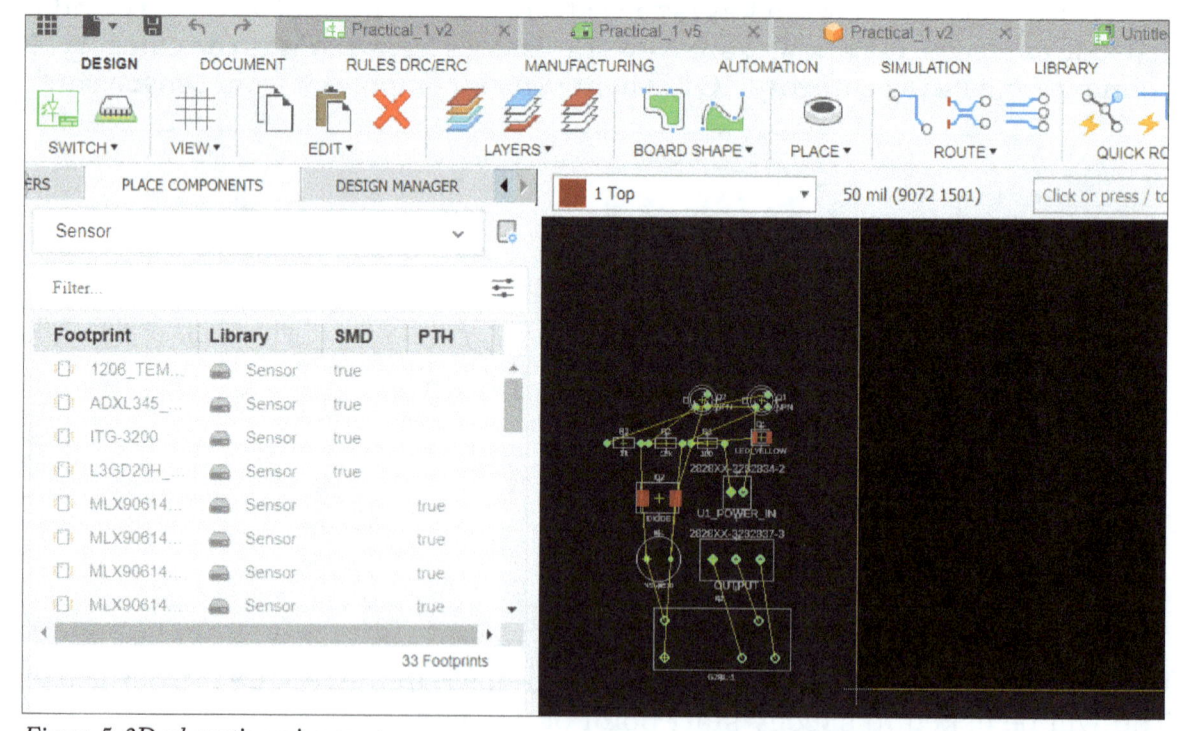

Figure-5. 2D schematic environment

- One by one, drag the components on black board at desired locations; refer to Figure-6. There are a few rules to be followed when placing components for routing at board like grouping components by types, obeying mechanical constraints, and so on; refer to the QR codes given next.

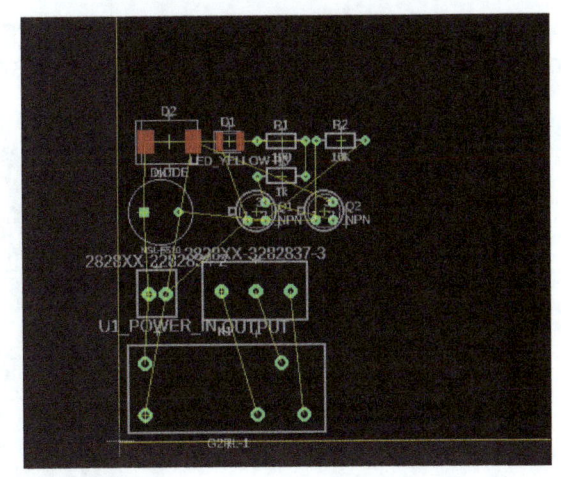

Figure-6. After rearranging components on board

- Drag the boundaries of board to adjust the size of PCB to minimum; refer to Figure-7.

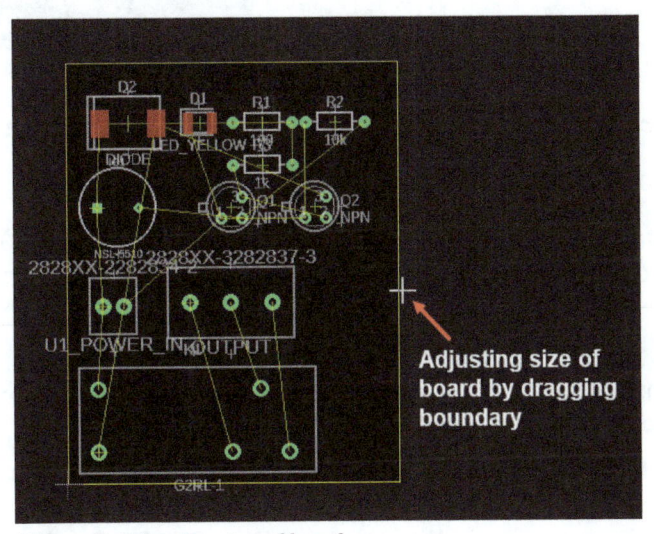

Figure-7. Adjusting size of board

- The next step after placing components on board is to create route wires. Although manual routing can be efficient by experienced PCB designers, you can still use Autorouter to speed up the routing process. Click on the **Autorouter** tool from the **QUICK ROUTE** drop-down in the **DESIGN** tab of the **Ribbon**. The **Autorouter Main Setup** dialog box will be displayed.
- Set desired parameters in the dialog box as discussed earlier and click on the **Continue** button. The **Routing Variants Dialog** will be displayed; refer to Figure-8.

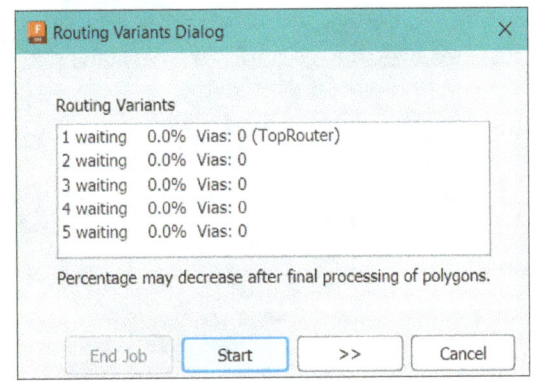

Figure-8. Routing Variants Dialog

- Click on the **Start** button from the **Dialog** to begin routing process. Once the process is complete, variants of routing will be displayed; refer to Figure-9.

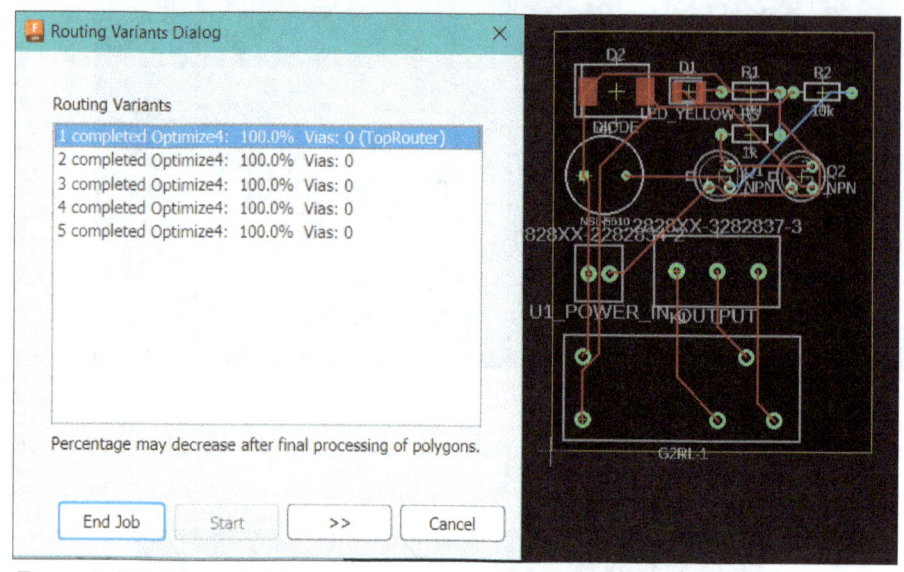

Figure-9. Routing variation

- After selecting desired variant, click on the **End Job** button. The routing will be generated.

You can perform desired placement manipulations on the component like moving component, rotating component, and so on. The route will be regenerated automatically.

Creating 3D PCB Design

- Click on the **Push to 3D PCB** tool from the **SWITCH** drop-down in the **DESIGN** tab of the **Ribbon**. The **PUSH TO 3D PCB** Dialog will be displayed.
- Select the **Most Detailed** option from the **Preset** drop-down to include all the 3D geometry objects of 2D PCB while generating the 3D PCB.
- Click on the **Push** button from the Dialog. The 3D PCB will be generated; refer to Figure-10.

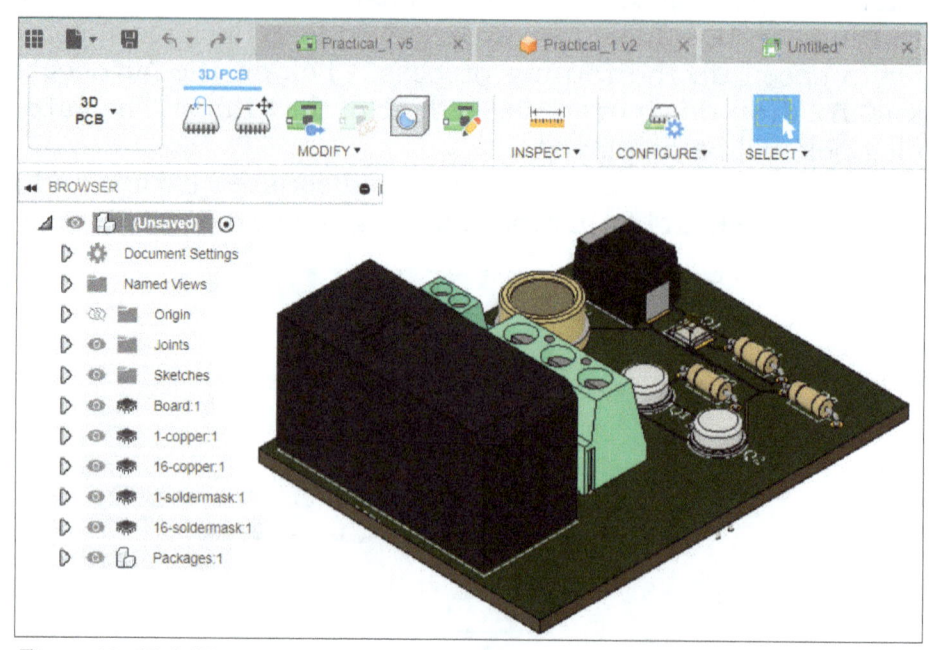

Figure-10. 3D PCB

Rearranging PCB for better design

- You can also perform move operations in 3D PCB using the **Move 3D PCB Component** tool, if needed. Make sure to push those changes in the 2D PCB and schematic as well. For example, in our 3D PCB model, the placement of connectors is not mechanically accessible to easily connect external wires; refer to Figure-11. Using the **Move 3D PCB Component** tool to rearrange the components so that you can decrease the size of PCB as well as place the connectors on outer side as shown in Figure-12.

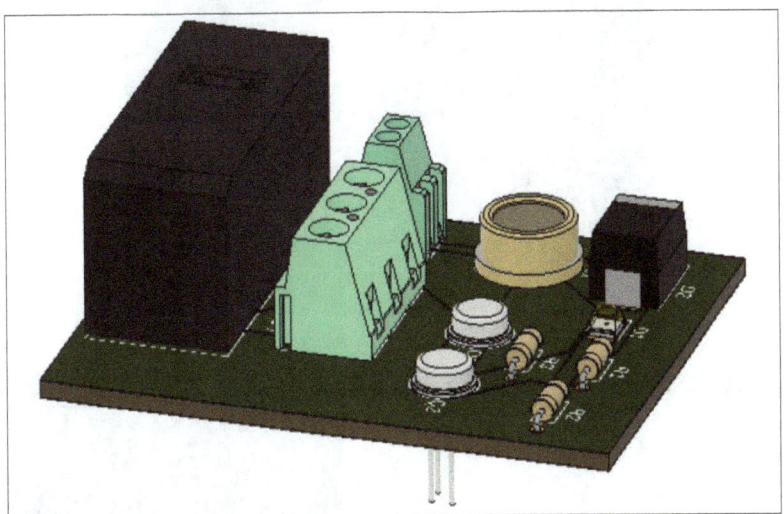

Figure-11. Connectors not accessible

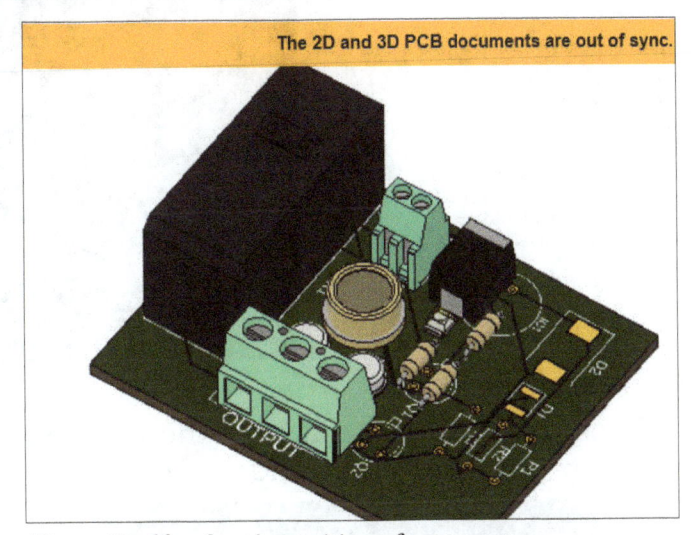

Figure-12. After changing positions of components

- After changing position, click on the **Push to 2D PCB** tool from the **MODIFY** drop-down in the **3D PCB** tab of **Ribbon** to push changes made in 3D PCB to 2D PCB. The changes will be reflected in the 2D PCB; refer to Figure-13. Note that the route wires in the PCB are now disorganized and you need to reroute the wires again.
- Use the **Unroute All** tool from the **UNROUTE** drop-down of **DESIGN** tab in the **Ribbon** to delete all the existing route wires and then use the **Autorouter** tool to generate automatic routing variations; refer to Figure-14. Select desired routing variation and click on the **End Job** button from the **Routing Variants Dialog** as discussed earlier.

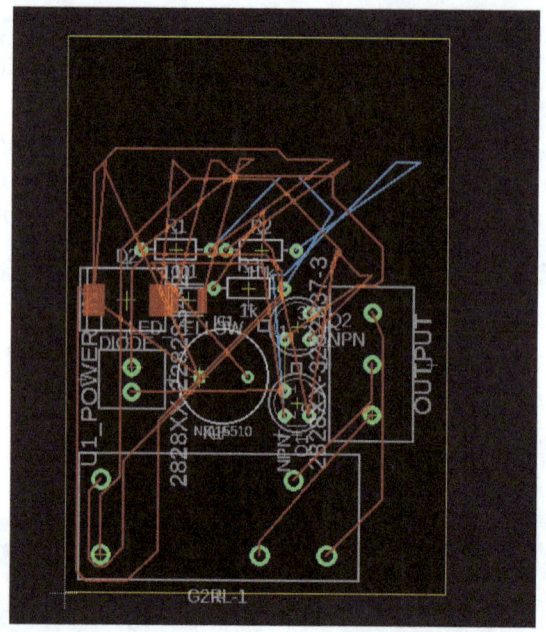

Figure-13. 2D PCB after modification

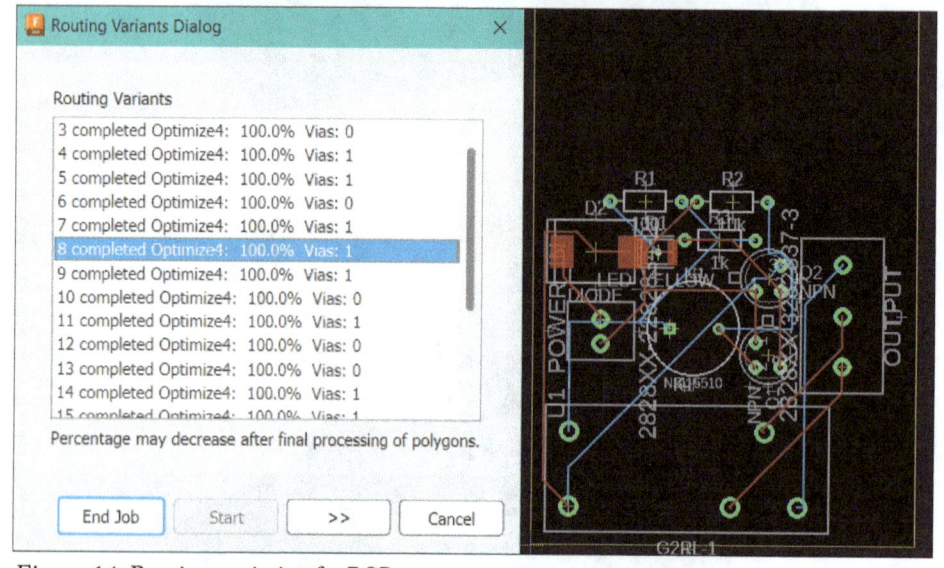

Figure-14. Routing variation for PCB

- Change the size of board to accommodate all the components while reducing wastage of PCB space by dragging the boundary of board; refer to Figure-15.

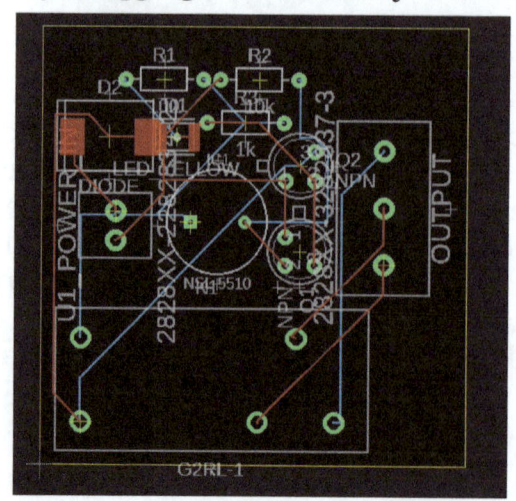

Figure-15. PCB after changing board size

Autodesk Fusion 360 PCB Black Book   7-9

- After performing changes, click on the **Push to 3D PCB** tool from the **SWITCH** drop-down of **Ribbon** in 2D PCB environment to update the 3D PCB accordingly. The **PUSH TO 3D PCB Dialog** will be displayed.
- Click on the **Push** button from **Dialog** to generate the updated model; refer to Figure-16.

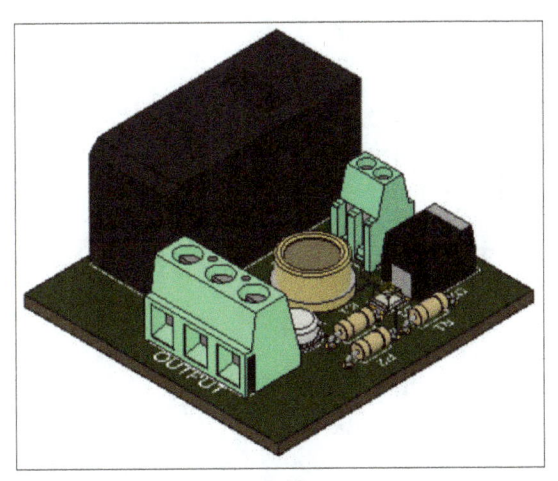

Figure-16. Updated 3D PCB

You can generate the manufacturing data of PCB from 2D PCB tab in application window using the **CAM Processor** tool of **MANUFACTURING** tab in the **Ribbon**; refer to Figure-17.

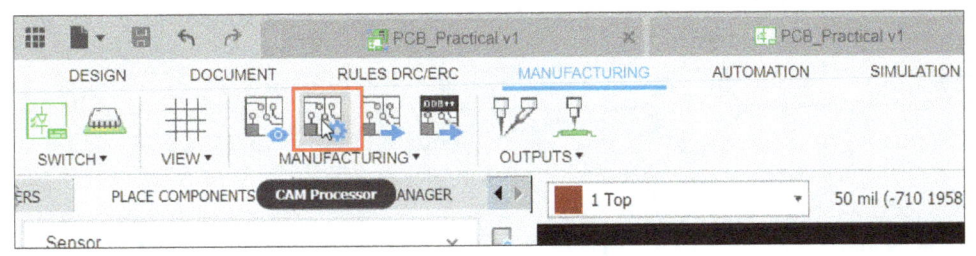

Figure-17. CAM Processor tool

PRACTICE SCHEMATICS

Create the Schematic drawing and manufacturable PCB models as per the drawings given next.

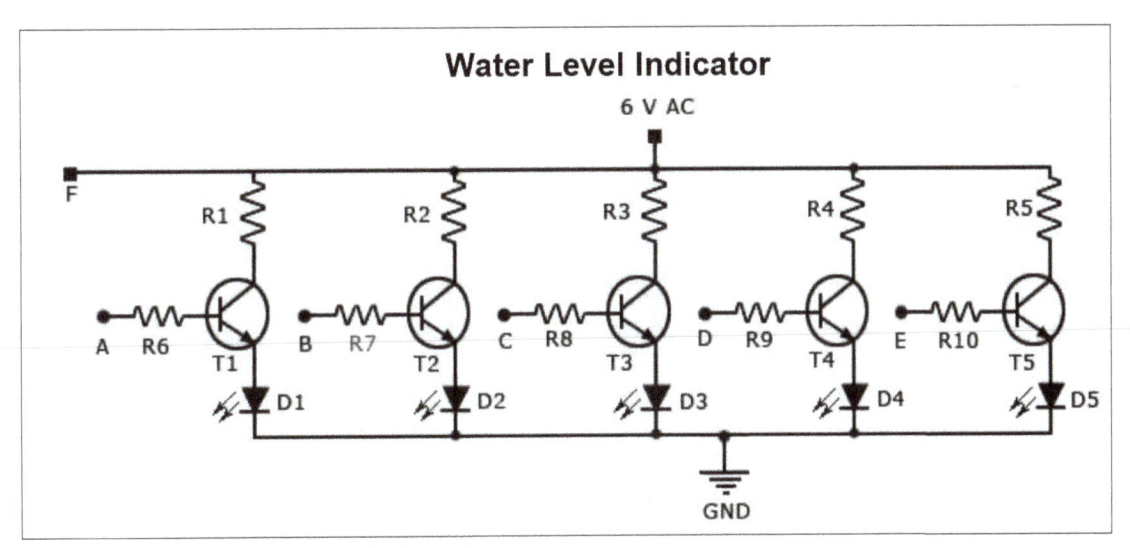

Figure-18. Water level indicator pcb schematic

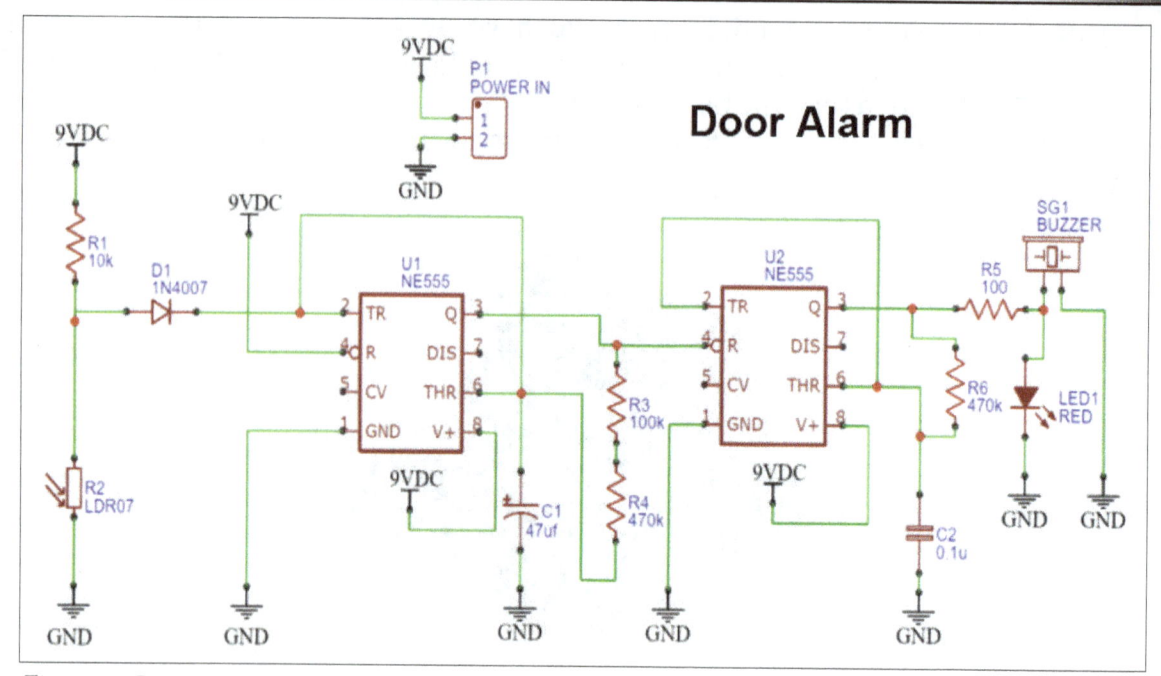

Figure-19. Door alarm schematic

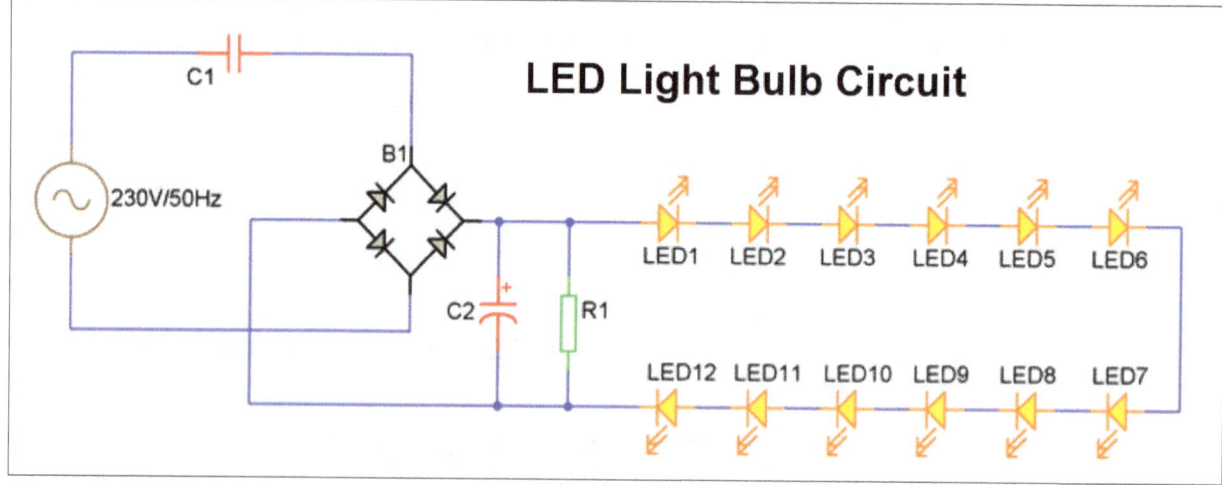

Figure-20. LED light bulb circuit

Index

A

AC Sweep simulation 3-26
Add Bus tool 2-29
Addlevel drop-down 6-25
Add local package option 6-26
Add new via type tool 5-6
Add next available layer pair button 5-5
Add Spice Model dialog box 3-15, 6-25
Add Spice Model tool 3-15
Add Version Description dialog box 1-16
Align tool 3-7
Analog Source Setup tool 3-17
API option 1-25
Arc tool 4-7
Arrange tool 3-7
Assembly and Drawing Outputs tool 6-19
Assembly Variant drop-down 2-11
Assembly Variant tool 4-2
Attribute tool 4-9
Autodesk Account tool 1-28
Automatic version on close check box 1-25
Automation 4-14
Autorouter tool 5-24

B

Bill of Materials tool 4-2
BOARD SHAPE drop-down 5-9
Break Out Bus tool 2-30
Break Out Pins tool 2-31
BROWSER 1-38

C

CAM Preview tool 6-13
CAM Processor tool 6-17
Capacitors 2-6
Capture Image tool 1-19
Change Active Units button 1-39
Change Keyboard Shortcut option 1-39
Change layer stack properties button 5-4
Change tool 3-9
Circle tool 4-7
Circuits 2-7
Circular Arrange button 3-8
Class tool 4-13

Clearance tab 5-7
Command Input box 2-20
CONNECT drop-down 2-24
Copy Format tool 3-9
Create Account button 1-3
Create a Team option 1-34
Create or join team tool 1-33

D

Data Panel 1-32
Data section 1-33
DC Sweep simulation 3-24
Default Orbit type drop-down 1-25
DESIGN MANAGER 2-10
Detect Board Shape check box 1-26
Diagnostic Log Files tool 1-31
Differential pair 5-13
Digital Source Setup tool 3-18
Diodes 2-3
Display Grid for Schematic check box 1-28
DISPLAY LAYERS MANAGER 2-12
Display Layers tool 5-3
Documentation Tools 4-2
Document Attributes tool 4-11
DRAW drop-down 4-4
DRC (Design Rule Check) 6-2
DRC tool 6-2

E

Edit material properties check box 5-6
Electro-magnetic Devices 2-7
Electronics option 1-26
Enable camera pivot check box 1-25
ERC tool 4-13
Export DXF button 6-15
Export Gerber 6-19
Export Image button 6-15
Export Libraries tool 4-17
Export multiple libraries radio button 4-17
Export ODB++ Output tool 6-20
Export tool 1-18
Extensions button 1-29
Extrude tool 6-9
Eye button 1-38

F

Fanout tool 5-22
File drop-down 1-6
File menu 1-8

Flip 3D PCB Component tool 6-4
Flip Board tool 5-3

G

Gate Swap tool 3-14
General node 1-24
Getting Started tool 1-31
Grid Settings tool 2-19
Group tool 2-19

H

Help drop-down 1-29
Hide all Polygon Pour fills tool 5-29
Hole (NPTH) tool 5-12

I

Import button 6-26
Import connected 3D Package check box 6-27
Info tool 2-18
Insert Derive tool 6-7
Insert into Current Design option 6-11
Insert Schematic tool 2-23
INSPECTOR Pane 2-16
Integrated Circuits (ICs) 2-3
In Use toggle button 3-20
Invite button 1-37
Invoke tool 3-11
IPC-D-356 Netlist tool 6-21

J

Joint tool 6-12
Junction tool 2-33

L

Label tool 2-33
Layer Stack Manager tool 5-3
Learn Fusion 360 tool 1-30
Library.io check box 3-20
Line tool 4-4
Loop Preserve button 2-25
Loop Remove button 2-25

M

Managers 2-10
MAP button 3-16
Mark Local Origin tool 2-19
Max points edit box 3-23
Meander tool 5-17
Merge into one library radio button 4-17

Milestone check 1-18
Mirror tool 3-6
Miter tool 3-2
Module tool 2-34
Move 3D PCB Component tool 6-5
Move tool 3-5
My Profile tool 1-28

N

Name tool 2-32
NC Drill 6-19
Net Breakout tool 2-28
Net Class drop-down 2-25
Net tool 2-24
New assembly variant dialog box 2-11
New design tool 1-6
New Design tool 6-7
New Drawing Template tool 1-9
New Drawing tool 1-8
New Electronics Design tool 1-8, 2-9
New Electronics Library tool 6-21
New Netclass option 2-26
New PCB tool 5-2
New Project button 1-15, 1-37
New Schematic tool 2-9

O

Offline cache time period (days) edit box 1-25
Open Library Manager button 3-19
Open Library Manager tool 4-15
Open on the Web button 1-37
Open tool 1-9
Operating Point option 3-23
Optimize tool 3-4, 3-14
Optoelectronic (Optronic) Devices 2-5
Outline Arc tool 5-10
Outline Circle tool 5-11
Outline Polyline tool 5-9
Outline Spline tool 5-10

P

PACKAGE GENERATOR 6-31
Pattern tool 2-22
PCBHole tool 6-6
People section 1-33
Phase Probe tool 3-22
Pin Array tool 6-29
Pin Swap tool 3-13
Pin tool 6-28

PLACE COMPONENTS MANAGER 2-14
Place Component tool 2-21, 5-11
PLACE drop-down 2-21
Polygon Cutout tool 5-27
Polygon Pour from Outline tool 5-29
Polygon Pour tool 5-26
Polygon Shape tool 4-8, 5-28
Port tool 2-35
Preferences tool 1-24
Print tool 4-3
Programmable Logic Devices (PLDs) 2-4
PTH Pad Array 6-30
PTH Pad tool 6-30
Push to 2D PCB tool 6-6
Push to 3D PCB tool 6-3

Q

QuickRoute Airwire tool 5-18
QUICK ROUTE drop-down 5-17
QuickRoute Guided tool 5-21
QuickRoute Multiple tool 5-20
QuickRoute Signal tool 5-19
QuickRoute Smooth tool 5-20

R

Recover Documents tool 1-11
Recovery time interval (min) edit box 1-25
Rectangle tool 4-7
Rectangular Arrange button 3-8
Rectangular Pattern button 2-22
Redo button 1-23
Re-fill Polygon Pours tool 5-29
Remove last layer pair tool 5-5
Remove Spice Model tool 3-16
Replace tool 3-12
Reposition Attributes tool 4-10
Reroute tool 5-31
Reset To Default Layout tool 1-23
Resistors 2-6
Reverse zoom direction check box 1-25
REWORK drop-down 3-2, 5-31
Rotate tool 3-6
Route Differential Pair tool 5-13
Route Manual tool 5-12
Route Multiple tool 5-14
Run Script tool 4-15
Run ULP tool 4-14

S

Save As Latest dialog box 1-18
Save As Latest tool 1-18
Save As tool 1-17
Save tool 1-14
Search Box 1-29
SELECTION FILTER Pane 2-17
Sensors 2-7
Share tool 1-21
Sheets tile 2-20
Shell tool 6-10
Show all Polygon Pour fills tool 5-29
Show Data Panel tool 1-23
Show/Hide Text Commands tool 1-23
Show tool 2-18
Signal tool 5-11
SIMULATE drop-down 3-15
Simulate tool 3-22, 3-23
Slice tool 3-3
SMD Pad Array 6-30
SMD PAD Dialog 6-30
SPICE 3-15
Split tool 3-3
Stop Command tool 2-19
SwapLevel edit box 6-25
Switches 2-7
Synchronize tool 4-12

T

Temperature edit box 3-23
Text tool 4-5
Toggle Single Layer View tool 5-3
Transducers 2-7
Transient simulation 3-27
Transistors 2-2

U

Undo tool 1-23
Unroute All tool 5-31
Unroute Incomplete tool 5-31
Unroute tool 5-30
Update design from all libraries tool 4-17
Update design from library tool 4-16
Upload tool 1-12
User Account drop-down menu 1-24
User language drop-down 1-24

V

Vacuum Tube (Valve) 2-5
VALIDATE tab 4-12
Value tool 3-10
Version Description edit box 1-18
Via tool 5-16
View Details On Web tool 1-22
View tool 1-22
View toolbar 2-17
V: Independent Voltage Source option 3-15
Voltage Probe tool 3-21

W

Work Offline tool 1-28

Z

Zoom In tool 2-18
Zoom Out tool 2-18
Zoom to Fit tool 2-19

Ethics of an Engineer

- Engineers shall hold paramount the safety, health and welfare of the public and shall strive to comply with the principles of sustainable development in the performance of their professional duties.

- Engineers shall perform services only in areas of their competence.

- Engineers shall issue public statements only in an objective and truthful manner.

- Engineers shall act in professional manners for each employer or client as faithful agents or trustees, and shall avoid conflicts of interest.

- Engineers shall build their professional reputation on the merit of their services and shall not compete unfairly with others.

- Engineers shall act in such a manner as to uphold and enhance the honor, integrity, and dignity of the engineering profession and shall act with zero-tolerance for bribery, fraud, and corruption.

- Engineers shall continue their professional development throughout their careers, and shall provide opportunities for the professional development of those engineers under their supervision.

OTHER BOOKS BY CADCAMCAE WORKS

Autodesk Revit 2024 Black Book
Autodesk Revit 2023 Black Book
Autodesk Revit 2022 Black Book

Autodesk Inventor 2024 Black Book
Autodesk Inventor 2023 Black Book
Autodesk Inventor 2022 Black Book

Autodesk Fusion 360 Black Book (V2.0.18477)
Autodesk Fusion 360 PCB Black Book (V2.0.15509)

AutoCAD Electrical 2024 Black Book
AutoCAD Electrical 2023 Black Book
AutoCAD Electrical 2022 Black Book
AutoCAD Electrical 2021 Black Book

SolidWorks 2024 Black Book
SolidWorks 2023 Black Book
SolidWorks 2022 Black Book

SolidWorks Simulation 2024 Black Book
SolidWorks Simulation 2023 Black Book
SolidWorks Simulation 2022 Black Book

SolidWorks Flow Simulation 2024 Black Book
SolidWorks Flow Simulation 2023 Black Book
SolidWorks Flow Simulation 2022 Black Book

SolidWorks CAM 2024 Black Book
SolidWorks CAM 2023 Black Book
SolidWorks CAM 2022 Black Book

SolidWorks Electrical 2024 Black Book
SolidWorks Electrical 2022 Black Book
SolidWorks Electrical 2021 Black Book

SolidWorks Workbook 2022

Mastercam 2023 for SolidWorks Black Book
Mastercam 2022 for SolidWorks Black Book
Mastercam 2017 for SolidWorks Black Book

Mastercam 2024 Black Book
Mastercam 2023 Black Book
Mastercam 2022 Black Book

Creo Parametric 10.0 Black Book
Creo Parametric 9.0 Black Book
Creo Parametric 8.0 Black Book
Creo Parametric 7.0 Black Book

Creo Manufacturing 10.0 Black Book
Creo Manufacturing 9.0 Black Book
Creo Manufacturing 4.0 Black Book

ETABS V21 Black Book
ETABS V20 Black Book
ETABS V19 Black Book
ETABS V18 Black Book

Basics of Autodesk Inventor Nastran 2024
Basics of Autodesk Inventor Nastran 2022
Basics of Autodesk Inventor Nastran 2020

Autodesk CFD 2023 Black Book
Autodesk CFD 2021 Black Book
Autodesk CFD 2018 Black Book

FreeCAD 0.21 Black Book
FreeCAD 0.20 Black Book
FreeCAD 0.19 Black Book
FreeCAD 0.18 Black Book

LibreCAD 2.2 Black Book